Physics Galaxy

# Lecture Notes on MECHANICS

JEE (Mains + Advance) | BITSAT | NEET | AIIMS

## Vol. I

*All Notes with Video Explanations!*

**Ashish Arora**
*Mentor & Founder*

**PHYSICSGALAXY.COM**
*World's largest encyclopedia of online video lectures on High School Physics*

G K Publications (P) Ltd

**CL MEDIA (P) LTD.**

**Edition 2019-20**



SBN : **978-93-8818-244-7**

Typeset by : *CL Media DTP Unit*

**Administrative and Production Offices**

Published by : **CL Media (P) Ltd.**
A-41, Lower Ground Floor,
Espire Building,
Mohan Cooperative Industrial Area,
Main Mathura Road,
New Delhi - 110044

Marketed by : **G.K. Publications (P) Ltd.**
A-41, Lower Ground Floor,
Espire Building,
Mohan Cooperative Industrial Area,
Main Mathura Road,
New Delhi - 110044

For product information :
Visit ***www.gkpublications.com*** or email to ***gkp@gkpublications.com***

***Dedicated***
***to***
***My Parents, Son, Daughter***
***and***
***My beloved wife***

In his teaching career since 1992 Ashish Arora personally mentored more than 10000 IITians and students who reached global heights in various career and profession chosen. It is his helping attitude toward students with which all his students remember him in life for his contribution in their success and keep connections with him live. Below is the list of some of the successful students in International Olympiad personally taught by him.

| | |
|---|---|
| **NAVNEET LOIWAL** | *International GOLD Medal in IPhO-2000 at LONDON*, Also secured **AIR-4** in **IIT JEE 2000**<br>PROUD FOR INDIA : Navneet Loiwal was the first Indian Student who won first International GOLD Medal for our country in International Physics Olympiad. |
| **DUNGRA RAM CHOUDHARY** | **AIR-1** in **IIT JEE 2002** |
| **HARSHIT CHOPRA** | *National Gold Medal in INPhO-2002* and got **AIR-2** in **IIT JEE-2002** |
| **KUNTAL LOYA** | A Girl Student got position **AIR-8** in **IIT JEE 2002** |
| **LUV KUMAR** | *National Gold Medal in INPhO-2003* and got **AIR-3** in **IIT JEE-2003** |
| **RAJHANS SAMDANI** | *National Gold Medal in INPhO-2003* and got **AIR-5** in **IIT JEE-2003** |
| **SHANTANU BHARDWAJ** | *International SILVER Medal in IPhO-2002 at INDONESIA* |
| **SHALEEN HARLALKA** | *International GOLD Medal in IPhO-2003 at CHINA* and got **AIR-46** in **IIT JEE-2003** |
| **TARUN GUPTA** | *National GOLD Medal in INPhO-2005* |
| **APEKSHA KHANDELWAL** | *National GOLD Medal in INPhO-2005* |
| **ABHINAV SINHA** | *Hon'ble Mension Award in APhO-2006* at *KAZAKHSTAN* |
| **RAMAN SHARMA** | *International GOLD Medal in IPhO-2007* at *IRAN* and got **AIR-20** in **IIT JEE-2007** |
| **PRATYUSH PANDEY** | *International SILVER Medal in IPhO-2007* at *IRAN* and got **AIR-85 in IIT JEE-2007** |
| **GARVIT JUNIWAL** | *International GOLD Medal in IPhO-2008* at *VIETNAM* and got **AIR-10** in **IIT JEE-2008** |
| **ANKIT PARASHAR** | *National GOLD Medal in INPhO-2008* |
| **HEMANT NOVAL** | *National GOLD Medal in INPhO-2008* and got **AIR-25** in **IIT JEE-2008** |
| **ABHISHEK MITRUKA** | *National GOLD Medal in INPhO-2009* |
| **SARTHAK KALANI** | *National GOLD Medal in INPhO-2009* |
| **ASTHA AGARWAL** | *International SILVER Medal in IJSO-2009* at *AZERBAIJAN* |
| **RAHUL GURNANI** | *International SILVER Medal in IJSO-2009* at *AZERBAIJAN* |
| **AYUSH SINGHAL** | *International SILVER Medal in IJSO-2009* at *AZERBAIJAN* |
| **MEHUL KUMAR** | *International SILVER Medal in IPhO-2010* at *CROATIA* and got **AIR-19** in **IIT JEE-2010** |
| **ABHIROOP BHATNAGAR** | *National GOLD Medal in INPhO-2010* |
| **AYUSH SHARMA** | *International Double GOLD Medal in IJSO-2010* at *NIGERIA* |
| **AASTHA AGRAWAL** | *Hon'ble Mension Award in APhO-2011* at *ISRAEL* and got **AIR-93** in **IIT JEE 2011** |
| **ABHISHEK BANSAL** | *National GOLD Medal in INPhO-2011* |
| **SAMYAK DAGA** | *National GOLD Medal in INPhO-2011* |
| **SHREY GOYAL** | *National GOLD Medal in INPhO-2012* and secured **AIR-24** in **IIT JEE 2012** |
| **RAHUL GURNANI** | *National GOLD Medal in INPhO-2012* |
| **JASPREET SINGH JHEETA** | *National GOLD Medal in INPhO-2012* |
| **DIVYANSHU MUND** | *National GOLD Medal in INPhO-2012* |
| **SHESHANSH AGARWAL** | *International SILVER Medal in IAO-2012 at KOREA* |
| **SWATI GUPTA** | *International SILVER Medal in IJSO-2012 at IRAN* |
| **PRATYUSH RAJPUT** | *International SILVER Medal in IJSO-2012 at IRAN* |
| **SHESHANSH AGARWAL** | *International BRONZE Medal in IOAA-2013 at GREECE* |
| **SHESHANSH AGARWAL** | *International GOLD Medal in IOAA-2014 at ROMANIA* |
| **SHESHANSH AGARWAL** | *International SILVER Medal in IPhO-2015 at INDIA* and secured **AIR-58** in **JEE(Advanced)-2015** |
| **VIDUSHI VARSHNEY** | *International SILVER Medal in IJSO-2015 to be held at SOUTH KOREA* |
| **AMAN BANSAL** | **AIR-1 in JEE Advanced 2016** |
| **KUNAL GOYAL** | **AIR-3 in JEE Advanced 2016** |
| **GOURAV DIDWANIA** | **AIR-9 in JEE Advanced 2016** |
| **DIVYANSH GARG** | *International SILVER Medal in IPhO-2016 at SWITZERLAND* |

# ABOUT THE AUTHOR

The complexities of Physics have given nightmares to many, but the homegrown genius of Jaipur-Ashish Arora has helped several students to live their dreams by decoding it.

Newton Law of Gravitation and Faraday's Magnetic force of attraction apply perfectly well with this unassuming genius. A Pied Piper of students, his webportal https://www.physicsgalaxy.com, The world's largest encyclopedia of video lectures on high school Physics possesses strong gravitational pull and magnetic attraction for students who want to make it big in life.

Ashish Arora, gifted with rare ability to train masterminds, has mentored over 10,000 IITians in his past 24 years of teaching sojourn including lots of students made it to Top 100 in IITJEE/JEE(Advance) including AIR-1 and many in Top-10. Apart from that, he has also groomed hundreds of students for cracking International Physics Olympiad. No wonder his student Navneet Loiwal brought laurel to the country by becoming the first Indian to win a Gold medal at the 2000 - International Physics Olympiad in London (UK).

His special ability to simplify the toughest of the Physics theorems and applications rates him as one among the best Physics teachers in the world. With this, Arora simply defies the logic that perfection comes with age. Even at 18 when he started teaching Physics while pursuing engineering, he was as engaging as he is now. Experience, besides graying his hair, has just widened his horizon.

Now after encountering all tribes of students - some brilliant and some not-so-intelligent - this celebrated teacher has embarked upon a noble mission to make the entire galaxy of Physics inform of his webportal PHYSICSGALAXY.COM to serve and help global students in the subject. Today students from more than 180 countries are connected with this webportal and his youtube channel 'Physics Galaxy'. Daily about more than 35000 video lectures are being watched and his pool of video lectures has cross and sutdents post their queries in INTERACT tab under different sections and topics of physics.

Physics Galaxy video lectures have already crossed 27 million views till now and growing further to set new benchmarks and all the video lectures of Physics Galaxy can be accessed through Physics Galaxy mobile application available on iOS and Android platform.

Dedicated to global students of middle and high school level, his website also www.physicsgalaxy.com has teaching sessions dubbed in American accent and subtitles in 87 languages.

# PREFACE

Physics is one of the most difficult subjects, for students of Science, because it requires good visualization power and high cognitive ability. To extend, it is very difficult to sail through in the present age of competition without mastering the concepts of Physics in depth. Scoring in Physics does not require hard work or rote learning but a simple understanding of physical situations which exist in the surroundings of our everyday life. Thus, in order to command the concepts of Physics, you should try to understand its core concepts in practical approach instead of merely depending on the theory.

It is against this backdrop that we have come up with ***Physics Galaxy*** series which include ***Lecture Notes*** which are designed for quick reference of video lectures available on the *YouTube channel of Physics Galaxy;* it can be easily accessed through ***www.physicsgalaxy.com*** or by the *mobile app of Physics Galaxy*.

The purpose of these *Lecture Notes* is to help students quickly revise the concepts in a systematic and coherent way after watching Video Lecture on any particular topic. In case of any confusion while understanding the solutions, students can simply scan the QR codes printed on each page of the book via 'Android' or 'iOS' device to quickly open the respective video lecture and understand the concept in a better way.

Furthermore, the handwritten format of the screenshots gives a better understanding of the concepts explained in the video lectures and helps in easily memorizing the concepts.

To sum up, by compiling screenshots of the video explanations, I have tried my best to make the entire content error free. However, owing to the nature of work, there is a slight possibility of inadvertent errors left untouched.

Therefore, I shall be grateful to the readers if they can point out errors and help me rectifying them for future editions. Feel free to give your valuable feedbacks at ***ashisharora@physicsgalaxy.com***.

***Date : August, 2018***

***Ashish Arora***

*PHYSICSGALAXY.COM*
*B-80, Model Town, Malviya Nagar, Jaipur-302017*
*e-mails: ashisharora@physicsgalaxy.com*
*ashash12345@gmail.com*

# CONTENTS


## 1 Motion in One Dimensions 1 – 24

### Lecture Notes Modules

- **Module-1:** Introduction
- **Module-2:** Instantaneous Velocity
- **Module-3:**Solved Example-1
- **Module-4:** Solved Example-2
- **Module-5:** Acceleration
- **Module-6:** Solved Example-3
- **Module-7:** Solved Example-4
- **Module-8:** Solved Example-5
- **Module-9:** Cases of Uniform Acceleration
- **Module-10:** Solved Example-6
- **Module-11:** Solved Example-7
- **Module-12:** Solved Example-8
- **Module-13:** Solved Example-9
- **Module-14:** Free Fall Motion
- **Module-15:** Solved Example-10
- **Module-16:** Concept of Relative Motion
- **Module-17:** Solved Example-11
- **Module-18:** Solved Example-12
- **Module-19:** Graphical Analysis of Motion
- **Module-20:** Displacement-Time Graphs
- **Module-21:** Velocity-Time Graphs
- **Module-22:** Acceleration-Time Graphs

## 2 Motion in Two Dimensions 25 – 37

### Lecture Notes Modules

- **Module-1:** Introduction
- **Module-2:** Solved Example-1
- **Module-3:** Solved Example-2
- **Module-4:** Solved Example-3
- **Module-5:** Trajectory of a 2D motion
- **Module-6:** Projectile Motion Under Gravity
- **Module-7:** Velocity of a Particle During Projectile
- **Module-8:** Position and Trajectory of Projectile
- **Module-9:** Solved Example-4
- **Module-10:** Solved Example-5
- **Module-11:** Solved Example-6
- **Module-12:** Solved Example-7
- **Module-13:** Projectile in Moving Reference Frame
- **Module-14:** Projectile on Inclined Plane

## 3 Concept of Relative Motion 39 – 51

### Lecture Notes Modules

- **Module-1:** Introduction to Relative Motion
- **Module-2:** Solved Example-1
- **Module-3:** Solved Example-2
- **Module-4:** Solved Example-3
- **Module-5:** Motion of a Body in Moving Ref Frame
- **Module-6:** Solved Example-4
- **Module-7:** Solved Example-5
- **Module-8:** Solved Example-6
- **Module-9:** Solved Example-7
- **Module-10:** Rain Fall Cases
- **Module-11:** Solved Example-8
- **Module-12:** Solved Example-9
- **Module-13:** Solved Example-10
- **Module-14:** Condition of Collision of Two Particle
- **Module-15:** Solved Example-11
- **Module-16:** Solved Example-12
- **Module-17:** Solved Example-13

## 4 Constrained Motion 53 – 63

### Lecture Notes Modules

- **Module-1:** Constrained Motion
- **Module-2:** Solved Example-1
- **Module-3:** Solved Example-2
- **Module-4:** Solved Example-3
- **Module-5:** Solved Example-4
- **Module-6:** Simple Pulley Constrained Motion
- **Module-7:** Solved Example-5
- **Module-8:** Solved Example-6
- **Module-9:** Solved Example-7
- **Module-10:** Wedge Constrained Motion
- **Module-11:** Solved Example-8
- **Module-12:** Solved Example-9
- **Module-13:** Solved Example-10

## 5 Newton's laws of Motion 65 – 88

### Lecture Notes Modules

- **Module-1:** Forces of Nature
- **Module-2:** Force can Change the State of Motion
- **Module-3:** Balanced & Unbalanced Forces
- **Module-4:** Newton's Laws of Motion
- **Module-5:** Free Body Diagram
- **Module-6:** Normal Reaction
- **Module-7:** Tension in String
- **Module-8:** Applications of Newton's Second Law
- **Module-9:** Solved Example-1
- **Module-10:** Solved Example-2
- **Module-11:** Solved Example-3
- **Module-12:** Solved Example-4
- **Module-13:** Solved Example-5
- **Module-14:** Concept of Weighing Machine
- **Module-15:** Spring Force
- **Module-16:** Concept of Spring Balance
- **Module-17:** Concept of Pseudo Force
- **Module-18:** Solved Example-6
- **Module-19:** Solved Example-7
- **Module-20:** Solved Example-8
- **Module-21:** Solved Example-9
- **Module-22:** Solved Example-10

## 6 Friction 89 – 98

### Lecture Notes Modules

- **Module-1:** Friction and its Classification
- **Module-2:** Types of Friction-Static Friction
- **Module-3:** Limiting & Kinetic Friction
- **Module-4:** Solved Example-1

- **Module-5:** Solved Example-2
- **Module-6:** Solved Example-3
- **Module-7:** Solved Example-4
- **Module-8:** Angle of Friction & Angle of Repose
- **Module-9:** Cases of Sliding on Rough Surface
- **Module-10:** Solved Example-6
- **Module-11:** Solved Example-7

## 7 Work Energy Power 99 – 121

### Lecture Notes Modules

- **Module-1:** Work & its Calculation
- **Module-2:** Work Done by a Variable Force
- **Module-3:** Solved Example-1
- **Module-4:** Solved Example-2
- **Module-5:** Energy & its Types
- **Module-6:** Conservative & Non-Conservative Forces
- **Module-7:** Potential Energy Stored in a Spring
- **Module-8:** Work-Energy Theorem
- **Module-9:** Relation in Force & Potential Energy
- **Module-10:** Solved Example-3
- **Module-11:** Solved Example-4
- **Module-12:** Solved Example-5
- **Module-13:** Work-Energy Equation for Rigid Bodies
- **Module-14:** Solved Example-6
- **Module-15:** Solved Example-7
- **Module-16:** Solved Example-8
- **Module-17:** Solved Example-9
- **Module-18:** Concept of Equilibrium
- **Module-19:** Power
- **Module-20:** Solved Example-10
- **Module-21:** Solved Example-11

## 8 Circular Motion 123 – 152

### Lecture Notes Modules

- **Module-1:** Circular Motion and Angular Properties
- **Module-2:** Relation in Linear and Angular Velocity
- **Module-3:** Angular Acceleration
- **Module-4:** Solved Example-1
- **Module-5:** Solved Example-2
- **Module-6:** Solved Example-3
- **Module-7:** Solved Example-4
- **Module-8:** Analysis of Acceleration in 2D Motion
- **Module-9:** Tangential and Normal Acceleration
- **Module-10:** Solved Example-5
- **Module-11:** Solved Example-6
- **Module-12:** Solved Example-7
- **Module-13:** Solved Example-8
- **Module-14:** Horizontal Circular Motion of a Conical Pendulum
- **Module-15:** Concept of Centrifugal Force
- **Module-16:** Turning of an Automobile on a Road
- **Module-17:** Solved Example-9
- **Module-18:** Solved Example-10
- **Module-19:** Solved Example-11
- **Module-20:** Solved Example-12
- **Module-21:** Vertical Circular Motion of a Pendulum Bob
- **Module-22:** Cases of Projection of a Pendulum Bob
- **Module-23:** Vertical Circular Motion of a Pendulum with a Rod
- **Module-24:** Circular Motion inside a Vertical Circular Track
- **Module-25:** Vertical Circular Motion inside a Tube
- **Module-26:** Vertical Circular Motion outside a Track
- **Module-27:** Solved Example-13
- **Module-28:** Solved Example-14
- **Module-29:** Solved Example-15
- **Module-30:** Solved Example-16
- **Module-31:** Solved Example-17

## 9 System of Particle-I 153 – 170

### Lecture Notes Modules

- **Module-1:** Centre of Mass
- **Module-2:** How to Locate Centre of Mass
- **Module-3:** Centre of Mass of Multiple Objects
- **Module-4:** Solved Example-1
- **Module-5:** Solved Example-2
- **Module-6:** Solved Example-3
- **Module-7:** Centre of Mass of Subtractive Objects
- **Module-8:** Centre of Mass of Continuous Extended Bodies
- **Module-9:** Centre of Mass of a Uniform Half Ring
- **Module-10:** Centre of Mass of a Uniform Half Disc
- **Module-11:** Centre of Mass of a Uniform Hollow Hemisphere
- **Module-12:** Centre of Mass of a Uniform Solid Hemisphere
- **Module-13:** Centre of Mass of a Uniform Hollow Cone
- **Module-14:** Centre of Mass of a Uniform Solid Cone
- **Module-15:** Solved Example-4
- **Module-16:** Solved Example-5
- **Module-17:** Solved Example-6
- **Module-18:** Displacement of Centre of Mass
- **Module-19:** Velocity and Acceleration of Centre of Mass
- **Module-20:** Solved Example-7
- **Module-21:** Solved Example-8

## 10 System of Particle-II 171 – 203

### Lecture Notes Modules

- **Module-1:** Conservation of Momentum
- **Module-2:** Solved Example-1
- **Module-3:** Solved Example-2
- **Module-4:** Solved Example-3
- **Module-5:** Effect of Internal Forces
- **Module-6:** Solved Example-4
- **Module-7:** Solved Example-5
- **Module-8:** Solved Example-6
- **Module-9:** Solved Example-7
- **Module-10:** Solved Example-8
- **Module-11:** Impulse & Conservation of Momentum
- **Module-12:** Solved Example-9
- **Module-13:** Solved Example-10
- **Module-14:** Solved Example-11
- **Module-15:** Collisions

- **Module-16:** Types of Collisions Based on Material Properties
- **Module-17:** Collision of Bodies of Different Materials
- **Module-18:** Coefficient of Restitution
- **Module-19:** Head-on Collisions
- **Module-20:** Solved Example-12
- **Module-21:** Partial Elastic Head-on Collision
- **Module-22:** Solved Example-13
- **Module-23:** Solved Example-14
- **Module-24:** Solved Example-15
- **Module-25:** Solved Example-16
- **Module-26:** Solved Example-17
- **Module-27:** Oblique Collision of a Ball on a Flat Surface
- **Module-28:** Oblique Collision in a 2D Plane
- **Module-29:** Solved Example-18
- **Module-30:** Solved Example-19
- **Module-31:** Solved Example-20
- **Module-32:** Mass Variation Cases
- **Module-33:** Super Elastic Collision

## 11 Rigid Body Dynamics 205 – 244

### Lecture Notes Modules

- **Module-1:** Rotational Motion
- **Module-2:** Moment of Inertia
- **Module-3:** Moment of Inertia of a Ring
- **Module-4:** Moment of Inertia of a Disc
- **Module-5:** Moment of Inertia of a Hollow Sphere
- **Module-6:** Moment of Inertia of a Solid Sphere
- **Module-7:** Moment of Inertia of a Hollow Cone
- **Module-8:** Moment of Inertia of a Solid Cone
- **Module-9:** Moment of Inertia of a Thin Rod
- **Module-10:** Axes Theorem-Perpendicular Axes Theorem
- **Module-11:** Axes Theorem-Parallel Axes Theorem
- **Module-12:** Radius of Gyration
- **Module-13:** Solved Example-1
- **Module-14:** Solved Example-2
- **Module-15:** Solved Example-3
- **Module-16:** Solved Example-4
- **Module-17:** Newton's Second Law in Rotational Motion
- **Module-18:** Solved Example-5
- **Module-19:** Solved Example-6
- **Module-20:** Solved Example-7
- **Module-21:** Solved Example-8
- **Module-22:** Solved Example-9
- **Module-23:** Angular Momentum
- **Module-24:** Conservation of Angular Momentum
- **Module-25:** Solved Example-10
- **Module-26:** Solved Example-11
- **Module-27:** Solved Example-12
- **Module-28:** Solved Example-13
- **Module-29:** Concept ofAngular Impulse
- **Module-30:** Kinetic Energy of a Rigid body in Rotational Motion
- **Module-31:** Rolling Motion and Pure Rolling of a body on Inclined Plane
- **Module-32:** Solved Example-14
- **Module-33:** Solved Example-15
- **Module-34:** Solved Example-16
- **Module-35:** Solved Example-17
- **Module-36:** Solved Example-18
- **Module-37:** Analysis of Rolling with Sliding
- **Module-38:** Solved Example-19
- **Module-39:** Solved Example-20
- **Module-40:** Work & Power in Rotational Motion
- **Module-41:** Solved Example-21
- **Module-42:** Solved Example-22

## 12 Fluid Statics 245 – 262

### Lecture Notes Modules

- **Module-1:** Introduction to Fluids
- **Module-2:** Pressure in a Fluid
- **Module-3:** Floatation of a Body due to Fluid Pressure
- **Module-4:** Fluid Pressure due to Gravity
- **Module-5:** Average Pressure on Side Walls of a Vessel
- **Module-6:** Force on Side Walls of a Random Shaped Vessel
- **Module-7:** Archimede's Principle
- **Module-8:** Buoyant Force on a Body at the Interface ofTwo Liquids
- **Module-9:** Apparent Weight ofa Body Submerged in a Liquid
- **Module-10:** Solved Example-1
- **Module-11:** Solved Example-2
- **Module-12:** Solved Example-3
- **Module-13:** Solved Example-4
- **Module-14:** Solved Example-5
- **Module-15:** Pascal's Principle of Pressure Distribution
- **Module-16:** Pressure at Different Levels in a Liquid
- **Module-17:** Mechanism of a Hydraulic Lift
- **Module-18:** Concept of Open tube Manometer
- **Module-19:** Pressure Distribution in an Accelerated Frame
- **Module-20:** Pressure Distribution in a Closed Accelerated Container
- **Module-21:** Pressure Distribution in a Rotating tube Filled with a Liquid
- **Module-22:** Solved Example-6
- **Module-23:** Solved Example-7
- **Module-24:** Solved Example-8
- **Module-25:** Solved Example-9

## 13 Fluid Dynamics 263 – 275

### Lecture Notes Modules

- **Module-1:** Introduction to Fluid Dynamics
- **Module-2:** Representation of Streamlines
- **Module-3:** Velocity Profile of a Flowing Fluid
- **Module-4:** Turbulent Flow
- **Module-5:** Equation of Continuity
- **Module-6:** Solved Example-1
- **Module-7:** Bernoulli's Theorem
- **Module-8:** Pressure-Velocity Trade-off

- **Module-9:** Solved Example-2
- **Module-10:** Solved Example-3
- **Module-11:** Solved Example-4
- **Module-12:** Solved Example-5
- **Module-13:** Working and Application of Pitot Tube
- **Module-14:** Working and Application of Venturimeter
- **Module-15:** Solved Example-6
- **Module-16:** Solved Example-7
- **Module-17:** Torecelli's Theorem
- **Module-18:** Solved Example-8
- **Module-19:** Solved Example-9
- **Module-20:** Cross-section of a Freely Falling Liquid
- **Module-21:** Reaction due to Ejection ofa Liquid
- **Module-22:** Solved Example-10
- **Module-23:** Solved Example-11

## 14 Surface Tension 277 – 291

### Lecture Notes Modules

- **Module-1:** Surface Tension
- **Module-2:** Significance of Surface Tension
- **Module-3:** Some Facts About Surface Tension
- **Module-4:** Solved Example-1
- **Module-5:** Solved Example-2
- **Module-6:** Concept of Surface Energy
- **Module-7:** Solved Example-3
- **Module-8:** Solved Example-4
- **Module-9:** Excess Pressure Inside a Liquid Drop
- **Module-10:** Excess Pressure Inside a Soap Bubble
- **Module-11:** Excess Pressure Inside a Surface Having Two Curvatures
- **Module-12:** Connecting Soap Bubbles of Different Radii
- **Module-13:** Angle of Contact
- **Module-14:** Shape of Liquid Meniscus in a Tube
- **Module-15:** CapillaryAction
- **Module-16:** Force Between Two Parallel Plates Due to Liquid Between Them
- **Module-17:** CapillaryAction with a Capillary of Insufficient Height
- **Module-18:** Solved Example-5
- **Module-19:** Solved Example-6
- **Module-20:** Solved Example-7
- **Module-21:** Solved Example-8
- **Module-22:** Solved Example-9

## 15 Viscosity 293 – 300

### Lecture Notes Modules

- **Module-1:** Viscosity
- **Module-2:** Some Facts about Viscosity
- **Module-3:** Flowing Fluid Relative to a Solid Surface in Contact
- **Module-4:** Stoke's Law and Terminal Velocity
- **Module-5:** Critical Velocity and Reynold's Number
- **Module-6:** Poisueille's Equation
- **Module-7:** Solved Example-1
- **Module-8:** Solved Example-2
- **Module-9:** Solved Example-3
- **Module-10:** Solved Example-4
- **Module-11:** Solved Example-5

## 16 Elasticity 301 – 316

### Lecture Notes Modules

- **Module-1:** Introduction to Elasticity
- **Module-2:** Rigid Body
- **Module-3:** Elastic and Partially Elastic Bodies
- **Module-4:** Inelastic or Plastic Bodies
- **Module-5:** Stress due to a Force
- **Module-6:** Types of Stress
- **Module-7:** Strain
- **Module-8:** Types of Strain
- **Module-9:** Relation Between Stress and Strain
- **Module-10:** Hooke's Law and Modulus of Elasticity
- **Module-11:** Types of Modulus of Elasticity
- **Module-12:** Solved Example-1
- **Module-13:** Solved Example-2
- **Module-14:** Solved Example-3
- **Module-15:** Solved Example-4
- **Module-16:** Solved Example-5
- **Module-17:** Solved Example-6
- **Module-18:** Some Facts about Elasticity
- **Module-19:** Elongation of Suspended Rod Under SelfWeight
- **Module-20:** Metal Wires of Rods as Equivalent Spring
- **Module-21:** Elastic Potential Energy Stored due to Elasticity
- **Module-22:** Solved Example-7
- **Module-23:** Solved Example-8

## 17 Gravitation Theory 317 – 349

### Lecture Notes Modules

- **Module-1:** Introduction to Gravitation
- **Module-2:** Limitations of Newton's Law of Gravitation
- **Module-3:** Vector Representation of Gravitational Force
- **Module-4:** Principle of Superposition of Forces
- **Module-5:** Solved Example-1
- **Module-6:** Solved Example-2
- **Module-7:** Solved Example-3
- **Module-8:** Gravitational Field
- **Module-9:** Gravitational Field Strength due to a Point
- Mass and Extended Bodies
- **Module-10:** Gravitational Field Strength due to a Uniform Rod
- **Module-11:** Gravitational Field Strength due to a Uniform Ring
- **Module-12:** Gravitational Field Strength due to a CircularArc
- **Module-13:** Gravitational Field Strength due to a Hollow Sphere

- **Module-14:** Gravitational Field Strength due to a Solid Sphere
- **Module-15:** Gravitational Field Strength due to a Long Cylinder
- **Module-16:** Solved Example-4
- **Module-17:** Solved Example-5
- **Module-18:** Solved Example-6
- **Module-19:** Solved Example-7
- **Module-20:** Gravitational Lines of Forces
- **Module-21:** Gravitational Field on Earth Surface
- **Module-22:** Value of g at a Height h Above Earth Surface
- **Module-23:** Value of g at a Height h Below Earth Surface
- **Module-24:** Effect of Earth's Rotation on Value of g
- **Module-25:** Effect of Shape of Earth on Value of g
- **Module-26:** Solved Example-8
- **Module-27:** Solved Example-9
- **Module-28:** Solved Example-10
- **Module-29:** Solved Example-11
- **Module-30:** Gravitational Interaction Energy
- **Module-31:** Gravitational Interaction Energy for a System of Particles
- **Module-32:** Solved Example-12
- **Module-33:** Solved Example-13
- **Module-34:** Gravitational Potential
- **Module-35:** Relation in g and V in Gravitational Field
- **Module-36:** Gravitational Potential due to a Point Mass
- **Module-37:** Gravitational Potential due to a Uniform Ring
- **Module-38:** Work Done in Gravitational Field
- **Module-39:** Gravitational Potential due to a Hollow Sphere
- **Module-40:** Gravitational Potential due to a Solid Sphere
- **Module-41:** Gravitational Potential Energy of a Body on Earth
- **Module-42:** Solved Example-14
- **Module-43:** Solved Example-15
- **Module-44:** Solved Example-16
- **Module-45:** Solved Example-17
- **Module-46:** Solved Example-18
- **Module-47:** Gravitational Self Energy of a Body
- **Module-48:** Gravitational Self Energy of a Uniform Hollow Sphere
- **Module-49:** Gravitational Self Energy of a Uniform Solid Sphere

## 18 Satellite Motion 351 – 367

### Lecture Notes Modules

- **Module-1:** Introduction to Satellite Motion
- **Module-2:** Energy of a Satellite in Circular Orbit
- **Module-3:** Solved Example-1
- **Module-4:** Solved Example-2
- **Module-5:** Solved Example-3
- **Module-6:** Bounded Motion and Escape Velocity
- **Module-7:** Escaping a Satellite
- **Module-8:** Solved Example-4
- **Module-9:** Solved Example-5
- **Module-10:** Motion of a Satellite in Elliptical Path
- **Module-11:** Conservation of Angular Momentum in Satellite Motion
- **Module-12:** Kepler's Laws of Planetary Motion-First Law
- **Module-13:** Kepler's Laws of Planetary Motion-Second Law
- **Module-14:** Kepler's Laws of Planetary Motion-Third Law
- **Module-15:** Projection of Spaceships and Satellites from Earth
- **Module-16:** Solved Example-6
- **Module-17:** Solved Example-7
- **Module-18:** Solved Example-8
- **Module-19:** Solved Example-9
- **Module-20:** Communication Satellites
- **Module-21:** Geostationary Satellite
- **Module-22:** Broadcasting Region of a Satellite
- **Module-23:** Solved Example-10
- **Module-24:** Solved Example-11
- **Module-25:** Solved Example-12

# Chapter 1
# Motion in One Dimensions

## Lecture Notes Modules


**Module-1:** Introduction
**Module-2:** Instantaneous Velocity
**Module-3:** Solved Example-1
**Module-4:** Solved Example-2
**Module-5:** Acceleration
**Module-6:** Solved Example-3
**Module-7:** Solved Example-4
**Module-8:** Solved Example-5
**Module-9:** Cases of Uniform Acceleration
**Module-10:** Solved Example-6
**Module-11:** Solved Example-7
**Module-12:** Solved Example-8
**Module-13:** Solved Example-9
**Module-14:** Free Fall Motion
**Module-15:** Solved Example-10
**Module-16:** Concept of Relative Motion
**Module-17:** Solved Example-11
**Module-18:** Solved Example-12
**Module-19:** Graphical Analysis of Motion
**Module-20:** Displacement-Time Graphs
**Module-21:** Velocity-Time Graphs
**Module-22:** Acceleration-Time Graphs

**Module-1: Introduction**

# Rectilinear Motion:

Motion of a body along a St. line

↓

Motion in One Dimension (1D)

# Refrence frame:

It is a region/zone from where observations are taken for motion parameters of a body.

[ displacement, velocity, accn, time ]

* It can be at rest or may be moving.

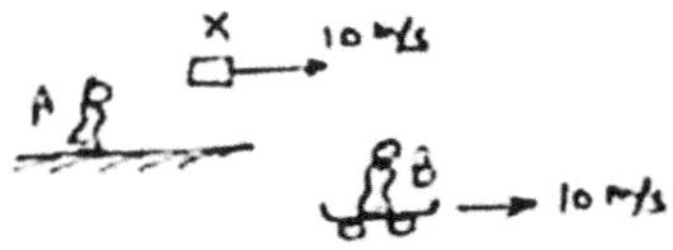


Scan for Video Explanation

Web Reference at www.physicsgalaxy.com

**Module-1: Introduction**

Ex:

ref frame → ground

Speed of Car w.r.to ground frame is $v_{CA} = 3$ m/s

ref. frame → trolley.

Speed of car w.r.to ref frame of trolley $v_{CB} = 2$ m/s

"A ref frame is a co-ordinate system with a clock for measurement of time. It may be at rest on ground or may be attached to a moving body/observer."

Scan for Video Explanation

Web Reference at www.physicsgalaxy.com

**Module-1: Introduction**

# Motion Parameters:

## Distance & Displacement:

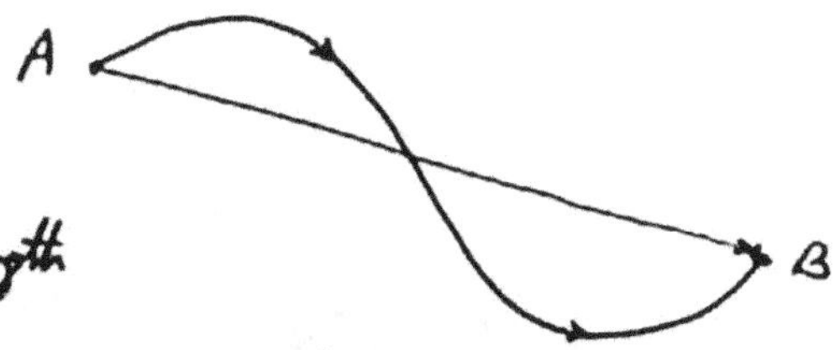


distance covered
↳ path length

Displacement
↳ $\vec{AB}$.

Rectilinear motion

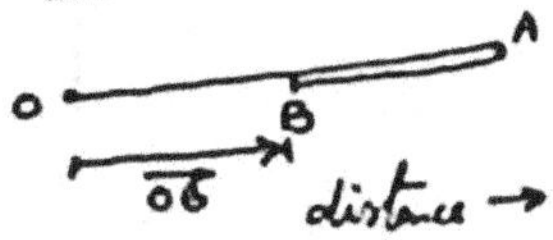


distance → OAB
displacement → $\vec{OB}$

## Speed & Velocity:

Speed → rate of change of position (magnitude)

Velocity → Rate of change of position (mag & dir)




**Module-1: Introduction**

## Average Speed & Average Velocity:

$$\text{Average Speed} = \frac{\text{Total distance travelled}}{\text{Time taken}}$$

$$\text{Average Velocity} = \frac{\text{Total displacement}}{\text{time taken}}$$

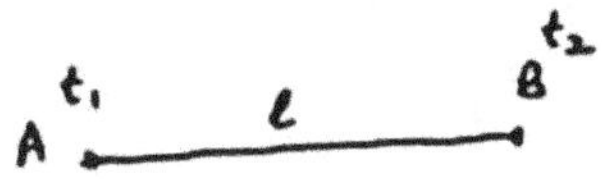


$$\text{Avg speed} = \frac{l}{t_2 - t_1} \leftarrow \text{Avg vel.}$$

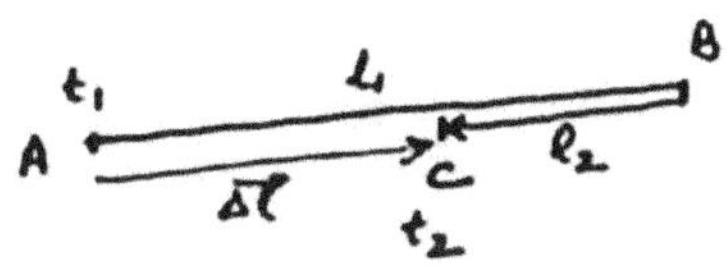


$$\text{Avg speed} = \frac{l_1 + l_2}{t_2 - t_1}$$

$$\text{Avg vel} = \frac{l_1 - l_2}{t_2 - t_1}$$

$\Bigg\} \neq \left(\frac{\Delta l}{\Delta t}\right)$ const

Web Reference at www.physicsgalaxy.com
Scan for Video Explanation

**Module-2: Instantaneous Velocity**

# Instataneous Velocity:

It is the velocity of a particle at an instant.

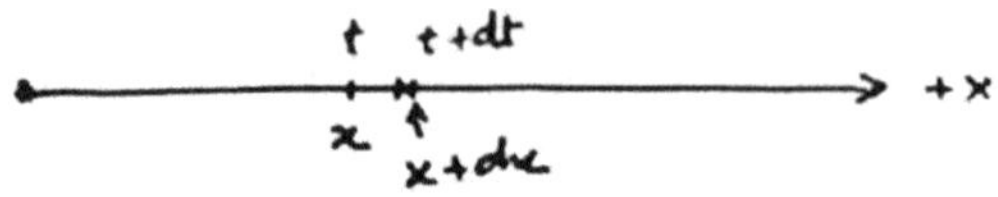


at the instant t instantaneous vel can be given

$$\vec{v} = \frac{d\vec{x}}{dt}$$ ← rate of change of position w.r.to time.

$y = f(x)$, $\left(\frac{dy}{dx}\right)$

Inst vel → differentiation of position w.r.to time

for a given motion if position $x = f(t)$ then inst vel is the slope of x v/s t curve.

Scan for Video Explanation

Web Reference at www.physicsgalaxy.com

**Module-2: Instantaneous Velocity**

**Ex :** If a particle moves such that its displacement is given as

$$x = 4t^4 - 3t^2 + 7t + 6$$

Find its velocity at t = 2 sec.

Sol$^n$.

$$v = \frac{dx}{dt} = 16t^3 - 6t + 7$$

$$= 16 \times 8 - 6 \times 2 + 7$$

$$= 128 - 12 + 7 = 123 \text{ m/s}$$

NOTE: As differentiation of displacement is velocity, on integrating velocity we get displacement.

Scan for Video Explanation

Web Reference at www.physicsgalaxy.com

**Module-3: Solved Example-1**

**Ex :** If velocity of a particle is given as

$$v = (3t^2 + 2) \text{ m/s}$$

Find the displacement covered by the particle from t = 1 to t = 2 sec.

Sol:

$$v = 3t^2 + 2$$

$$\frac{dx}{dt} = 3t^2 + 2$$

integrating both sides

$$\int_{x_1}^{x_2} dx = \int_{t=1}^{t=2} (3t^2 + 2)\, dt$$

$$x_2 - x_1 = \left[t^3 + 2t\right]_1^2$$

$$x_2 - x_1 = (8 + 4) - (1 + 2)$$

$$= 12 - 3 = 9\text{m Ans.}$$

Web Reference at www.physicsgalaxy.com | Scan for Video Explanation

**Module-4: Solved Example-2**

**Ex :** A body moves at velocity v = 2t + 4 m/s. Find the average velocity of body from t = 1s to t = 2s. Initially at t = 0 body starts from origin. ($x = 0$)

Sol: let at $t_1 = 1$ sec $x = x_1$

& at $t_2 = 2$ sec $x = x_2$

avg vel $v = \frac{x_2 - x_1}{t_2 - t_1} = \frac{x_2 - x_1}{1} = x_2 - x_1$

$v_{avg} = 7$ m/s Ans.

$$\frac{dx}{dt} = 2t + 4$$

$$\int_{x_1}^{x_2} dx = \int_1^2 (2t + 4)\, dt$$

$$x_2 - x_1 = \left[t^2 + 4t\right]_1^2 = (4 + 8) - (1 + 4)$$

$$= 12 - 5 = 7$$

Web Reference at www.physicsgalaxy.com | Scan for Video Explanation

**Module-5: Acceleration**

# Acceleration : → Rate of change of velocity

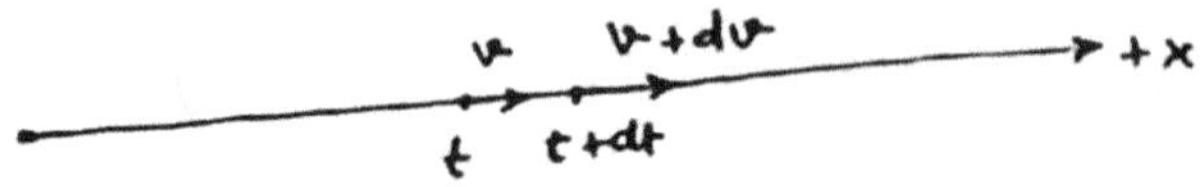


acc$^n$ at time t can be given as $\boxed{\vec{a} = \frac{d\vec{v}}{dt}}$ → the derivative of $\vec{v}$ or slope of v v/s t curve.

disp $\xrightarrow{d}$ vel $\xrightarrow{d}$ acc$^n$ (reverse: $\int$, $\int$)

$$a = \frac{dv}{dt} = \frac{d^2x}{dt^2}$$

$$a = \frac{dv}{dt} \times \frac{dx}{dx} = v\frac{dv}{dx}$$

acc$^n$ = inst vel × slope of v v/s x curve.

Scan for Video Explanation

Web Reference at www.physicsgalaxy.com

**Module-5: Acceleration**

$$a = \frac{dv}{dt} \quad —(1) \qquad \text{OR} \qquad a = v\frac{dv}{dx} \quad —(2)$$

If a is given as a f$^n$ of time → we use eq$^n$ (1)

If a is given as a f$^n$ of disp → we use eq$^n$ (2)

If a is given in terms of v → either eq$^n$ (1) or eq$^n$ (2)

**Ex:** If for a part its disp is $x = 3t^3 + 2t^2 + 5t + 6$ m find acc$^n$ of part at t = 1 sec

**Sol$^n$:** Inst vel $v = \frac{dx}{dt} = 9t^2 + 4t + 5$ m/s

acc$^n$ $a = \frac{dv}{dt} = 18t + 4$ m/s$^2$

at t = 1 sec $a = 18 + 4 = 22$ m/s$^2$ Ans.

Scan for Video Explanation

Web Reference at www.physicsgalaxy.com

**Module-5: Acceleration**

# Average Acceleration:

For a motion of particle between two instants its avg accⁿ can be given as-

$$\langle \bar{a}_{avg} \rangle = \frac{\text{Total change in vel}}{\text{Time Taken}} = \frac{\Delta \vec{v}}{\Delta t} = \frac{\vec{v}_2 - \vec{v}_1}{t_2 - t_1}$$

Ex. If a truck moving at 4 m/s toward south takes a turn toward east & its vel changes to 3 m/s. If turning time is 10 sec, find the avg accⁿ of truck.

Solⁿ: $\vec{v}_1 = 4\ m/s \downarrow$

$\vec{v}_2 = 3\ m/s \rightarrow$

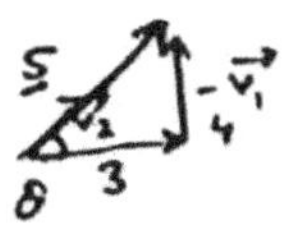


$|\vec{a}_{avg}| = \frac{|\vec{v}_2 - \vec{v}_1|}{10} = \frac{5}{10} = 0.5\ m/s^2$ in dir 53° N of E.

$\tan\theta = 4/3$

$\theta = 53°$

Web Reference at www.physicsgalaxy.com | Scan for Video Explanation

**Module-6: Solved Example-3**

**Ex :** If a car accelerats with acceleration depends on displacement as

$$a = 6x^2 + 2x + 5$$

Find velocity of particles when it is at x = 3m. Given that initially it starts from x = 1 with velocity 2 m/s.

Solⁿ:

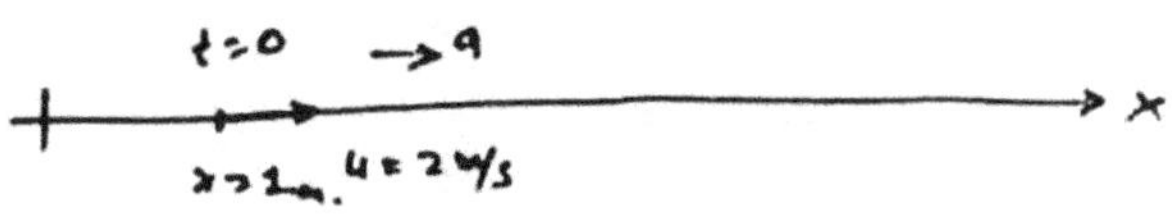


We use $a = v\frac{dv}{dx} = 6x^2 + 2x + 5$

$$\int_2^v v\,dv = \int_1^3 (6x^2 + 2x + 5)\,dx$$

$$\left[\frac{v^2}{2}\right]_2^v = \left[2x^3 + x^2 + 5x\right]_1^3$$

$$\frac{v^2}{2} - 2 = (2\times 27 + 9 + 15) - (2 + 1 + 5)$$

$$v^2 = (78 - 8 + 2)\times 2$$

$v^2 = (72)\times 2$

$= 144$

$v = \sqrt{144} = 12\ m/s.$

Web Reference at www.physicsgalaxy.com | Scan for Video Explanation

**Module-7: Solved Example-4**

Ex : A particle moves with acceleration given as

$$a = 8t + 6 \text{ m/s}^2$$

Find displacement of particles at t = 3 sec if at t = 0 it starts from origin with initial speed 1 m/s.

Sol$^n$:

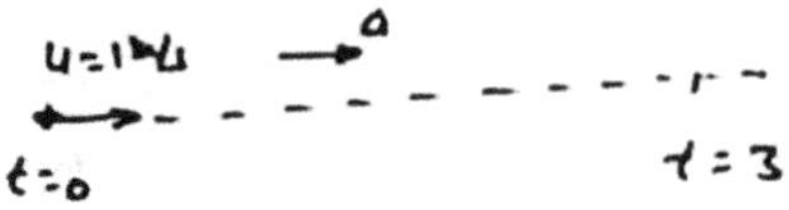


we use $\frac{dv}{dt} = 8t + 6$

$$\int_1^v dv = \int_0^t (8t + 6)\, dt$$

$$v - 1 = \frac{8t^2}{2} + 6t \Rightarrow v = 4t^2 + 6t + 1 \text{ m/s}$$

$$\frac{dx}{dt} = 4t^2 + 6t + 1$$

$$\int_0^x dx = \int_0^t (4t^2 + 6t + 1)\, dt$$

$$x = \frac{4t^3}{3} + 3t^2 + t$$

at t = 3 sec

$$x = \frac{4(3)^3}{3} + 3(3)^2 + 3$$

$$= 36 + 27 + 3 = 66 \text{ m. Ans.}$$

Scan for Video Explanation

Web Reference at www.physicsgalaxy.com

**Module-8: Solved Example-5**

Ex : Instantaneous velocity of a particle is given as

$$v = 5e^t \text{ m/s}$$

Find the average acceleration and average velocity of particle between instants $t_1$ = 2 sec and $t_2$ = 4 sec.

(Given at t = 0, x = 0)

Sol$^n$:

$$\text{avg acc}^n = \frac{\vec{v}_2 - \vec{v}_1}{t_2 - t_1} = \frac{\Delta\vec{v}}{\Delta t} = \frac{5e^4 - 5e^2}{4 - 2}$$

$$= \frac{5}{2}e^2(e^2 - 1) \text{ m/s}^2 \text{ Ans.}$$

$$\text{Avg vel} = \frac{(x_2 - x_1)}{t_2 - t_1}$$

$$\frac{dx}{dt} = 5e^t$$

$$\int_0^{x_1} dx = \int_0^2 5e^t\, dt \longrightarrow x_1 = 5[e^t]_0^2 = 5e^2 - 5$$

$$\int_0^{x_2} dx = \int_0^4 5e^t\, dt \longrightarrow x_2 = 5[e^t]_0^4 = 5e^4 - 5$$

$$v_{avg} = \frac{(5e^4 - 5) - (5e^2 - 5)}{4 - 2}$$

$$= \frac{5}{2}e^2(e^2 - 1) \text{ m/s Ans.}$$

Scan for Video Explanation

Web Reference at www.physicsgalaxy.com

**Module-9: Cases of Uniform Acceleration**

# Analysis of Motion with Uniform Acceleration:

When a particle moves at constant accⁿ a & starts with initial velocity u, the vel of particle after time t can be obtained as-

$$a = \frac{dv}{dt}$$

$$\int_u^v dv = \int_{t=0}^{t} a\,dt$$

$$v - u = at$$

$$\boxed{v = u + at} \quad \text{—(1)}$$

First Speed Equation for constantly accelerated motion.

Web Reference at www.physicsgalaxy.com | Scan for Video Explanation

**Module-9: Cases of Uniform Acceleration**

Further we use

$$v = u + at$$

$$\frac{dx}{dt} = u + at$$

$$\int_0^s dx = \int_0^t (u + at)\,dt$$

$$s = ut + \frac{at^2}{2}$$

$$\boxed{s = ut + \tfrac{1}{2}at^2} \quad \text{—(2)}$$

Second Speed eqⁿ for uniformly accelerated motion.

Web Reference at www.physicsgalaxy.com | Scan for Video Explanation

**Module-9: Cases of Uniform Acceleration**

Similarly if we use $a = v\frac{dv}{dx}$

$$\int_u^v v\,dv = \int_0^s a\,dx$$

$$\left[\frac{v^2}{2}\right]_u^v = as$$

$$\frac{v^2 - u^2}{2} = as$$

$$\boxed{v^2 = u^2 + 2as} \quad \text{—(3)}$$

Third Speed Eqn for Uniformly accelerated motion.

Scan for Video Explanation

Web Reference at www.physicsgalaxy.com

**Module-9: Cases of Uniform Acceleration**

# Displacement of a particle during nth second for uniformly accelerated motion:

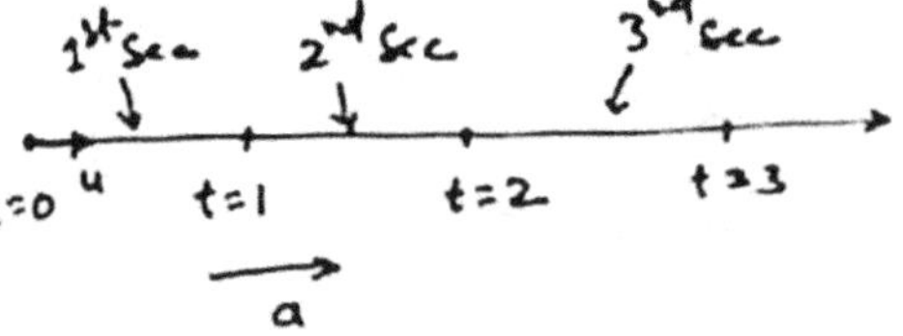


nth Second ↓ t=n-1 to t=n

disp. covered in nth Second of motion

$$\overrightarrow{\Delta S_n} = \vec{S}_n - \vec{S}_{n-1}$$

$$= \left[un + \tfrac{1}{2}an^2\right] - \left[u(n-1) + \tfrac{1}{2}a(n-1)^2\right]$$

$$= un + \tfrac{1}{2}an^2 - un + u - \tfrac{1}{2}an^2 - \tfrac{1}{2}a + an$$

$$\boxed{\Delta S_n = u + \tfrac{1}{2}a(2n-1)}$$

Scan for Video Explanation

Web Reference at www.physicsgalaxy.com

**Module-9: Cases of Uniform Acceleration**

NOTE: All the Speed Equations are Vector Equations & the motion parameters $\vec{s}, \vec{v}, \vec{u}, \vec{a}$ are taken with appropriate signs acc to the ref frame chosen.

(1) $\vec{v} = \vec{u} + \vec{a}t$

(2) $\vec{s} = \vec{u}t + \frac{1}{2}\vec{a}t^2$

(3) $v^2 = u^2 + 2\vec{a}.\vec{s}$

(4) $\vec{\Delta s_n} = \vec{u} + \frac{\vec{a}}{2}(2n-1)$

Web Reference at www.physicsgalaxy.com | Scan for Video Explanation

**Module-10: Solved Example-6**

**Ex :** A car is moving at 20 m/s. The driver applies the brakes which causes a retardation of 4 m/s$^2$. Find the distance travelled by car before it stops, if reaction time of driver is 0.4 sec.

Sol: Reaction time : The time taken by a human to execute the action after one sees a need of it.

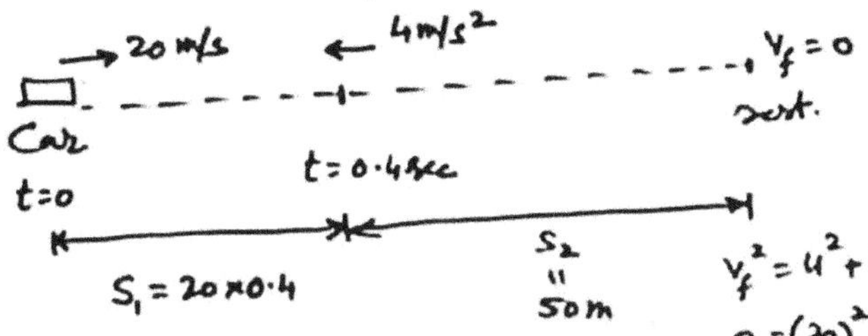


$S_1 = 20 \times 0.4 = 8$ m.

$v_f^2 = u^2 + 2aS$

$0 = (20)^2 - 8S_2$

$S_2 = \frac{400}{8} = 50$ m

$S_1 + S_2 = 8 + 50 = 58$ m Ans.

Web Reference at www.physicsgalaxy.com | Scan for Video Explanation

**Module-11: Solved Example-7**

**Ex :** A particle starts at initial speed 17 m/s and retards at 2 m/s$^2$. Find the distance covered by the particle in 9$^{th}$ second of motion.

$t = 8$ to $t = 9$

Sol$^n$:

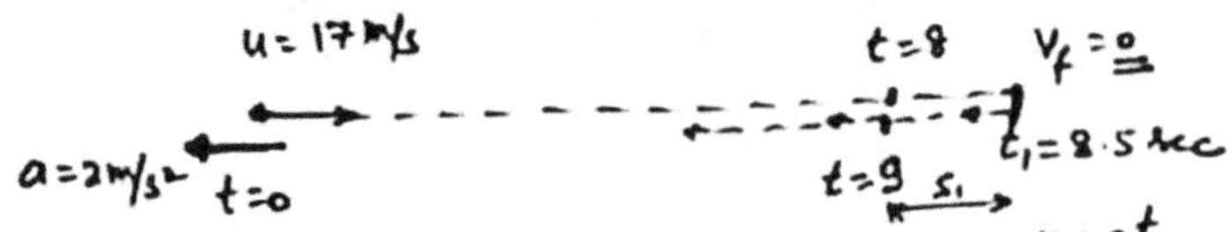


$v_f = u + at$

$0 = 17 - 2t_1$

$t_1 = \frac{17}{2} = 8.5$ sec

disp of part in $n^{th}$ sec

✓ $\overrightarrow{\Delta S_n} = u + \frac{a}{2}(2n-1)$

✓ $\overrightarrow{\Delta S_9} = 17 + \left(\frac{-2}{2}\right)(2\times 9 - 1)$

$= 17 - 17 = 0$ ✓

$s_1 = \frac{1}{2}at^2$

$= \frac{1}{2}\times 2\times(0.5)^2$

$= 0.25$ m

distance covered in 9$^{th}$ sec $= 2s_1 = 0.5$ m Ans.

Scan for Video Explanation

Web Reference at www.physicsgalaxy.com

**Module-12: Solved Example-8**

**Ex :** A car and a truck starts from same point along same direction, car with initial speed 2 m/s & acceleration 2 m/s$^2$ and truck with a constant speed 20 m/s$^2$. Find the time after which car will overtake the truck.

Sol$^n$:

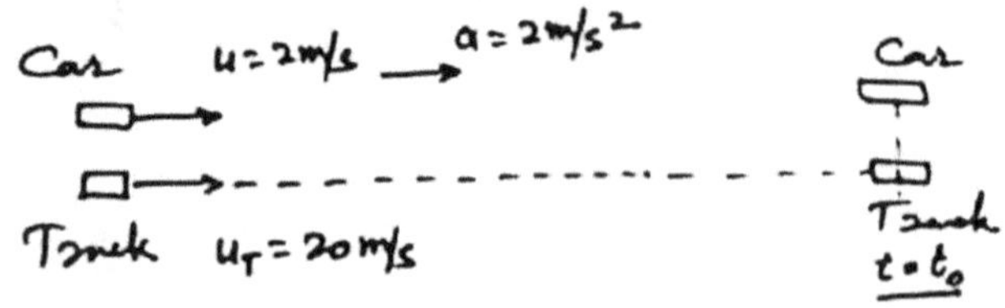


$t = 18$ sec

displacements Truck $S_T = 20t_0$

Car $S_C = 2t_0 + \frac{1}{2}(2)t_0^2$

for overtaking $S_T = S_C$

$20t_0 = 2t_0 + t_0^2$

$t_0 = 18$ sec Ans.

Scan for Video Explanation

Web Reference at www.physicsgalaxy.com

**Module-13: Solved Example-9**

**Ex :** **Find the ratio of distances travelled in 1st, 2nd, & 3rd second by a particle started from rest with uniform acceleration.**

Sol:

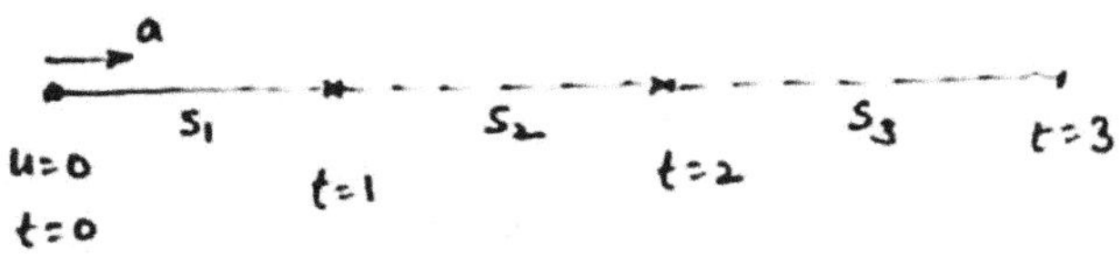


$s_1 : s_2 : s_3 = ?$

$$\vec{\Delta S_n} = \vec{u} + \frac{\vec{a}}{2}(2n-1)$$

$$s_1 = \frac{a}{2}(2-1) = a/2$$

$$s_2 = \frac{a}{2}(3) = 3a/2$$

$$s_3 = \frac{a}{2}(5) = 5a/2$$

$s_1 : s_2 : s_3 = 1 : 3 : 5$ Ans.

Web Reference at www.physicsgalaxy.com | Scan for Video Explanation

**Module-14: Free Fall Motion**

# Free Fall Motion:

Motion of a body moving freely under Earth's gravity is called free fall motion

under free fall acc$^n$ = $g \downarrow$

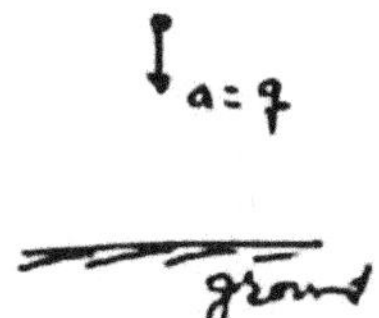


if we take downward dir +ve $a = +g$

$$v = u + gt$$
$$s = ut + \frac{1}{2}gt^2$$
$$v^2 = u^2 + 2gh$$
$$\Delta h_n = u + \frac{g}{2}(2n-1)$$

if we take upward dir +ve $a = -g$.

$$v = u - gt$$
$$s = ut - \frac{1}{2}gt^2$$
$$v^2 = u^2 - 2gh$$
$$\Delta h_n = u - g/2(2n-1)$$

Web Reference at www.physicsgalaxy.com | Scan for Video Explanation

**Module-14: Free Fall Motion**

**Ex :** A boy throws a ball up at speed u. Find :
(a) time to reach max height.
(b) max height attained.

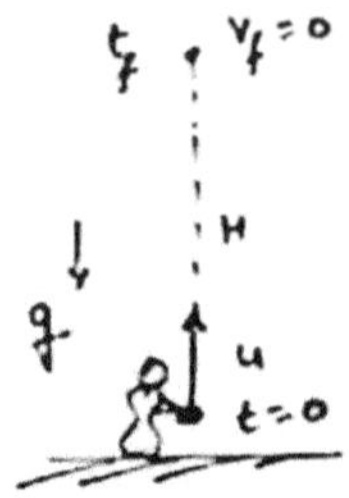


Sol$^n$:

We use $v = u + at$

$0 = u + (-g)t_f$

(a) $\boxed{t_f = u/g}$ Ans.

We use $v^2 = u^2 + 2as$

$0 = u^2 + 2(-g)H$

(b) $\boxed{H = \frac{u^2}{2g}}$ Ans.

Total time of flight

$\boxed{T = 2t_f = \frac{2u}{g}}$

V. imp.

Scan for Video Explanation

Web Reference at www.physicsgalaxy.com

**Module-15: Solved Example-10**

**Ex :** A stone is released from rest from a height above the ground. It covers a distance 6 m in last second of its motion. Find the height from which it was dropped.

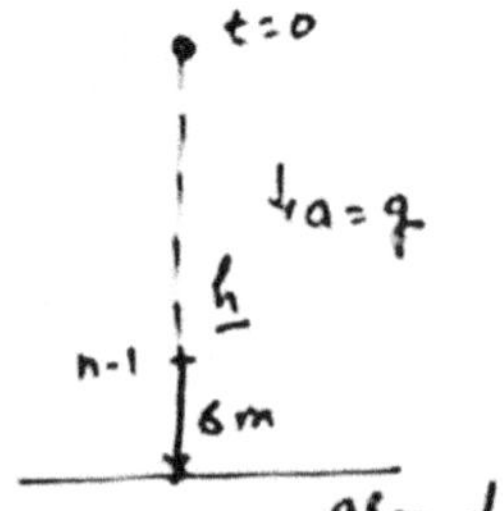


$(g = 10\ m/s^2)$

Sol$^n$: Let stone falls for n secs.

⇒ in $n^{th}$ sec.

$\Delta S_n = u + \frac{a}{2}(2n-1)$

$\Delta S_n = 6 = \frac{g}{2}(2n-1)$

$2n - 1 = 6/5$

$n = \frac{11}{10}$ sec.

Total height

$h = \frac{1}{2}gt^2$

$= \frac{1}{2} \times 10 \times \left(\frac{11}{10}\right)^2 = \frac{121}{20}$ m Ans.

Scan for Video Explanation

Web Reference at www.physicsgalaxy.com

**Module-16: Concept of Relative Motion**

# Concept of Relative Motion :

Relative Motion → observation w.r.to a given reference.

Case-I: When bodies move in same dir

A $v_A$ → B $v_B$ →

$$\boxed{v_{BA} = v_B - v_A}$$

↑ relative vel of body B w.r.to body A.

Case-II: When bodies move in opp. dir

A $v_A$ → ← $v_B$ B

$$\boxed{v_{BA} = v_B + v_A}$$

→ Same concept of Relative motion can be applied to acc^n of bodies also.



Scan for Video Explanation

**Module-16: Concept of Relative Motion**

**Ex :** In the situation shown in figure if both the boys throw balls at same time, find the time when the balls will collide.

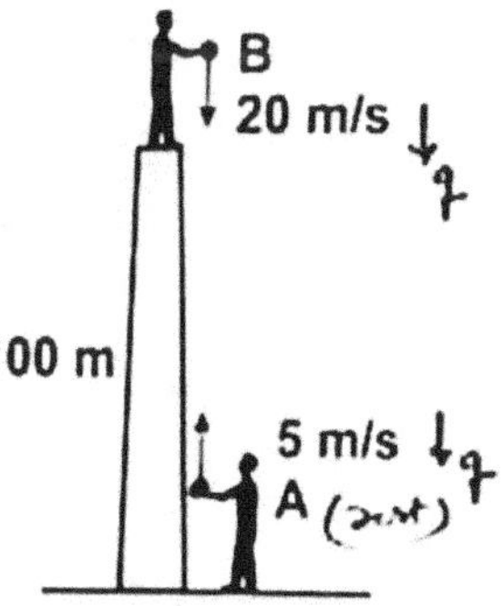


Sol^n:

relative vel of B w.r.to A

$v_{BA} = 20 + 5 = 25$ m/s.

$a_{BA} = g - g = 0$

time when B will meet A is $t = \frac{\text{distance}}{\text{vel}} = \frac{100}{25} = 4$ sec. Ans.



Scan for Video Explanation

**Module-16: Concept of Relative Motion**

**Ex :** Two cars moving toward each other at velocities $v_A$ and $v_B$ are at a separation d. At this instant both are retarded with acceleration $a_A$ and $a_B$ respectively. Find minimum value of d such that no collision occurs.

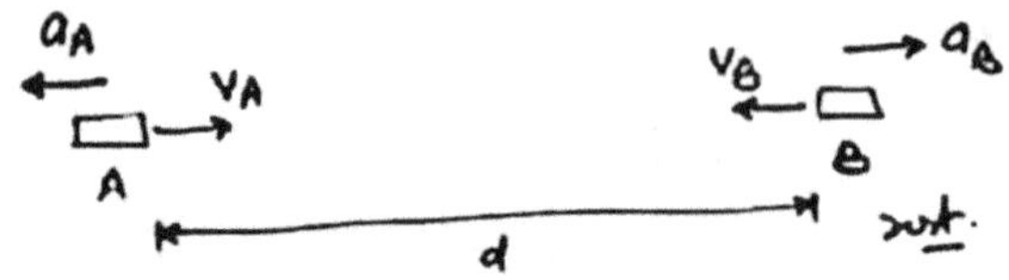


Sol$^n$:

$$v_{AB} = v_A + v_B$$
$$a_{AB} = -(a_A + a_B)$$

$$v^2 = u^2 + 2aS$$
$$0 = (v_A + v_B)^2 - 2(a_A + a_B)d$$
$$d = \frac{(v_A + v_B)^2}{2(a_A + a_B)}$$
(min)

Scan for Video Explanation

Web Reference at www.physicsgalaxy.com

**Module-17: Solved Example-11**

**Ex :** A balloon is rising up at 20 m/s and a stone is released from it, when it is at a height 100 m above ground. Find the time after which the stone will hit the ground.

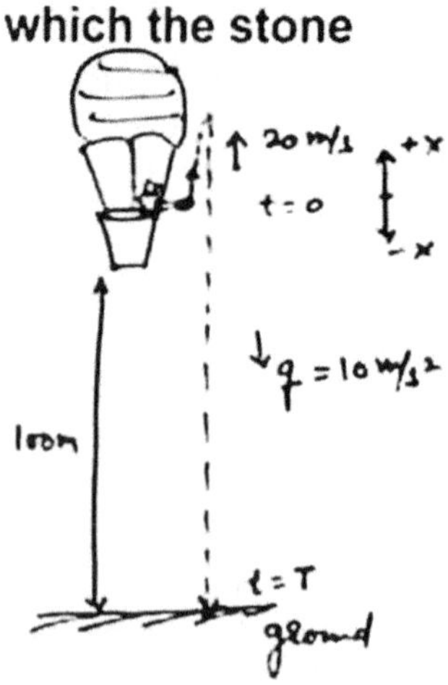


Sol$^n$:

$$u = +20 \text{ m/s}$$
$$h = -100 \text{ m}$$
$$a = -10 \text{ m/s}^2$$

we use
$$h = uT + \frac{1}{2}aT^2$$
$$-100 = 20T - 5T^2$$
$$T^2 - 4T - 20 = 0$$
$$T = \frac{4 \pm \sqrt{16+80}}{2} = 2 \pm \frac{1}{2}\sqrt{96}$$
$$= 2 \pm \sqrt{24} \text{ sec}$$
$$T = 2 + \sqrt{24} \text{ sec} \quad \text{Ans.}$$

Scan for Video Explanation

Web Reference at www.physicsgalaxy.com

**Module-18: Solved Example-12**

**Ex :** **Inside a lift accelerating up at 2 m/s$^2$ a loose bolt falls from its ceiling. If height of lift is 2m find the speed of bolt with which it hits the floor. Initial speed of lift is 5 m/s in upward direction.**

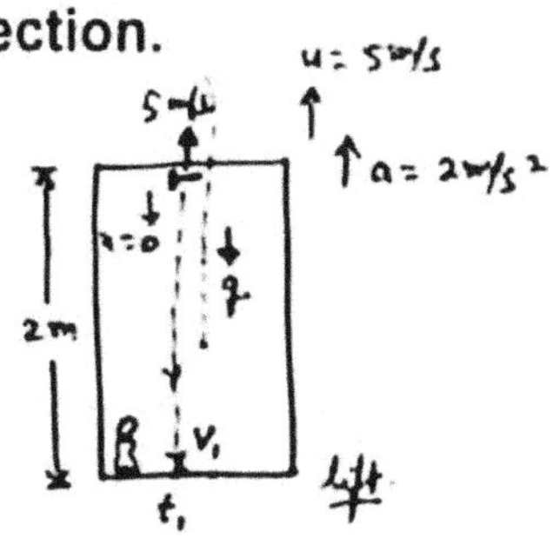


Sol$^n$:

Initial speed of bolt w.r.to lift is

$u = 0$

acc$^n$ of bolt w.r.to lift

$a_{BL} = 10 + 2 = 12\ m/s^2$

in reference of lift $h = ut + \frac{1}{2}at^2$

$2 = \frac{1}{2} \times 12 \times t_1^2 \Rightarrow t_1^2 = \frac{1}{3} \Rightarrow t_1 = \frac{1}{\sqrt{3}}$ sec

vel of bolt as seen from lift $v_1 = u + at = 12 \times \frac{1}{\sqrt{3}} = 4\sqrt{3}\ m/s$

net vel w.r.to ground. $v_2 = u + at = 5 - 10\left(\frac{1}{\sqrt{3}}\right) = 5 - \frac{10}{\sqrt{3}}\ m/s$



**Module-19: Graphical Analysis of Motion**

# Graphical Analysis of Motion:

Rectilinear motion → 3 types of graphs are used in motion analysis.

(1) Displacement v/s time → distance vs time

(2) Velocity v/s time → speed v/s time

(3) Acc$^n$ v/s time

## Module-19: Graphical Analysis of Motion

# Displacement v/s Time Graphs:

Position v/s time

Slope at pt P

$$\frac{dx}{dt} = \tan\theta = v_P$$

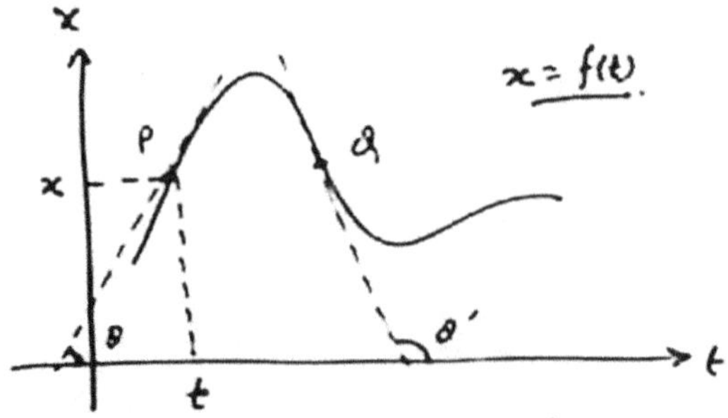


$v_P = \tan\theta = +ve.$ $v_Q = \tan\theta' = -ve.$

Part. Situation

A part moves → rest↑ accelerates, uniformly, retards↑ rest

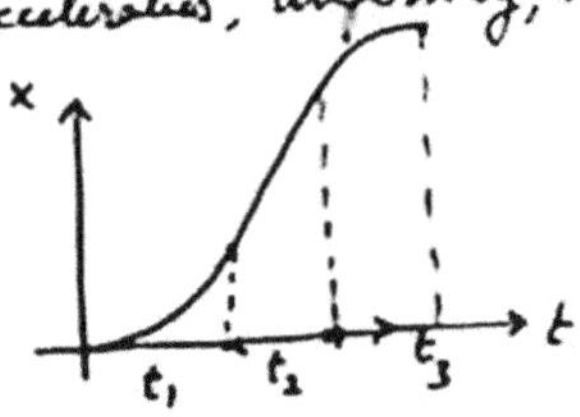


Scan for Video Explanation

Web Reference at www.physicsgalaxy.com

## Module-19: Graphical Analysis of Motion

Exp.

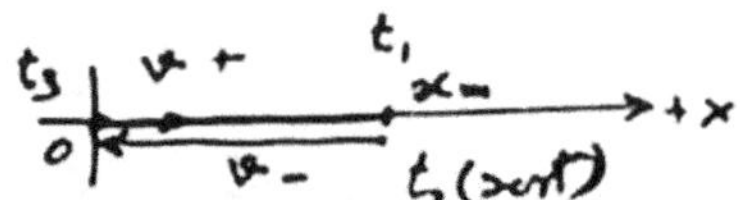


Particle

Pos-time graph:

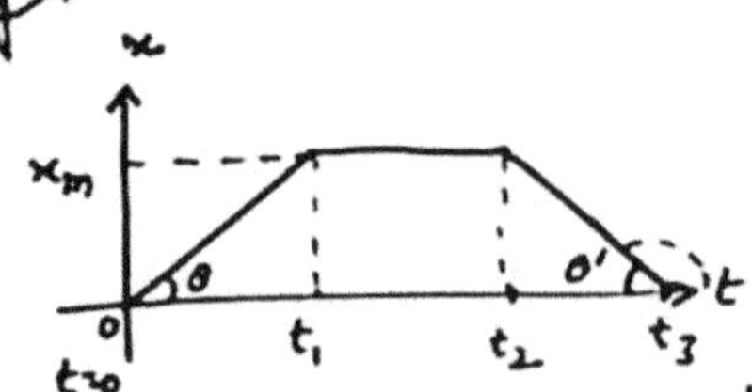


$\tan\theta = v$

$\tan\theta' = v$

distance-time graph.

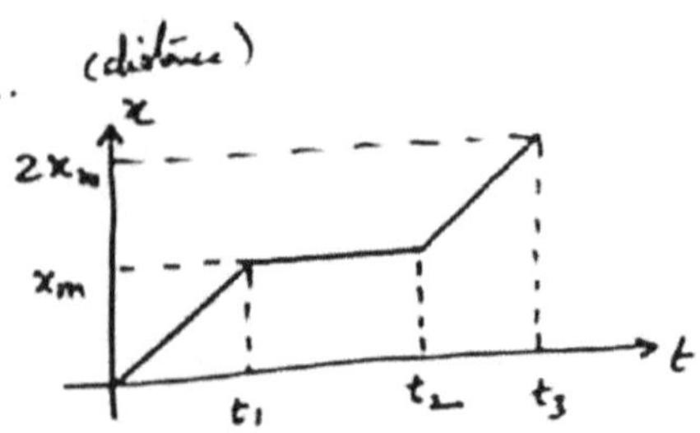


Scan for Video Explanation

Web Reference at www.physicsgalaxy.com

**Module-19: Graphical Analysis of Motion**

# Velocity-time Graphs:

$$\frac{dv}{dt} = \tan\theta = a_x$$

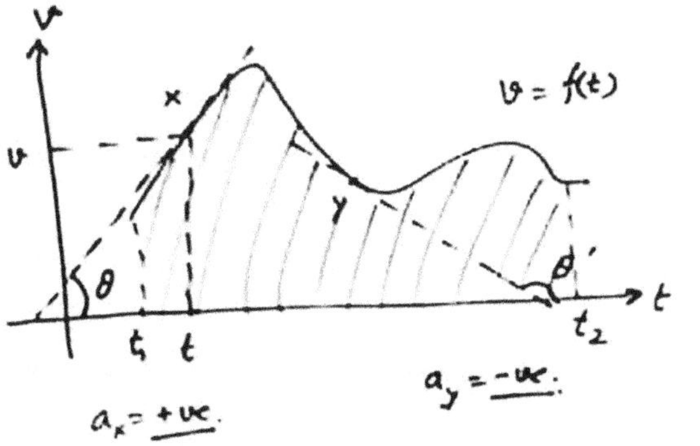


Area under the curve

$$\text{Area} = \int_{t_1}^{t_2} v\,dt \rightarrow \text{displacement of particle from } t = t_1 \text{ to } t_2.$$

Web Reference at www.physicsgalaxy.com | Scan for Video Explanation

**Module-19: Graphical Analysis of Motion**

Motion Situation:

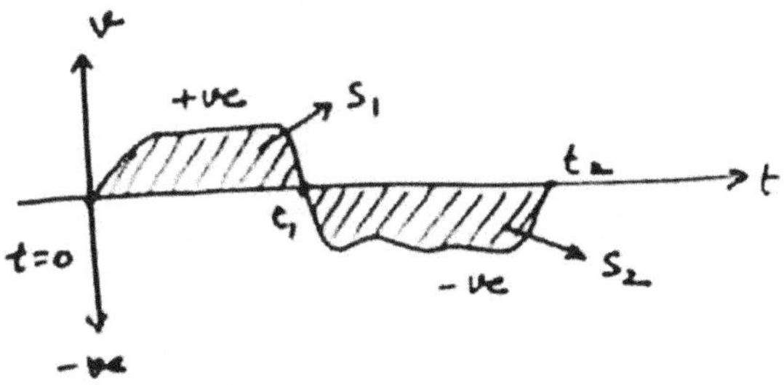


Total displacement of particle from t = 0 to $t_2$ is

$$\text{disp} = S_1 - S_2$$

$$\text{distance travelled} = S_1 + S_2$$

# Accⁿ time graphs:

$$A = \int_{t_1}^{t_2} a\,dt \rightarrow \text{net change in vel of part from } t_1 \text{ to } t_2.$$

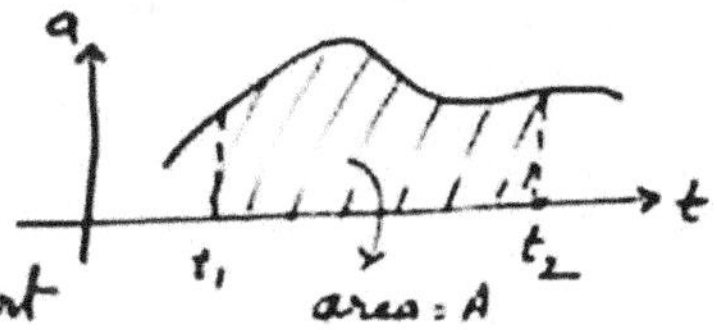


Web Reference at www.physicsgalaxy.com | Scan for Video Explanation

**Module-20: Displacement-Time Graphs**

# Various Cases of Displacement v/s Time Graphs:

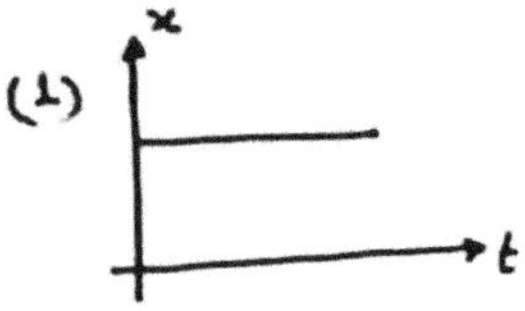


$v = 0$

↓

particle at rest.

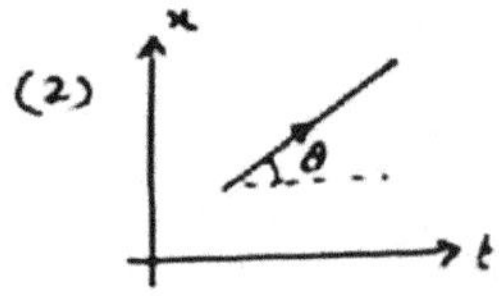


$v = \tan\theta$ +ve

↓

a particle moving away from Origin at Constant vel.

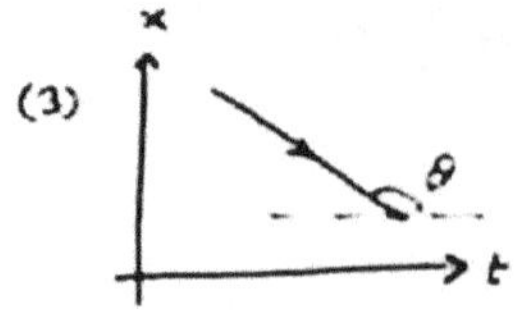


$v = \tan\theta =$ −ve.

↓

a particle moving toward origin at Constant vel.

Scan for Video Explanation

Web Reference at www.physicsgalaxy.com

**Module-20: Displacement-Time Graphs**

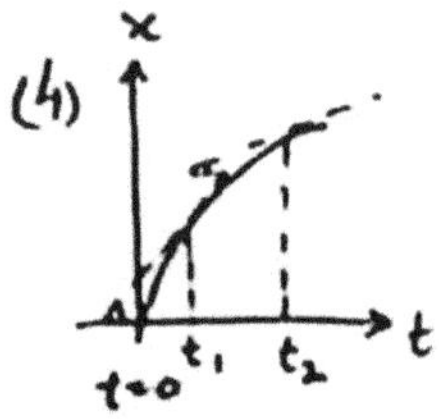


$\vec{v} = \tan\theta \downarrow$ +ve

$\vec{a}$ −ve.

↓

a particle is moving with +ve vel & −ve acc$^n$

(retarded motion)

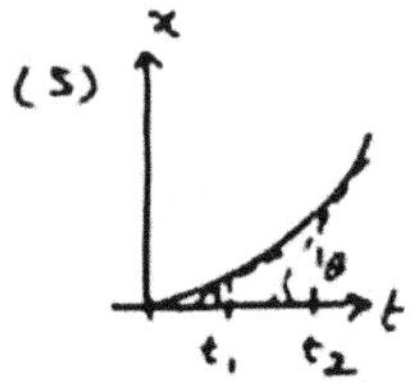


$\vec{v} = \tan\theta \uparrow$ +ve

$\vec{a} =$ +ve

↓

a particle is moving with +ve vel & +ve acc$^n$

(accelerated motion)

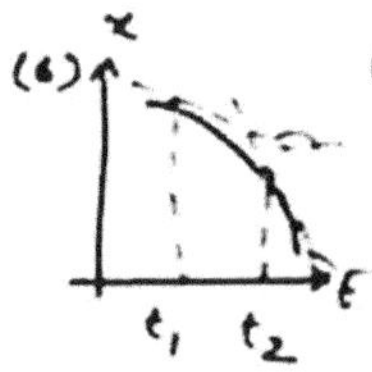


$\vec{v} = \tan\theta \uparrow$ −ve

$\vec{a} =$ −ve

↓

particle moving with −ve vel & −ve acc$^n$

(accelerated motion)

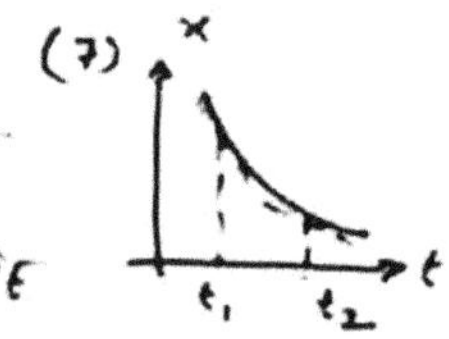


$\vec{v} = \tan\theta \downarrow$ −ve

$\vec{a} =$ +ve

↓

part. moving with −ve vel & +ve acc$^n$

(Retarded motion)

Scan for Video Explanation

Web Reference at www.physicsgalaxy.com

**Module-20: Displacement-Time Graphs**

**Ex :** **A boy on top of a tower of height H throws a ball up with initial speed u, taking upward direction positive draw the displacement v/s time curve for ball's motion till it hits the ground.**

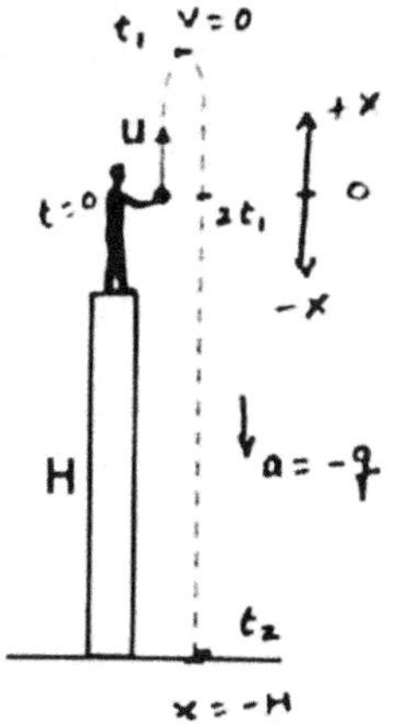

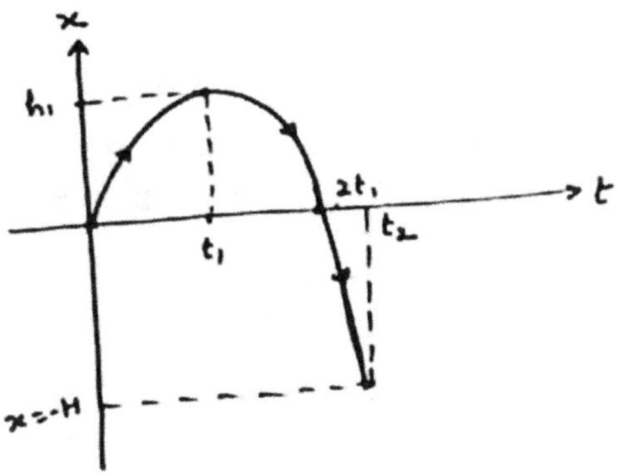

Web Reference at www.physicsgalaxy.com

Scan for Video Explanation

**Module-20: Displacement-Time Graphs**

**Ex :** **A car moves at speed $v_0$. If at $t = 0$ driver sees a need to apply brakes and applies the brakes at a reaction time $t_R$ and finally comes to rest in a distance s after brakes are applied. Draw position time curve for car's motion.**

Sol:

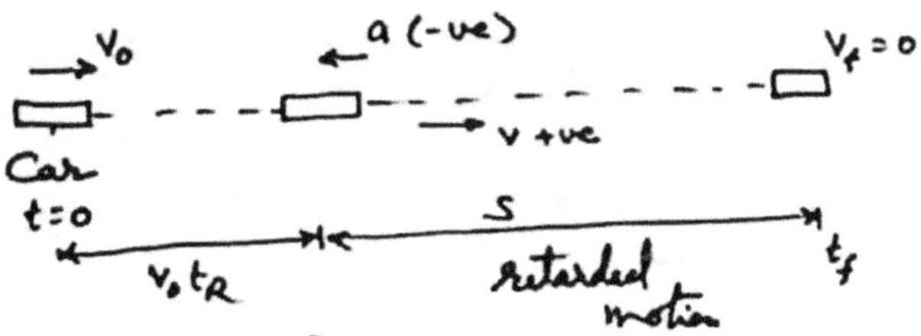

$\tan\theta = v_0$

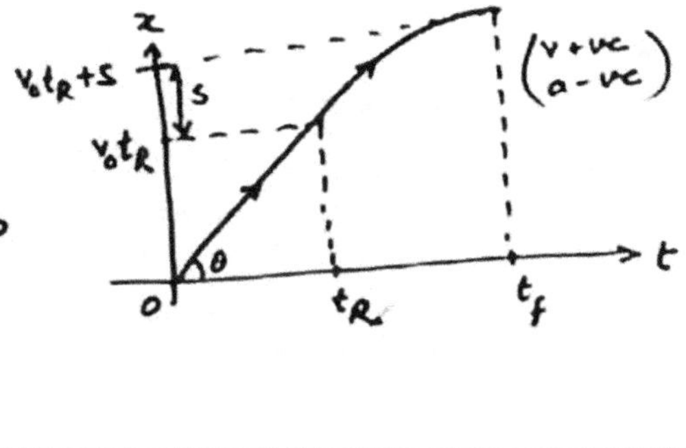

Web Reference at www.physicsgalaxy.com

Scan for Video Explanation

**Module-21: Velocity-Time Graphs**

# Velocity v/s Time Graphs:

slope → Acceleration
Area → Displacement

(1)

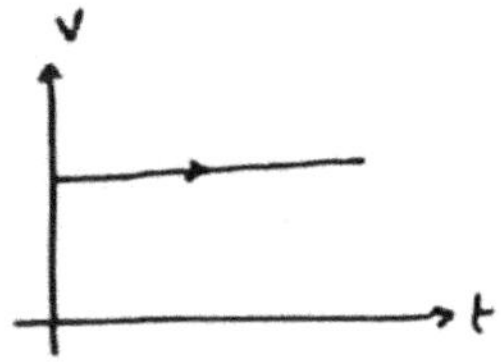

⇒ $a = 0$, uniform motion (const vel)

(2)

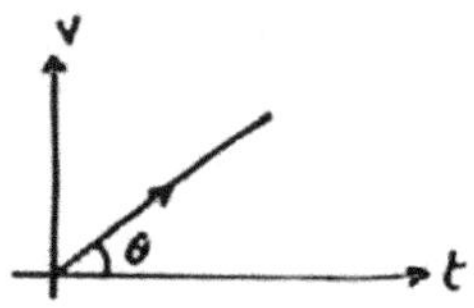

⇒ $a$ = Constant (+ve)
increasing speed
$\vec{a}$ +ve ⊕ $\vec{v}$ +ve

(3)

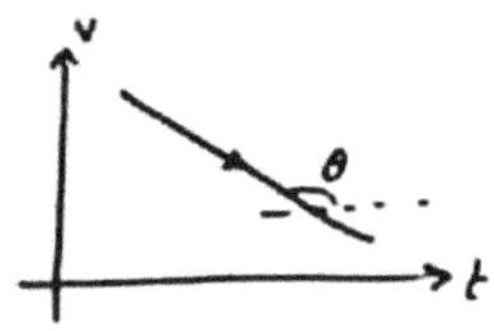

⇒ $a$ = Constant (−ve)
decreasing speed
$\vec{a}$ −ve ⊕ $\vec{v}$ +ve

Scan for Video Explanation

Web Reference at www.physicsgalaxy.com

**Module-21: Velocity-Time Graphs**

(4)

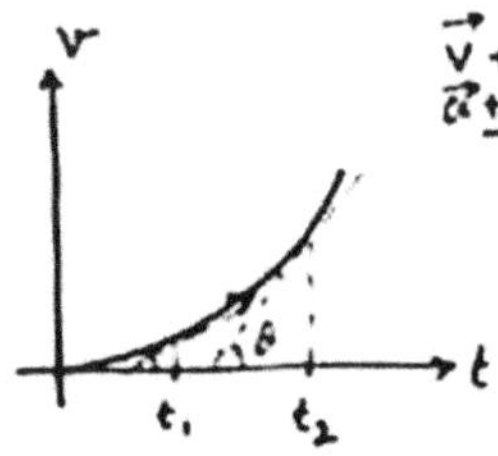

$\vec{v}$ +ve
$\vec{a}$ +ve

⇒ variable acc$^n$
→ increasing acc$^n$
→ increasing speed

(5)

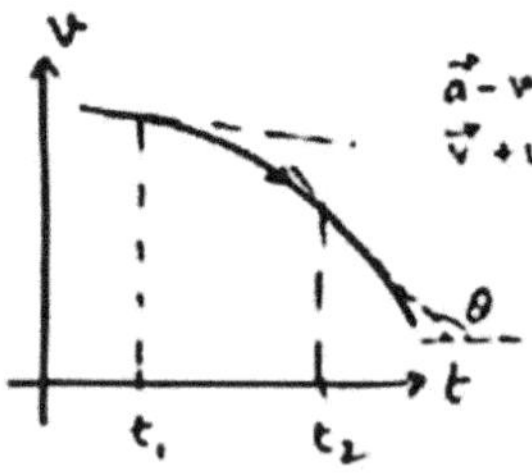

$\vec{a}$ −ve
$\vec{v}$ +ve

⇒ variable acc$^n$
→ increasing acc$^n$
→ decreasing speed.

(6)

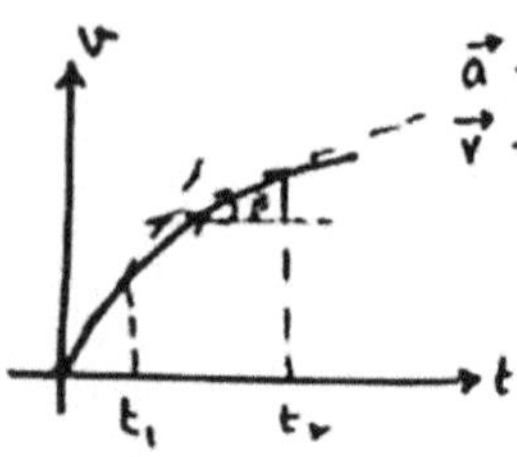

$\vec{a}$ +ve
$\vec{v}$ +ve

⇒ variable acc$^n$
→ decreasing acc$^n$
→ increasing speed.

Scan for Video Explanation

Web Reference at www.physicsgalaxy.com

**Module-21: Velocity-Time Graphs**

**Ex :** **A boy on top of a tower of height H throws a ball up with initial speed u, taking upward direction positive draw the velocity v/s time curve for ball's motion till it hits the ground.**

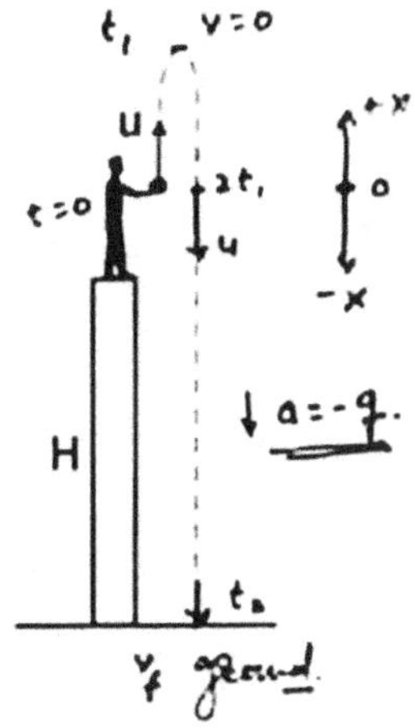


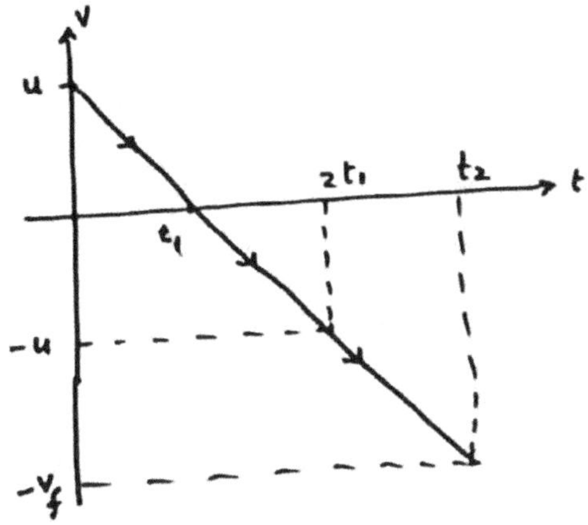




**Module-21: Velocity-Time Graphs**

**Ex :** **A ball is dropped at t = 0 from a height H as shown, it collides elastically with floor and rebounds to same height. Draw velocity v/s time graph for ball's motion taking downward direction positive.**

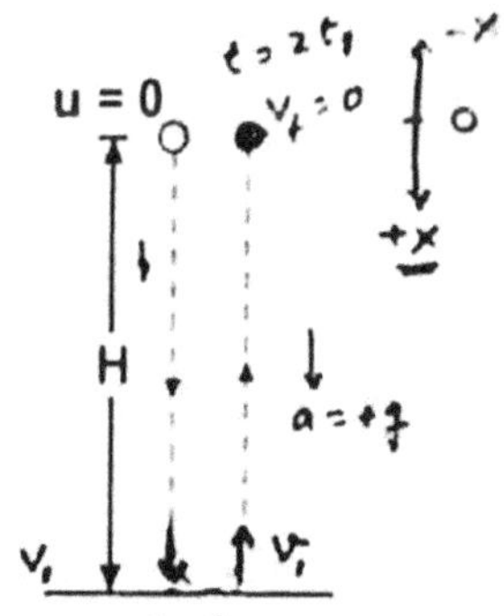


Sol: elastic collision → no energy is lost.

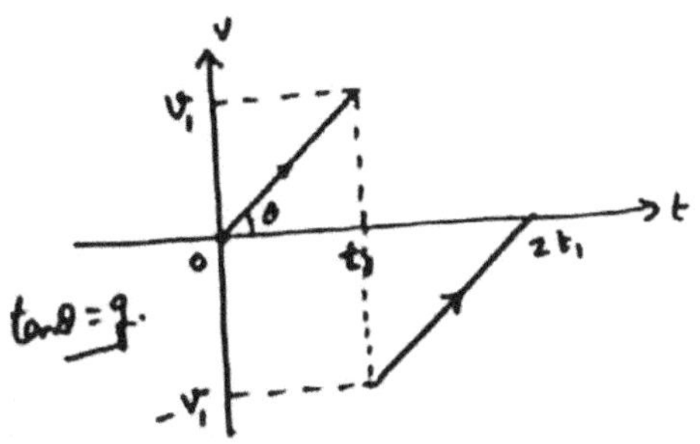


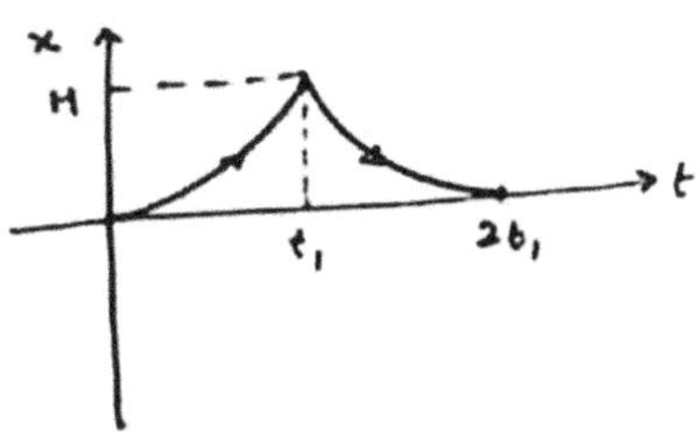

**Module-22: Acceleration-Time Graphs**

# Acceleration v/s Time Graphs:

$\frac{da}{dt} \rightarrow$ Rate of change of accn.

Area = change in velocity

Area $= \vec{V}_B - \vec{V}_A$

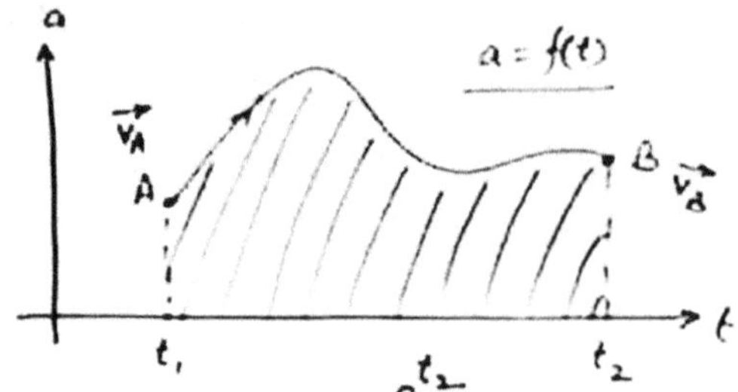

Area $= \int_{t_1}^{t_2} a\,dt$

Ex: A part starts at 5 m/s & its accn varies according to the graph shown find the vel of part at t = 10 sec

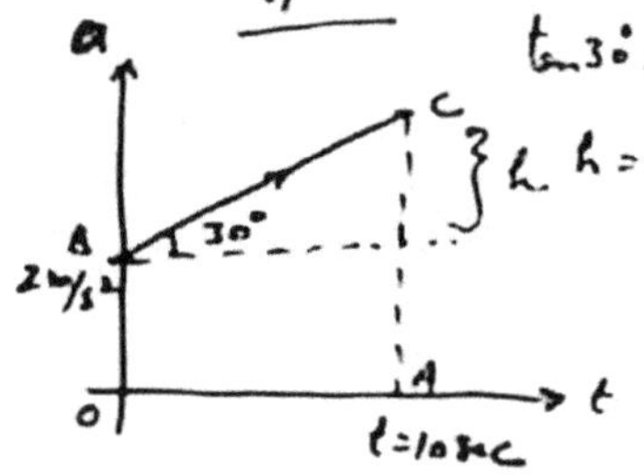

$\tan 30° = \frac{h}{10}$, $h = 10/\sqrt{3}$

Soln: $V_{10sec} = V_0 +$ Area OACB

$V_{10sec} = 5 + 20 + \frac{50}{\sqrt{3}}$

$= 25 + \frac{50}{\sqrt{3}}$ m/s Ans.

Area OACB $= 20 + \frac{1}{2} \times 10 \times \frac{10}{\sqrt{3}}$

$= 20 + \frac{50}{\sqrt{3}}$

Scan for Video Explanation

# Chapter 2
# Motion in Two Dimensions

## Lecture Notes Modules


**Module-1:** Introduction
**Module-2:** Solved Example-1
**Module-3:** Solved Example-2
**Module-4:** Solved Example-3
**Module-5:** Trajectory of a 2D motion
**Module-6:** Projectile Motion Under Gravity
**Module-7:** Velocity of a Particle During Projectile
**Module-8:** Position and Trajectory of Projectile
**Module-9:** Solved Example-4
**Module-10:** Solved Example-5
**Module-11:** Solved Example-6
**Module-12:** Solved Example-7
**Module-13:** Projectile in Moving Reference Frame
**Module-14:** Projectile on Inclined Plane

**Module-1: Introduction**

# Motion in Two Dimensions: Motion in a Plane.

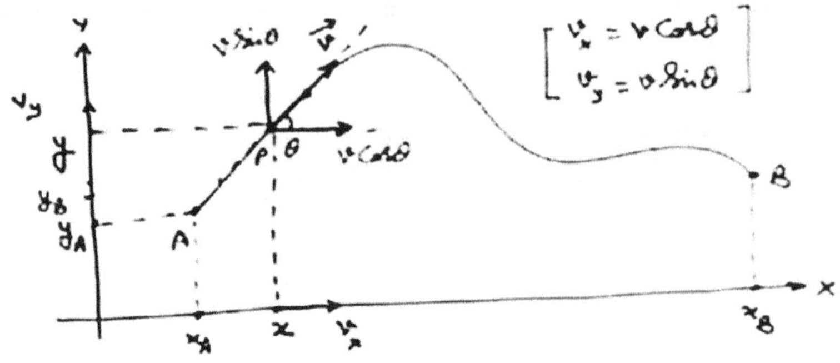


$$\begin{bmatrix} v_x = v\cos\theta \\ v_y = v\sin\theta \end{bmatrix}$$

$$\vec{v} = v_x\hat{i} + v_y\hat{j}$$

$$\begin{bmatrix} v = \sqrt{v_x^2 + v_y^2} \\ \tan\theta = \frac{v_y}{v_x} \quad \text{or} \quad \theta = \tan^{-1}\left(\frac{v_y}{v_x}\right) \end{bmatrix}$$

NOTE: A 2D motion can be analyzed by splitting it into two ⊥ 1D motions with $t$ as a common variable.

Scan for Video Explanation



**Module-1: Introduction**

In 2D motion all the analysis of 1D motion can be applied along x and Y directions simultaneously.

along x-dir $\quad v_x = \frac{dx}{dt} \quad$ & $\quad a_x = \frac{dv_x}{dt} = v_x\frac{dv_x}{dx}$

along Y-dir $\quad v_y = \frac{dy}{dt} \quad$ & $\quad a_y = \frac{dv_y}{dt} = v_y\frac{dv_y}{dy}$.

Ex: A particle moves in a 2D motion such that its position depends on time as -

$$x = 4t^2 + 2t + 3$$
$$y = 5t + 7$$

Find the vel of particle at $t = 2$ s.

Sol$^n$: $v_x = \frac{dx}{dt} = 8t + 2$ , $v_y = \frac{dy}{dt} = 5$

$t = 2$ sec: $v_x = 18$ m/s , $v_y = 5$ m/s

$$\left.\begin{array}{l} \vec{v} = 18\hat{i} + 5\hat{j} \\ v = \sqrt{(18)^2 + (5)^2} \\ \theta = \tan^{-1}(5/18) \end{array}\right\} \text{Ans.}$$

Scan for Video Explanation

Web Reference at www.physicsgalaxy.com

**Module-2: Solved Example-1**

**Ex :** A particle starts in a 2D motion such that its position vector at any instant can be given as

$$\vec{r} = (2t^3 + 4t)\,\hat{i} + (3t^2 + 7)\,\hat{j}$$

Find the velocity and acceleration of particle at t = 1 sec.

Sol$^n$:

$x = 2t^3 + 4t \longrightarrow v_x = 6t^2 + 4 \longrightarrow a_x = 12t$

$y = 3t^2 + 7 \longrightarrow v_y = 6t \longrightarrow a_y = 6$

t = 1 sec  $\vec{v} = 10\hat{i} + 6\hat{j}$  $\vec{a} = 12\hat{i} + 6\hat{j}$

$v = \sqrt{100 + 36}$  $a = \sqrt{144 + 36}$

$= \sqrt{136}$ m/s  $= \sqrt{180}$ m/s$^2$

Ans.

Web Reference at www.physicsgalaxy.com | Scan for Video Explanation

**Module-3: Solved Example-2**

**Ex :** A particle starts from origin such that its velocity along Y direction remains constant at 2 m/s and in X direction it is imparted an acceleration which varies with time as $a_X = 3t + 2$ m/s$^2$. Find the position vector of particle at t = 2 sec.

Sol$^n$:

$u_y = 2$ m/s  $y = u_y t$

$a_x = 3t + 2$ m/s  $= 2 \times 2 = 4$ m.

in x dir$^n$ $\frac{dv_x}{dt} = 3t + 2$

$\int_0^{v_x} dv_x = \int_0^t (3t+2)\,dt$

$v_x = \frac{3t^2}{2} + 2t$

$\frac{dx}{dt} = \frac{3t^2}{2} + 2t$

$\int_0^x dx = \int_0^t \left(\frac{3t^2}{2} + 2t\right) dt$

$x = \frac{t^3}{2} + t^2$

at t = 2  $x = 4 + 4 = 8$ m.

Position at t = 2 sec is (8, 4)

$\vec{r} = 8\hat{i} + 4\hat{j}$ Ans.

Web Reference at www.physicsgalaxy.com | Scan for Video Explanation

**Module-4: Solved Example-3**

**Ex :** A particle is moving in 2D such that in x and y directions its velocities are given as

$$V_x = 5t^2 + 7t + 3$$

$$V_y = 2t^3$$

Find the angle between direction of velocity and acceleration at t = 1 sec.

Sol$^n$: at t = 1 sec $\vec{v} = 15\hat{i} + 2\hat{j}$ $\hat{v} = \frac{\vec{v}}{v} = \frac{15\hat{i} + 2\hat{j}}{\sqrt{229}}$

$$\vec{a} = \frac{d\vec{v}}{dt} = (10t + 7)\hat{i} + (6t^2)\hat{j}$$

at t = 1 sec $\vec{a} = 17\hat{i} + 6\hat{j}$ $\hat{a} = \frac{\vec{a}}{a} = \frac{17\hat{i} + 6\hat{j}}{\sqrt{325}}$

If $\theta$ is the angle between $\vec{v}$ & $\vec{a}$ at t = 1 sec

$$\theta = \cos^{-1}(\hat{v} \cdot \hat{a}) = \text{-----} \quad \text{Ans.}$$

Scan for Video Explanation

Web Reference at www.physicsgalaxy.com

**Module-5: Trajectory of a 2D motion**

# Trajectory of a Two Dimensional Motion:

path of Motion → Trajectory

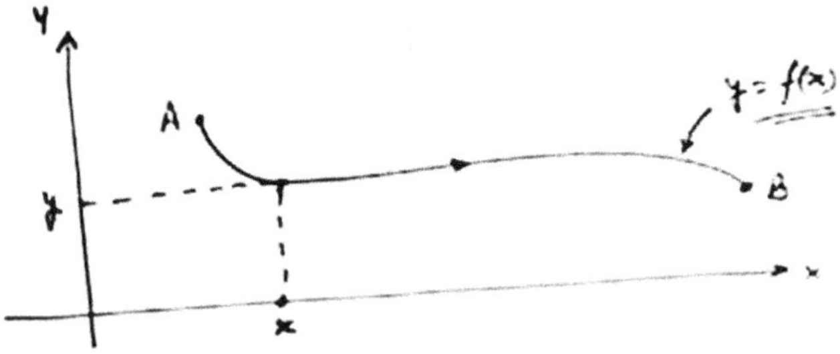


Equation of Trajectory: $\left[\begin{array}{l} x = f_1(t) \quad \text{— (1)} \\ y = f_2(t) \quad \text{— (2)} \end{array}\right.$

eliminating t from (1) & (2) we get $y = f(x)$

Scan for Video Explanation

Web Reference at www.physicsgalaxy.com

**Module-5: Trajectory of a 2D motion**

**Ex :** If a particle travels in 2D with velocity given as

$$\vec{V} = 4t\hat{i} + 5\hat{j}$$

Find the path of motion of particle and its equation of trajectory if it starts from origin.

Sol$^n$:

$$\vec{v} = 4t\hat{i} + 5\hat{j}$$

$$v_x = 4t \text{ m/s}$$

$$\int_0^x dx = \int_0^t 4t\,dt$$

$$x = 2t^2 \quad -(1)$$

$$v_y = 5 \text{ m/s}$$

$$y = 5t \quad -(2)$$

$$t = y/5$$

from eq$^n$(1) $x = 2(y/5)^2$

$x = 2y^2/25$ parabolic path.

eq$^n$ of traj of part. $\rightarrow$ $y = 5\sqrt{x/2}$ Ans.

Web Reference at www.physicsgalaxy.com | Scan for Video Explanation

**Module-5: Trajectory of a 2D motion**

# Distance Travelled by a particle in 2D Motion:

↓

length of path.

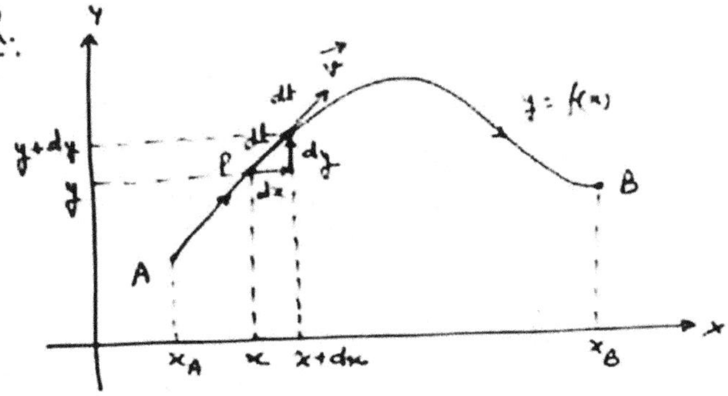


length of path

$$l_{AB} = \int dl$$

$$= \int \sqrt{dx^2 + dy^2}$$

$$= \int \sqrt{1 + \left(\frac{dy}{dx}\right)^2}\,.\,dx$$

$$\boxed{l_{AB} = \int_{x_A}^{x_B} \sqrt{1 + \left(\frac{dy}{dx}\right)^2}\,dx}$$

$$l_{AB} = \int_{x_A}^{x_B} \sqrt{1 + [f'(x)]^2}\,.\,dx.$$

Web Reference at www.physicsgalaxy.com | Scan for Video Explanation

**Module-6: Projectile Motion Under Gravity**

# Projectile Motion:

Motion of a body after its projection is called Projectile motion

projection → to impart some initial velocity

# Projectile Motion under gravity:

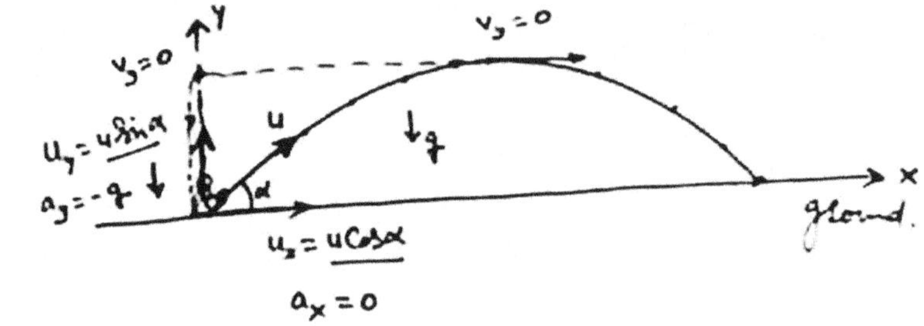


Scan for Video Explanation

Web Reference at www.physicsgalaxy.com

**Module-6: Projectile Motion Under Gravity**

# Std. Parameters of Projectile Motion on Horizontal Plane:

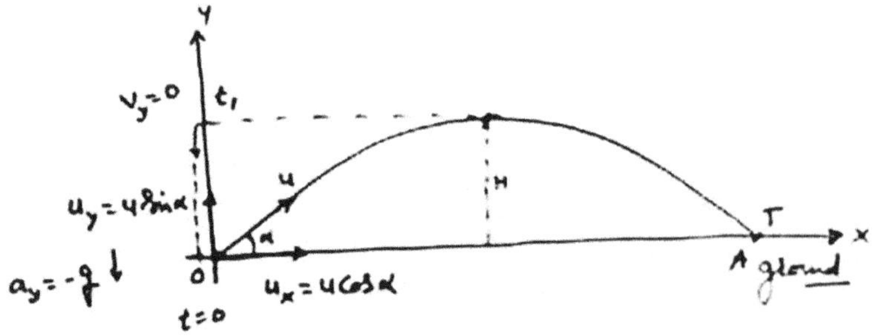


* Time to reach max height → $v_y = u_y - gt_1$

$$0 = u\sin\alpha - gt_1 \Rightarrow \boxed{t_1 = \frac{u\sin\alpha}{g}}$$

* Time of flight → $\boxed{T = 2t_1 = \frac{2u\sin\alpha}{g}}$

* max. height attained by projectile → $h = ut - \frac{1}{2}gt^2$

$$H = u\sin\alpha\left(\frac{u\sin\alpha}{g}\right) - \frac{1}{2}g\frac{u^2\sin^2\alpha}{g^2}$$

$v^2 = u^2 - 2gh$ → $\boxed{H = \frac{u^2\sin^2\alpha}{2g}}$

Scan for Video Explanation

Web Reference at www.physicsgalaxy.com

**Module-6: Projectile Motion Under Gravity**

# Range of Projectile on Horizontal Plane:

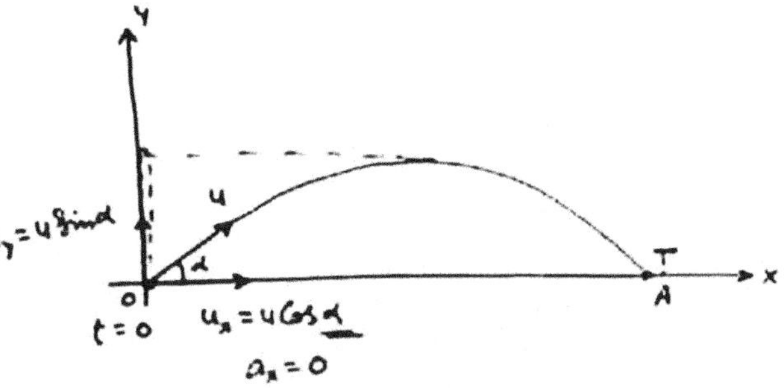


time of flight

$$T = \frac{2u\sin\alpha}{g}$$

Range of projectile = OA

$$R = u\cos\alpha \times T = u\cos\alpha \times \frac{2u\sin\alpha}{g}$$

$$\boxed{R = \frac{u^2\sin 2\alpha}{g}}$$

**Web Reference at www.physicsgalaxy.com** **Scan for Video Explanation**

**Module-6: Projectile Motion Under Gravity**

# Range Analysis of Projectile Motion:

$$R = \frac{u^2\sin 2\alpha}{g}$$

$$(\sin 2\alpha)_{max} = 1$$

$$2\alpha = 90^\circ$$

$$\alpha = 45^\circ$$

$$R_{max} = \frac{u^2}{g}$$

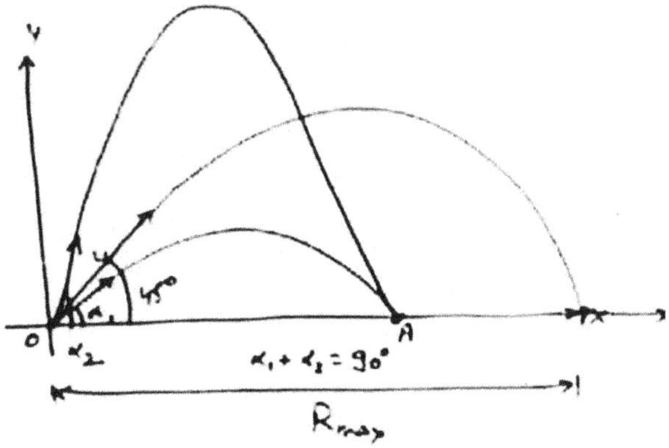


at proj angles $\alpha \neq 45^\circ$ always there exist two angles for which the range of proj on horizontal ground are equal. These angle are Complimentary angle such that $\alpha_1 + \alpha_2 = 90^\circ$

**Web Reference at www.physicsgalaxy.com** **Scan for Video Explanation**

**Module-7: Velocity of a Particle During Projectile**

# Velocity of a Particle during Projectile:

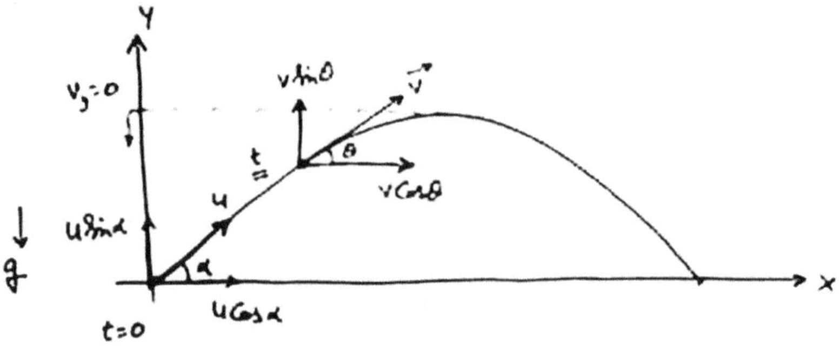


Initial vel vector $\vec{u} = (u\cos\alpha)\hat{i} + (u\sin\alpha)\hat{j}$

along x dir. $v\cos\theta = u\cos\alpha$ — (1)

along y dir $v\sin\theta = u\sin\alpha - gt$ — (2) $[v_y = u_y - gt]$

vel after time t of projectile $\boxed{\vec{v} = (u\cos\alpha)\hat{i} + (u\sin\alpha - gt)\hat{j}}$

Scan for Video Explanation

Web Reference at www.physicsgalaxy.com

**Module-8: Position and Trajectory of Projectile**

# Position of Particle and Trajectory during Projectile Motion:

Position of part at time t

$$\begin{bmatrix} x = u\cos\alpha . t & — (1) \\ y = u\sin\alpha t - \frac{1}{2}gt^2 & — (2) \end{bmatrix}$$

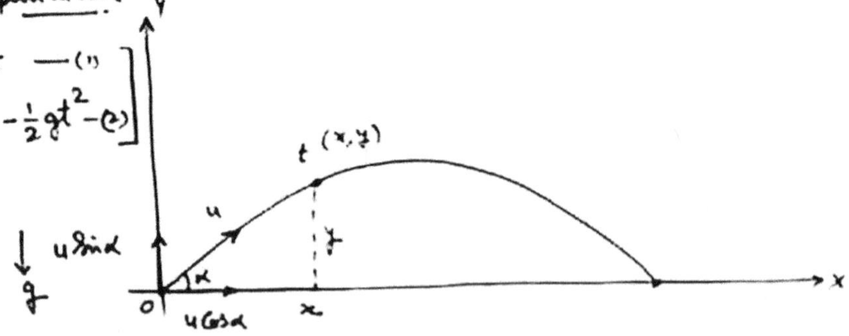


Equation of Trajectory from eq (1) $t = \dfrac{x}{u\cos\alpha}$

from eqn (2) $y = u\sin\alpha\left(\dfrac{x}{u\cos\alpha}\right) - \dfrac{1}{2}g\left(\dfrac{x}{u\cos\alpha}\right)^2$

$$\boxed{y = x\tan\alpha - \frac{gx^2}{2u^2\cos^2\alpha}}$$ v. imp

Scan for Video Explanation

Web Reference at www.physicsgalaxy.com

**Module-8: Position and Trajectory of Projectile**

# Equation of Trajectory for a Horizontally Thrown Projectile :

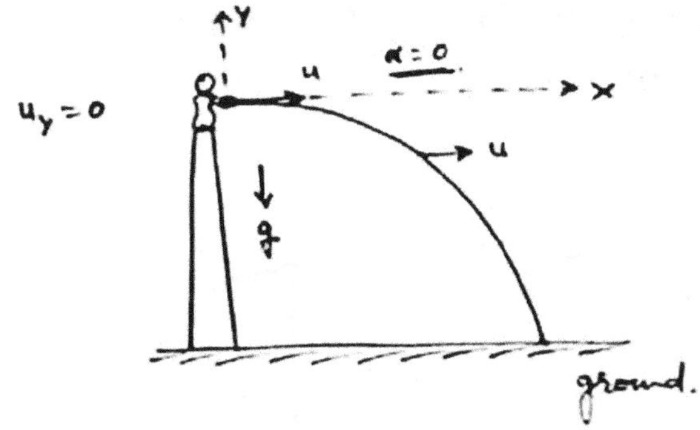


trajectory eqⁿ: $y = x\tan\alpha - \dfrac{gx^2}{2u^2\cos^2\alpha}$

$\alpha = 0$ $\boxed{y = -\dfrac{gx^2}{2u^2}}$ v. imp.



**Module-9: Solved Example-4**

**Ex :** If initial velocity of a particle projected from ground is $\vec{u} = 6\hat{i} + 8\hat{j}$ m/s, find the horizontal range of projectile.

Solⁿ:

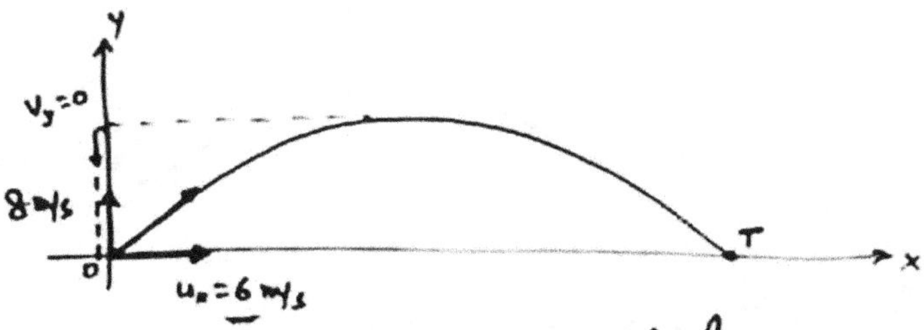


$a = -g$
$= -10$ m/s²

Const
$a_x = 0$

Horizontal Range $R = 6T$

$T = \dfrac{2u_y}{g} = \dfrac{2\times 8}{10} = 1.6$ sec

$R = 6 \times 1.6$
$= 9.6$ m Ans.

**Module-10: Solved Example-5**

**Ex :** **In the situation shown, a player kicks the ball at 45° with 20 m/s. The distance of a 3 m high Goal post is 25 m from the player, find whether there will be a goal or not.**

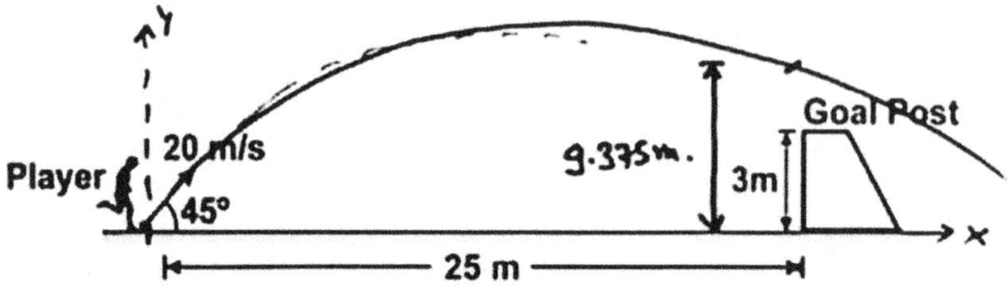


Sol$^n$: Eqn of traj $y = x\tan 45° - \dfrac{gx^2}{2(20)^2\cos^2 45°}$ NO GOAL!

$$y = x - \frac{gx^2}{400} = x - \frac{x^2}{40}\Big|_{x=25}$$

$$y = 25 - \frac{625}{40} = 25 - 15.625 = 9.375\,m.$$

Scan for Video Explanation

Web Reference at www.physicsgalaxy.com

**Module-11: Solved Example-6**

**Ex :** **A particle is thrown from ground at an angle 30° with horizontal at 40 m/s. Find the time duration for which it will be at a height more them 10 m above the ground.**

Sol$^n$:

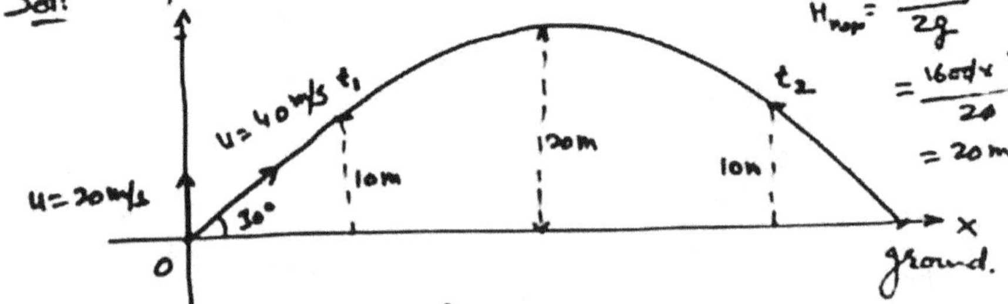


$$H_{max} = \frac{u^2\sin^2\theta}{2g} = \frac{1600 \times 1/4}{20} = 20\,m.$$

in y dir. $h = u_y t - \frac{1}{2}gt^2$ $\Delta t = t_2 - t_1$

$$10 = 20t - 5t^2$$

$$t^2 - 4t + 2 = 0 \Rightarrow t = \frac{4 \pm \sqrt{16-8}}{2} = 2 \pm \sqrt{2}\,sec$$

$t_1 = 2 - \sqrt{2}$ sec

$t_2 = 2 + \sqrt{2}$ sec

$\Delta t = t_2 - t_1 = 2\sqrt{2}$ sec Ans.

Scan for Video Explanation

Web Reference at www.physicsgalaxy.com

**Module-12: Solved Example-7**

**Ex : A body is projected horizontally from top of a high tower at speed 20 m/s. Find the displacement of body at t = 10 sec.**

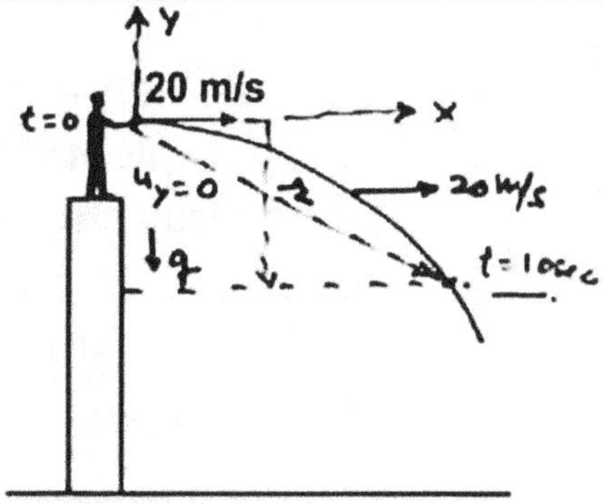


Solⁿ: Position of body at t=10sec

$u_x = 20\,m/s$ $\quad x = u_x t$

$= 20\times 10 = 200\,m.$

$u_y = 0 \quad a_y = g \quad y = \frac{1}{2}gt^2$

$y = \frac{1}{2}\times 10\times 100 = 500\,m.$

$r = \sqrt{x^2+y^2} = \sqrt{(200)^2+(500)^2}$

$= 100\sqrt{4+25} = 100\sqrt{29}\,m$ Ans.



**Module-13: Projectile in Moving Reference Frame**

# Projectile from a moving reference frame:

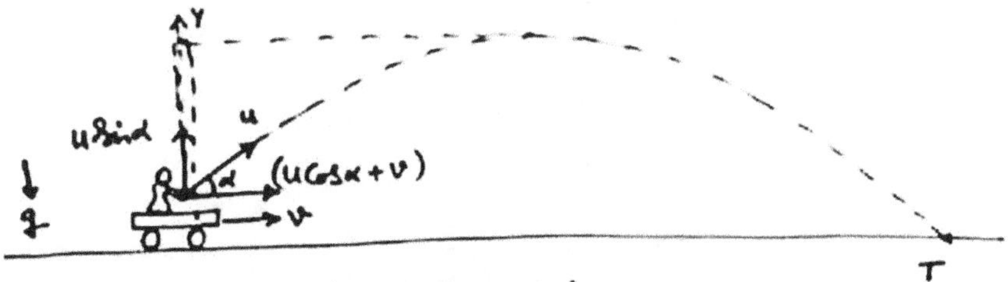


Initial vel Components of ball's motion

$u_x = u\cos\alpha + v$

$u_y = u\sin\alpha$

Time of flight $\quad T = \frac{2u_y}{g} = \frac{2u\sin\alpha}{g}$

Horizontal Range $\quad R = u_x \times T$

$= (u\cos\alpha + v)\cdot\left(\frac{2u\sin\alpha}{g}\right)$

**Module-13: Projectile in Moving Reference Frame**

**Ex :** A boy in an automobile throws a ball at 40 m/s at an angle 45° from horizontal as shown. Just after throw, boy jumps and start running in same direction to catch the ball. Find the speed at which boy should run to catch the ball and the time and position at which he will catch the ball.

Sol$^n$:

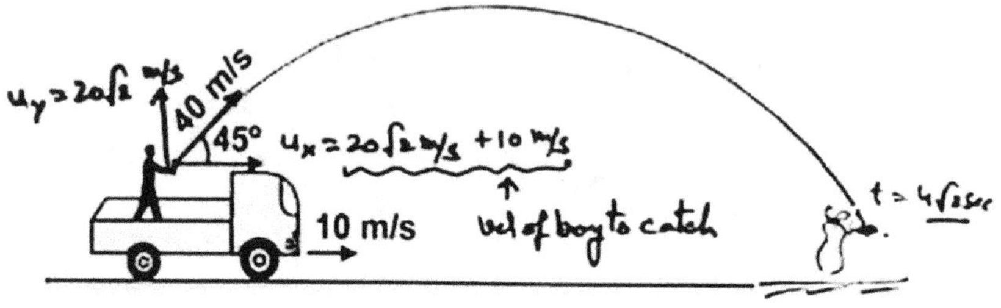


Time of flight $T = \frac{2u_y}{g} = \frac{2\times 20\sqrt{2}}{10} = 4\sqrt{2}$ sec Ans.

Range $R = u_x T = (20\sqrt{2}+10)\times 4\sqrt{2} = 160 + 40\sqrt{2}$ m. Ans.

Scan for Video Explanation

Web Reference at www.physicsgalaxy.com

**Module-14: Projectile on Inclined Plane**

# Projectile Motion on Inclined Plane:

Case-I: Projection up the Plane

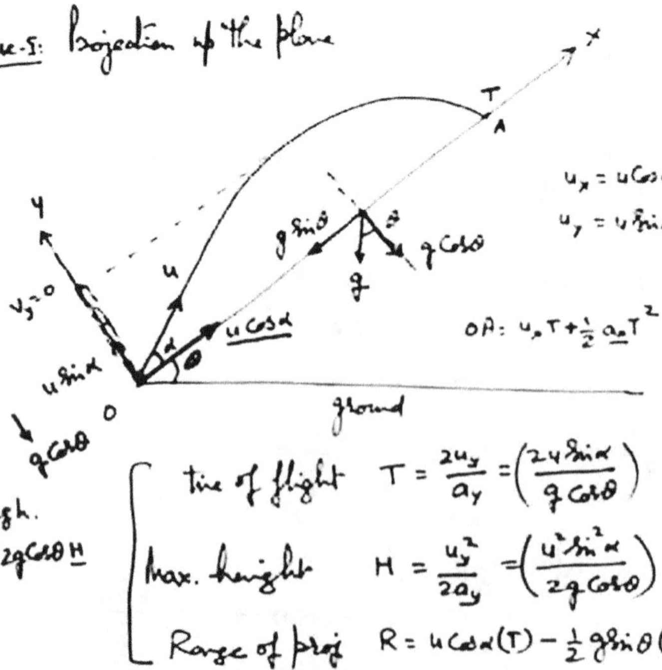


$u_x = u\cos\alpha \quad a_x = -g\sin\theta$

$u_y = u\sin\alpha \quad a_y = -g\cos\theta$

$OA: u_xT + \frac{1}{2}a_xT^2$

$[v_y = u_y - a_y T]$

$v_y^2 = u_y^2 - 2gh.$

$0 = u_y^2 - 2g\cos\theta\, H$

Time of flight $T = \frac{2u_y}{a_y} = \left(\frac{2u\sin\alpha}{g\cos\theta}\right)$

Max. height $H = \frac{u_y^2}{2a_y} = \left(\frac{u^2\sin^2\alpha}{2g\cos\theta}\right)$

Range of proj $R = u\cos\alpha(T) - \frac{1}{2}g\sin\theta(T)^2 = \ldots$

Scan for Video Explanation

Web Reference at www.physicsgalaxy.com

**Module-14: Projectile on Inclined Plane**

Case-II: Projection down the Plane

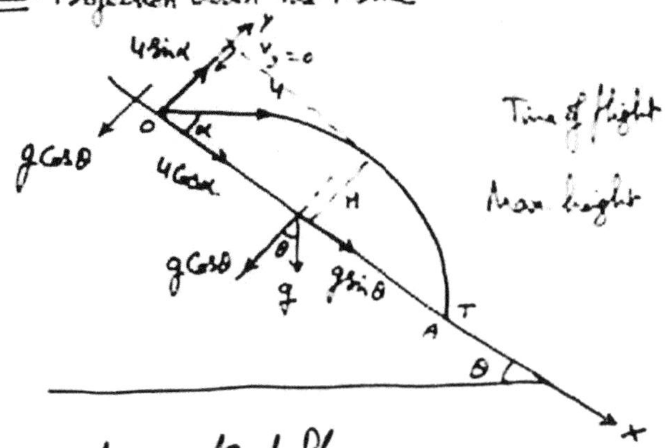


Time of flight $T = \frac{2u_y}{a_y} = \frac{2u\sin\alpha}{g\cos\theta}$

Max. height $H = \frac{u_y^2}{2a_y} = \frac{u^2\sin^2\alpha}{2g\cos\theta}$

Range along inclined Plane.

$$R = u\cos\alpha(T) + \frac{1}{2}g\sin\theta(T)^2$$

$= - - - \checkmark$



**Module-14: Projectile on Inclined Plane**

\# Maximum Range of Projectile on inclined Plane:

$$R = u_x T \pm \frac{1}{2}a_x T^2 \qquad a_x = g\sin\theta.$$

R will be max when $\frac{dR}{d\alpha} = 0$

↓

max Range along inclined Plane. $R_{max} = \frac{u^2}{g(1 \pm \sin\theta)}$

θ → angle of inc.

+ sign → proj up the plane.
− sign → proj down the plane.

$\theta = 0$ → horiz. ground. $R_{max} = \frac{u^2}{g}$. $\theta = 0$, $\alpha = 45°$

# Chapter 3
# Concept of Relative Motion

## Lecture Notes Modules


**Module-1:** Introduction to Relative Motion
**Module-2:** Solved Example-1
**Module-3:** Solved Example-2
**Module-4:** Solved Example-3
**Module-5:** Motion of a Body in Moving Ref Frame
**Module-6:** Solved Example-4
**Module-7:** Solved Example-5
**Module-8:** Solved Example-6
**Module-9:** Solved Example-7
**Module-10:** Rain Fall Cases
**Module-11:** Solved Example-8
**Module-12:** Solved Example-9
**Module-13:** Solved Example-10
**Module-14:** Condition of Collision of Two Particle
**Module-15:** Solved Example-11
**Module-16:** Solved Example-12
**Module-17:** Solved Example-13

**Module-1: Introduction to Relative Motion**

# Relative Motion:

Motion of a body as seen from a reference frame which may be at rest or in motion is called relative motion of the body w.r.to the frame.

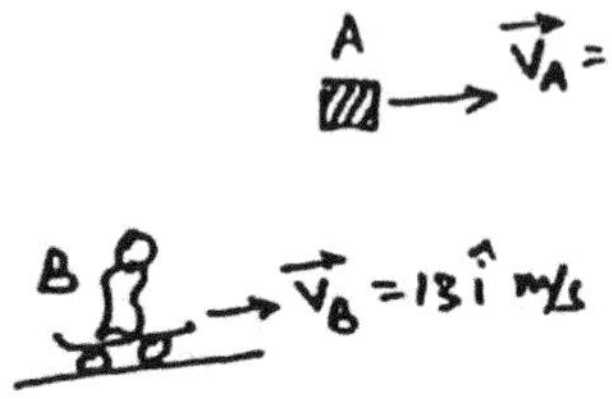


→ relative velocity of A relative to B is

$$\boxed{\vec{V}_{AB} = \vec{V}_A - \vec{V}_B}$$
$$= 20\hat{i} - 13\hat{i}$$
$$= 7\hat{i} \text{ m/s Ans.}$$

Scan for Video Explanation

Web Reference at www.physicsgalaxy.com

**Module-1: Introduction to Relative Motion**

Relative Motion → It is the motion as seen by some one.

velocity of B relative to A $\vec{V}_{BA} = \vec{V}_B - \vec{V}_A$

$$= (-V_B) - (V_A)$$
$$= -(V_A + V_B)$$

in same dir of two bodies
relative speed $= |V_1 - V_2|$.

in opp dir motion of two bodies
relative speed $= V_1 + V_2$.

Scan for Video Explanation

Web Reference at www.physicsgalaxy.com

**Module-1: Introduction to Relative Motion**

**Ex :** On a straight road cyclist are riding one after the other at a separation of 30 m at a speed of 10 m/s. In the same direction boys are running at a separation of 20 m at speed 5 m/s. If an old man is walking in opposite direction, find his speed so that he will meet each cycle and boy together.

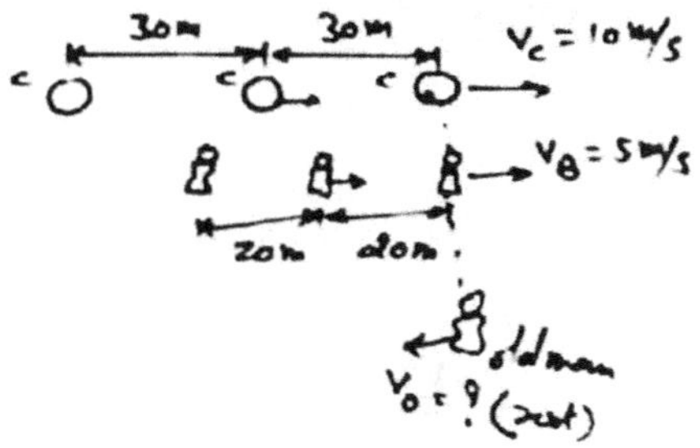


Sol$^n$:

vel of cyclist w.r.t. old man

$V_{co} = 10 + V_0$

vel of boy w.r.t. old man

$V_{BO} = 5 + V_0$

time for boy & cyc to reach old man are equal $\Rightarrow$ $\frac{3d}{10+V_0} = \frac{2d}{5+V_0}$

$15 + 3V_0 = 20 + 2V_0$

$V_0 = 5$ m/s Ans.

Web Reference at www.physicsgalaxy.com | Scan for Video Explanation

**Module-2: Solved Example-1**

**Ex :** A thief is running away on a straight road at 9 m/s. A policeman chases him on a motor cycle at 10 m/s. If the separation between them is 100 m, how long will it take for policeman to catch the thief ?

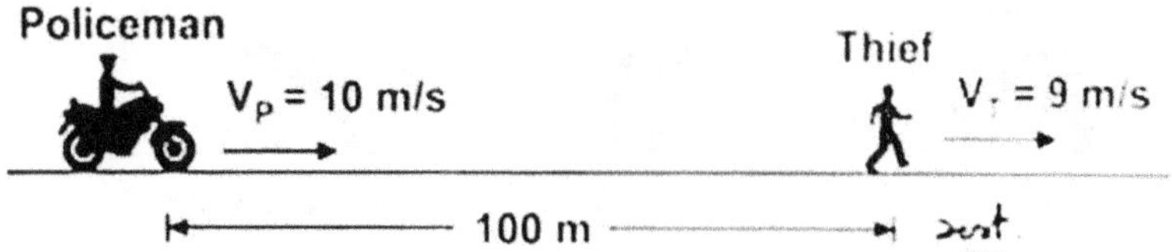


Sol$^n$: vel of policeman w.r.t. thief $\vec{V}_{PT} = \vec{V}_P - \vec{V}_T$

$= 10 - 9 = 1$ m/s

time to catch $= \frac{100}{1} = 100$ sec Ans.

Web Reference at www.physicsgalaxy.com | Scan for Video Explanation

**Module-3: Solved Example-2**

**Ex :** A bus is moving at 10 m/s on a straight road. A scooterist 80 m behind moving at 5 m/s wishes to overtake the bus in 100 sec. Find the acceleration required for scooterist to do this.

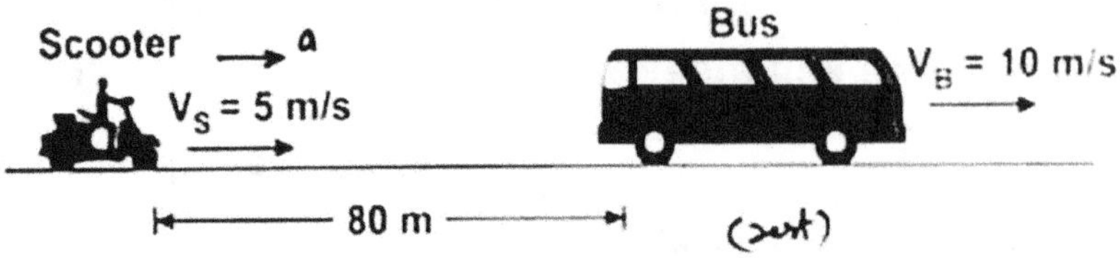


Sol$^n$: vel of Scooter w.r.to Bus is $v_{SB} = v_S - v_B$

$= 5 - 10 = -5$ m/s

to overtake the Bus we are

$$S = ut + \frac{1}{2}at^2$$

$$80 = (-5)(100) + \frac{1}{2}a(100)^2$$

$$80 = -500 + 5000a$$

$$a = \frac{580}{5000} = \frac{116}{1000} = 0.116 \text{ m/s}^2$$ Ans.

Scan for Video Explanation

Web Reference at www.physicsgalaxy.com

**Module-4: Solved Example-3**

**Ex :** Three elephants A, B and C are moving along a straight line as shown with constant speeds. When B catches C the separation between A and C becomes 3d. Find speed of B.

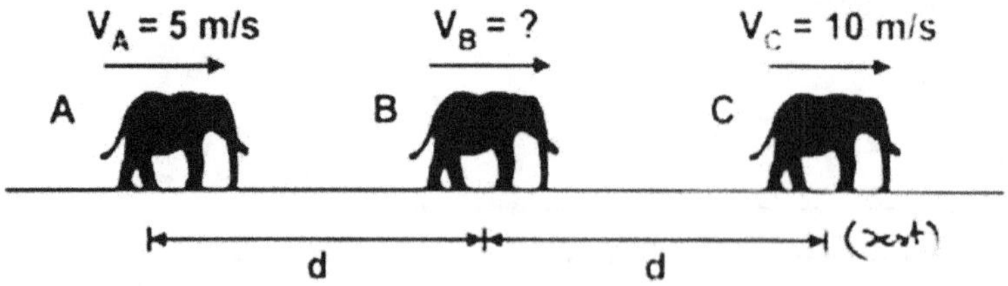


Sol$^n$: vel of B w.r.to C $\vec{V}_{BC} = \vec{V}_B - \vec{V}_C = V_B - 10$

vel of A w.r.to C $\vec{V}_{AC} = \vec{V}_A - \vec{V}_C = 5 - 10 = -5$ m/s

time for B to catch C is $t = \left(\frac{d}{V_B - 10}\right)$

$$d = 5 \times \frac{d}{V_B - 10} \Rightarrow V_B = 15 \text{ m/s}$$ Ans.

Scan for Video Explanation

Web Reference at www.physicsgalaxy.com

**Module-5: Motion of a Body in Moving Ref Frame**

# Motion of a body in a moving reference frame:

Basic Ex:

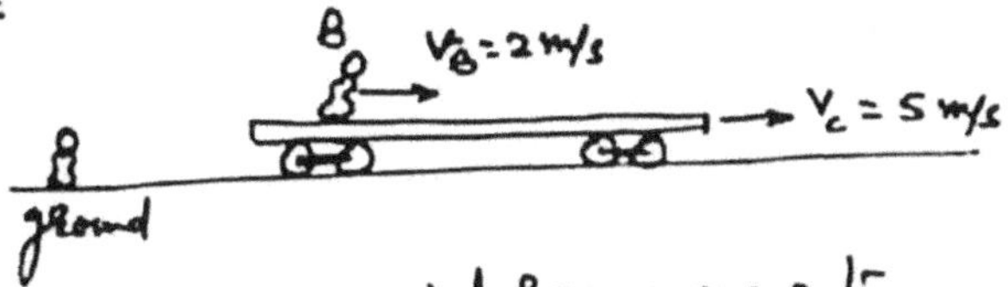


vel of swimmer w.r.to ground $V_{BG} = V_B + V_C$

$= 2 + 5 = 7$ m/s

Ex:

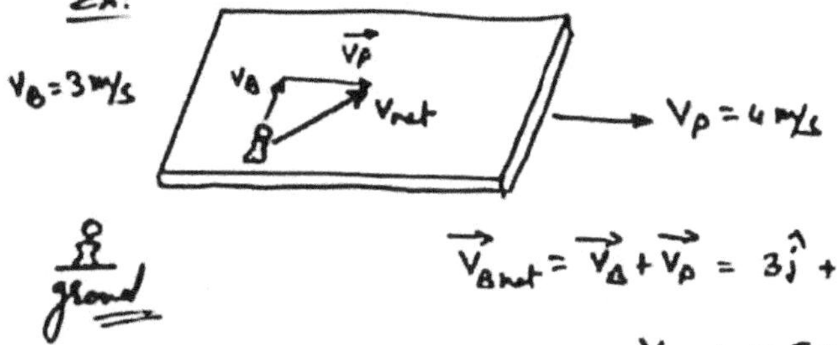


$\vec{V_P} = 4\hat{i}$

$\vec{V_B} = 3\hat{j}$

$\vec{V_{Bnet}} = \vec{V_B} + \vec{V_P} = 3\hat{j} + 4\hat{i}$

$V_{Bnet} = 5$ m/s $\angle \tan^{-1} 3/4 = 5 \angle 37°$ m/s

Web Reference at www.physicsgalaxy.com | Scan for Video Explanation

**Module-5: Motion of a Body in Moving Ref Frame**

# Motion of a Swimmer in a River:

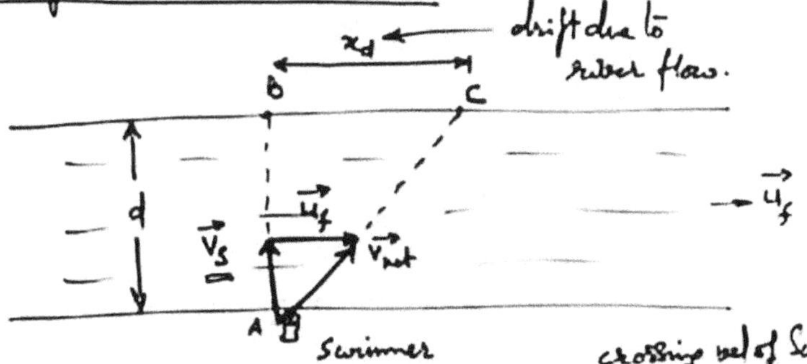


$\vec{V_S} \rightarrow$ vel of swimmer relative to water

crossing vel of Swimmer $= V_S$

Drift vel of Swimmer $= u_f$

net vel of swimmer w.r.to ground is

$\vec{V_{Snet}} = \vec{V_S} + \vec{u_f}$

time to cross river $t = \frac{d}{V_S}$

$V_{Snet} = \sqrt{V_S^2 + u_f^2}$

Drift $x_d = u_f \times t = \left(u_f \frac{d}{V_S}\right)$

$t = \frac{d}{V_S} = \frac{AC}{\sqrt{V_S^2 + u_f^2}}$

Web Reference at www.physicsgalaxy.com | Scan for Video Explanation

**Module-5: Motion of a Body in Moving Ref Frame**

# Downstream swimming in a flowing River:

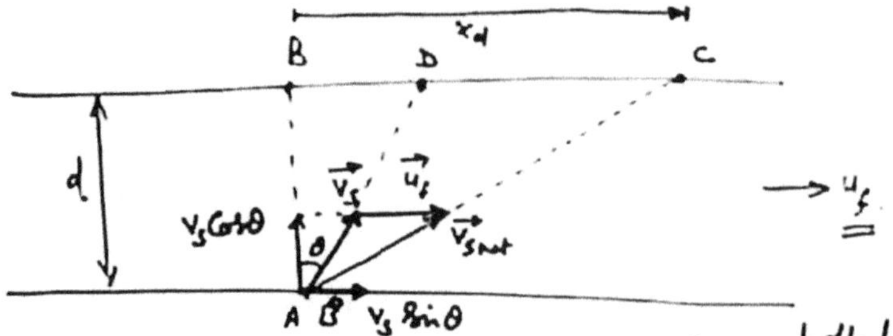

$DC \rightarrow$ drift due to river flow.

crossing vel of Swimmer $v_c = v_s \cos\theta$

Total drift vel of Swimmer $v_D = v_s \sin\theta + u_f$

$DC = u_f \cdot \dfrac{d}{v_s \cos\theta}$

time to cross river $t = \dfrac{d}{v_s \cos\theta}$

Total drift $x_d = (v_s \sin\theta + u_f) \cdot \dfrac{d}{v_s \cos\theta}$.

Scan for Video Explanation

Web Reference at www.physicsgalaxy.com

**Module-5: Motion of a Body in Moving Ref Frame**

# Upstream Swimming in a flowing River:

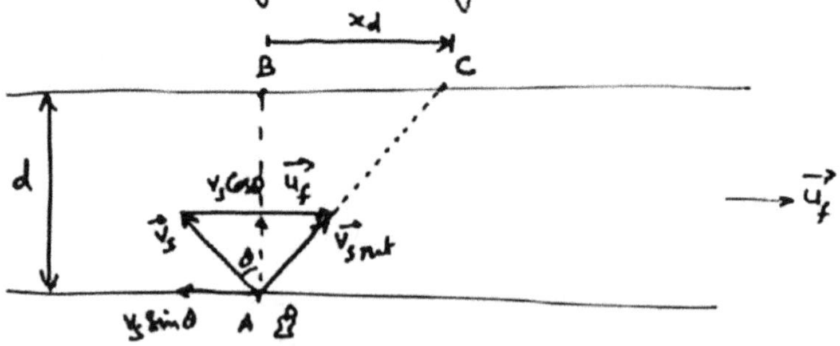

crossing vel $v_c = v_s \cos\theta$

Total drift vel $v_D = u_f - v_s \sin\theta$

time to cross river $t = \dfrac{d}{v_s \cos\theta} = \dfrac{AC}{v_{s\,net}}$

Total drift $x_d = (u_f - v_s \sin\theta) \cdot \dfrac{d}{v_s \cos\theta}$.

Scan for Video Explanation

Web Reference at www.physicsgalaxy.com

**Module-6: Solved Example-4**

**Ex :** A swimmer wishes to cross a river flowing at 3 kph in minimum time. Find the direction in which he should swim to achieve this if he can swim at 5 kph relative to water.

Sol$^n$:

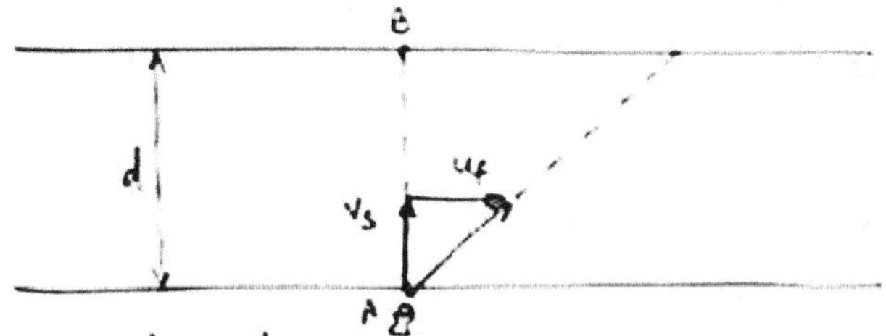


time to cross $= \frac{d}{V_c} = \frac{d}{V_s}$

NOTE: To minimize time to cross the river swimmer must swim in the dir perpendicular to river current. Ans.



**Module-7: Solved Example-5**

**Ex :** If a swimmer wishes to cross a river in the shortest path find the direction in which he should swim to achieve this. The velocity of river current is 4 kph and the swimmer can swim relative to water at 5 kph.

Sol$^n$:

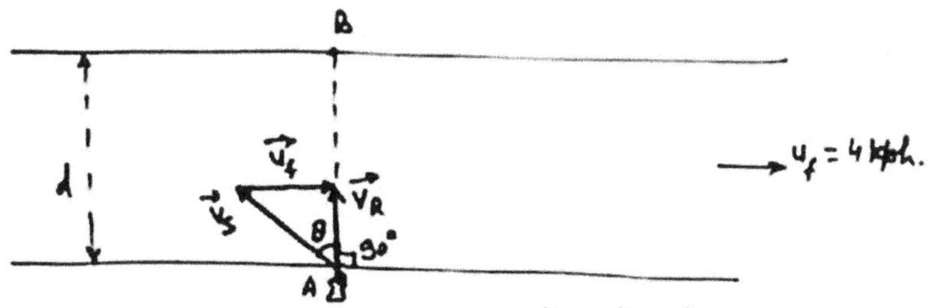


$\vec{V_R} = \vec{V_s} + \vec{u_f}$

$\sin\theta = \frac{u_f}{V_s} = \frac{4}{5}$

$V_R = \sqrt{V_s^2 - u_f^2}$

$= \sqrt{25 - 16} = \sqrt{9}$

$\theta = 53°$

$= 3$ kph.

at an angle 143° from the dir of river current. Ans.

**Module-8: Solved Example-6**

**Ex :** An Aeroplane which flies at speed $V_p$ tracks a horizontal square of side $l$ as seen from ground. Given that wind exist in direction parallel to one side of the square at speed $V_w$. Find the time taken by plane to complete one round of the square.

**Sol$^n$:**

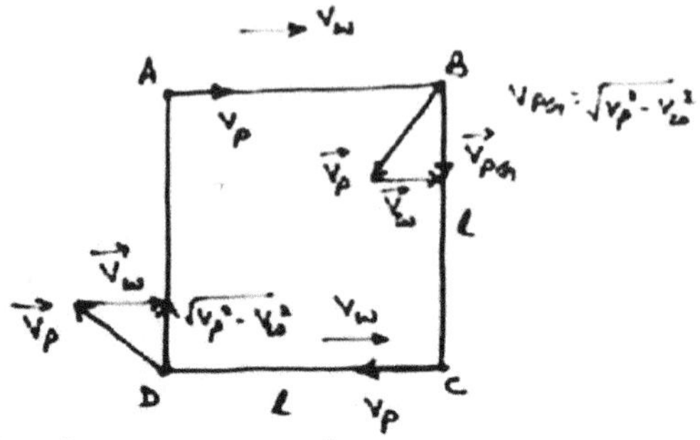


$$t_{AB} = \frac{l}{V_p + V_w}$$

$$t_{BC} = \frac{l}{\sqrt{V_p^2 - V_w^2}}$$

$$t_{CD} = \frac{l}{V_p - V_w}$$

$$t_{DA} = \frac{l}{\sqrt{V_p^2 - V_w^2}}$$

time for one complete round is

$$T = \frac{l}{V_p + V_w} + \frac{2l}{\sqrt{V_p^2 - V_w^2}} + \frac{l}{V_p - V_w} \quad \text{Ans.}$$

Scan for Video Explanation

Web Reference at www.physicsgalaxy.com

**Module-9: Solved Example-7**

**Ex :** A river flows at 20 m/s in which a swimmer jumps from one bank of river who can swim at 10 m/s relative to water. Find the direction from normal to bank of river at which he should swim so as to minimize his drift along river flow.

**Sol$^n$:**

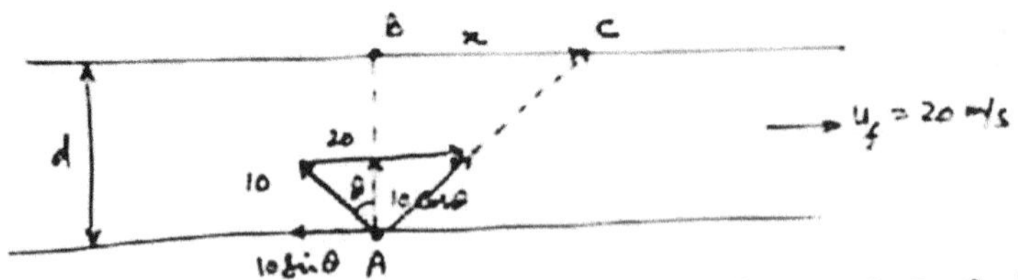


$V_c = 10\cos\theta$

$V_D = 20 - 10\sin\theta$

drift $x = (20 - 10\sin\theta)\dfrac{d}{10\cos\theta} = (2\sec\theta - \tan\theta)d$

$x$ will be min when $\dfrac{dx}{d\theta} = 0$ ; $\dfrac{dx}{d\theta} = 2\sec\theta\tan\theta - \sec^2\theta = 0$

$\sec\theta[2\tan\theta - \sec\theta] = 0$

$2\tan\theta = \sec\theta$

$\underline{\theta = 30^\circ}$ Ans. $\Leftarrow$ $\sin\theta = 1/2$

Scan for Video Explanation

Web Reference at www.physicsgalaxy.com

**Module-10: Rain Fall Cases**

# Observation of Rain Fall:

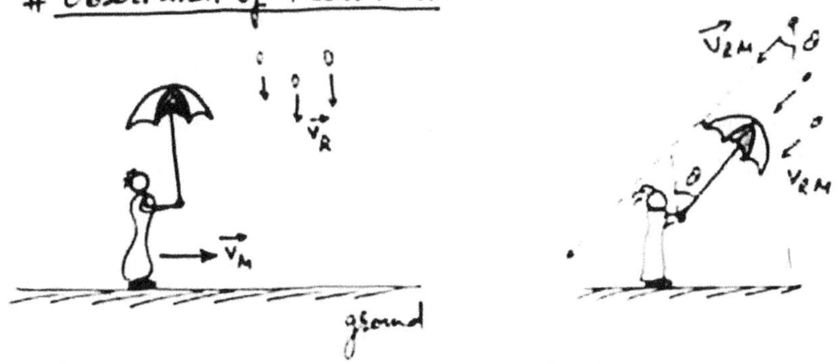


vel of rain drops w.r.to man $\vec{V}_{RM} = \vec{V}_R - \vec{V}_M$

$\tan\theta = \left(\frac{V_M}{V_R}\right)$

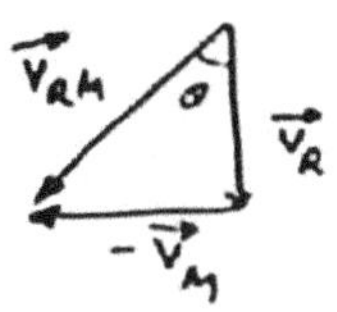




**Module-10: Rain Fall Cases**

**Ex :** A man is standing on ground with an umbrella hat as shown. Suddenly rain fall starts at speed $V_R$ = 20 m/s at an angle 30° with the vertical. Find the speed and direction in which man should start running so as to save himself.

Sol$^n$

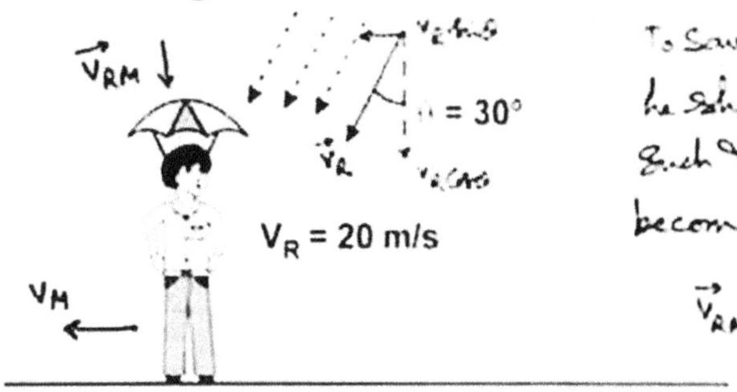


To save the man he should run such that $\vec{V}_{RM}$ becomes ↓ (vert down)

$\vec{V}_{RM} = \vec{V}_R - \vec{V}_M$

$\vec{V}_M = \vec{V}_R - \vec{V}_{RM}$

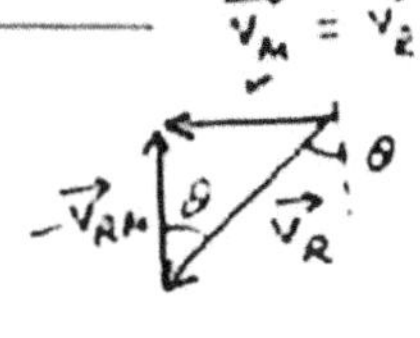


$V_M = V_R \sin\theta$

$= 20 \sin 30°$

$= 10$ m/s Ans.

**Module-11: Solved Example-8**

**Ex :** A truck shown in figure is travelling on horizontal road. At its back there are some cotton bags placed as shown in figure. Rain starts falling in vertical direction at speed 10 m/s. Find the velocity at which truck should move so that cotton bags will not get wet.

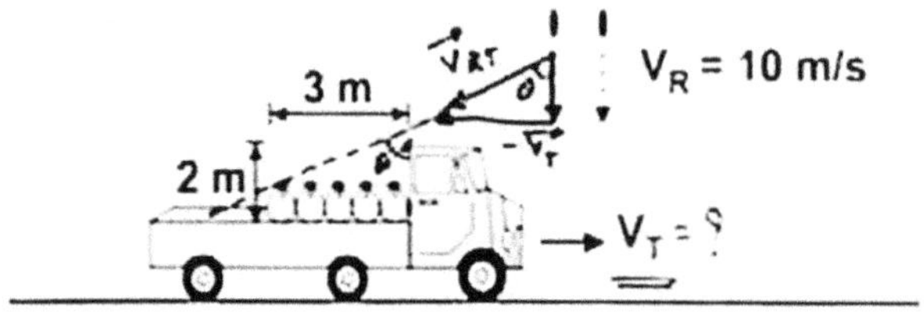


**Sol$^n$:** $\vec{V}_{RT} = \vec{V}_R - \vec{V}_T$

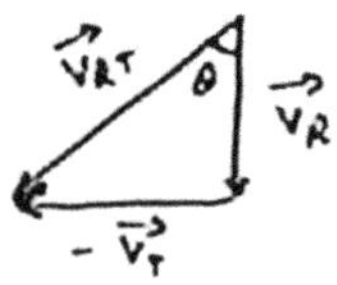

$\theta = \tan^{-1}(3/2)$

$\tan\theta = \frac{V_T}{V_R} = \frac{3}{2}$

$V_T = \frac{3}{2} \times 10 = 15 \text{ m/s}$ Ans.

Scan for Video Explanation

Web Reference at www.physicsgalaxy.com

**Module-12: Solved Example-9**

**Ex :** A pipe which can rotate in vertical plane is mounted on a cart as shown. The cart travels along the horizontal path at 2 m/s. At what angle α to the horizontal the pipe be placed so that rain drops move parallel to the walls of pipe, if rain is falling vertically at 6 m/s.

**Sol$^n$:**

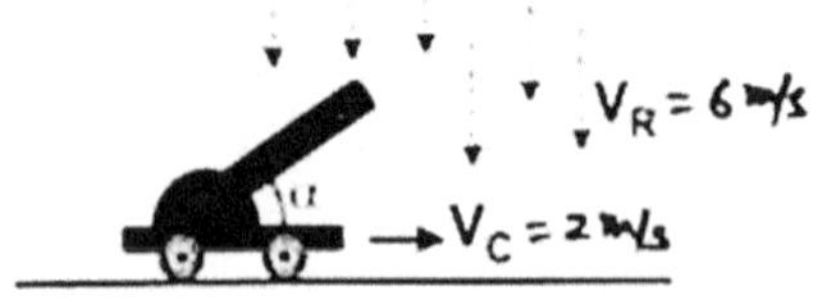


$\vec{V}_{RC} = \vec{V}_R - \vec{V}_C$

$\tan\alpha = \frac{V_R}{V_C} = \frac{6}{2}$

$\alpha = \tan^{-1}(3)$ Ans.

Scan for Video Explanation

Web Reference at www.physicsgalaxy.com

**Module-13: Solved Example-10**

**Ex :** A man wearing a hat of extended length 12 cm is running in rain falling vertically down at 10 m/s. Find the maximum speed with which he can run so that rain drops does not fall on his face as shown.

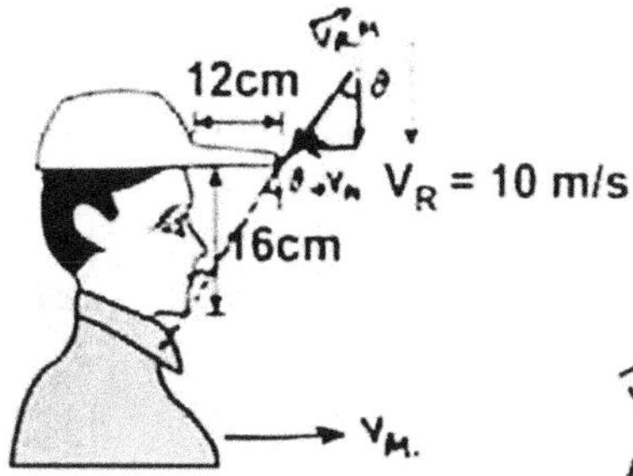


$\tan\theta = \frac{V_M}{V_R} = \frac{12}{16} = \frac{3}{4}$

Sol$^n$: $\vec{V}_{RM} = \vec{V}_R - \vec{V}_M$

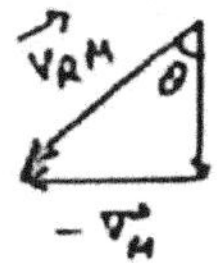


$V_M = \frac{3}{4} \cdot V_R$

$V_M = \frac{3}{4} \times 10$
$= 7.5$ m/s

Ans.



**Module-14: Condition of Collision of Two Particle**

# Condition of Collision of two particles in Space :

Vel of A w.r.to B.

$\vec{V}_{AB} = \vec{V}_A - \vec{V}_B$

we consider B at rest.

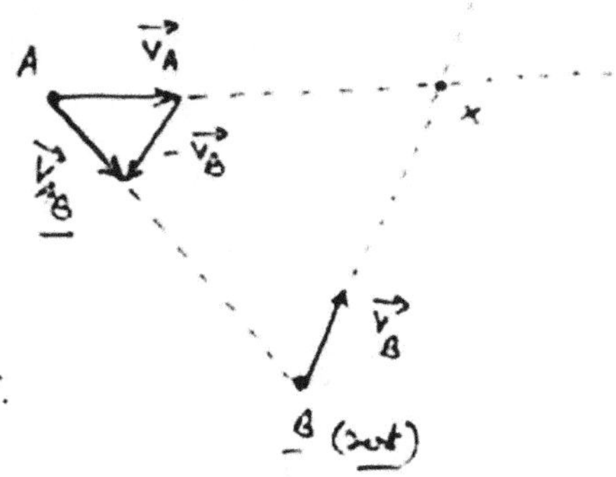


NOTE: For A to hit B, its relative velocity w.r.to B must be directed along line joining A to B.

## Module-14: Condition of Collision of Two Particle

Recall Ex:

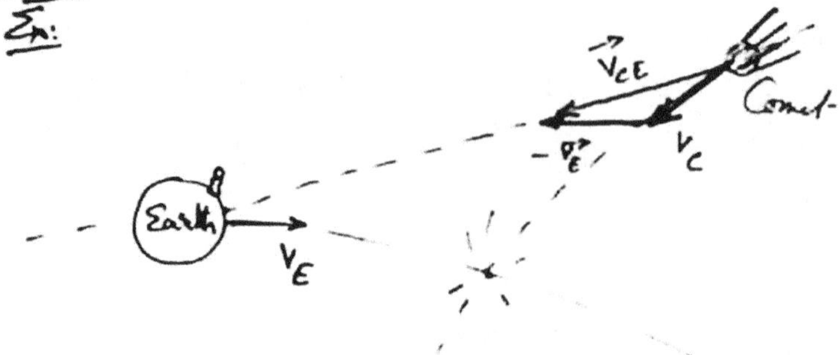


vel of Comet relative to Earth $\vec{V}_{CE} = \vec{V}_C - \vec{V}_E$

Scan for Video Explanation

Web Reference at www.physicsgalaxy.com

## Module-15: Solved Example-11

**Ex :** A car is going toward east at 8 m/s. To the passengers in car, a train appears to be moving in north direction at 15 m/s. Find the actual velocity of train.

Sol:

Car $\longrightarrow$ $\vec{V}_C = 8$ m/s

vel of Train w.r.to Car $\vec{V}_{TC} = 15$ m/s (↑)

$\vec{V}_{TC} = \vec{V}_T - \vec{V}_C$

$\vec{V}_T = \vec{V}_{TC} + \vec{V}_C$

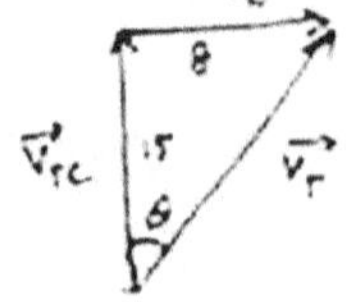


$\theta = \tan^{-1}(8/15)$

$V_T = \sqrt{(15)^2 + (8)^2}$
$= \sqrt{225 + 64} = \sqrt{289}$
$= 17$ m/s

$\vec{V}_T = 17$ m/s at $\tan^{-1}(8/15)$ E of N.

Ans.

Scan for Video Explanation

Web Reference at www.physicsgalaxy.com

**Module-16: Solved Example-12**

**Ex :** A helicopter is flying toward south at 50 kph. A train is moving with same speed toward east. Find velocity of helicopter as seen by passengers in train.

**Soln:**

$\vec{V}_H = 50$ kph ↓ $\quad \vec{V}_H = -50\hat{j}$

$\vec{V}_T = 50$ kph → $\quad \vec{V}_T = 50\hat{i}$

$\vec{V}_{HT} = \vec{V}_H - \vec{V}_T$

$= -50\hat{j} - 50\hat{i}$

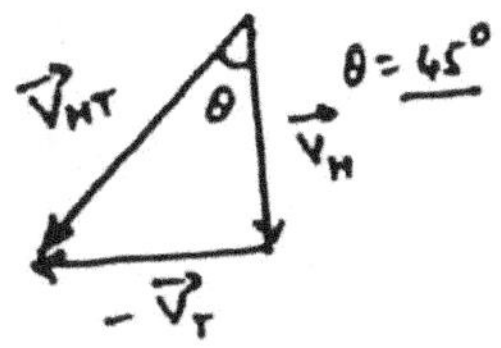


$V_{HT} = 50\sqrt{2}$ m/s at $\theta = \tan^{-1}(1) = 45°$ along SW dir.

$V_{HT} = 50\sqrt{2}$ m/s at 45° in SW dir.

Ans.

Web Reference at www.physicsgalaxy.com | Scan for Video Explanation

**Module-17: Solved Example-13**

**Ex :** An airplane flies toward west at 240 kph and after 1 Hr it is at a position which is 300 km west and 80 km south from its starting point. Find the wind speed and its direction.

**Soln:**

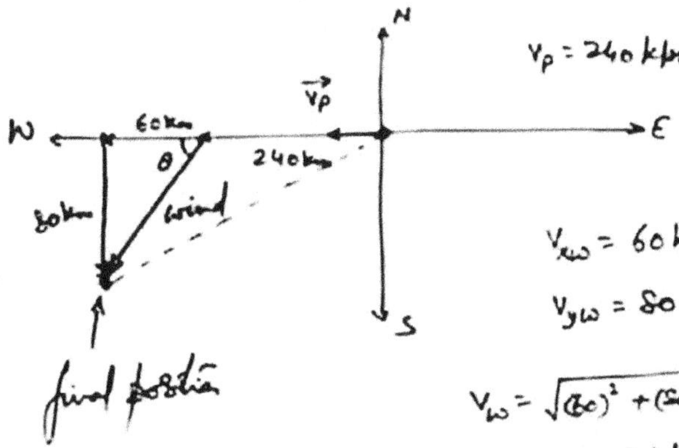


$V_P = 240$ kph.

$\tan\theta = \frac{80}{60} = \frac{4}{3}$

$\theta = 53°$

$V_{xw} = 60$ kph (west)

$V_{yw} = 80$ kph (south)

$V_w = \sqrt{(60)^2 + (80)^2}$

$V_w = 100$ kph.

at 53° South of West

Ans.

Web Reference at www.physicsgalaxy.com | Scan for Video Explanation

# Chapter 4
# Constrained Motion

## Lecture Notes Modules


**Module-1:** Constrained Motion
**Module-2:** Solved Example-1
**Module-3:** Solved Example-2
**Module-4:** Solved Example-3
**Module-5:** Solved Example-4
**Module-6:** Simple Pulley Constrained Motion
**Module-7:** Solved Example-5
**Module-8:** Solved Example-6
**Module-9:** Solved Example-7
**Module-10:** Wedge Constrained Motion
**Module-11:** Solved Example-8
**Module-12:** Solved Example-9
**Module-13:** Solved Example-10

**Module-1: Constrained Motion**

# Constrained Motion:

Motion of a body restricted under some physical conditions or governed by some mathematical rules is called Constrained Motion

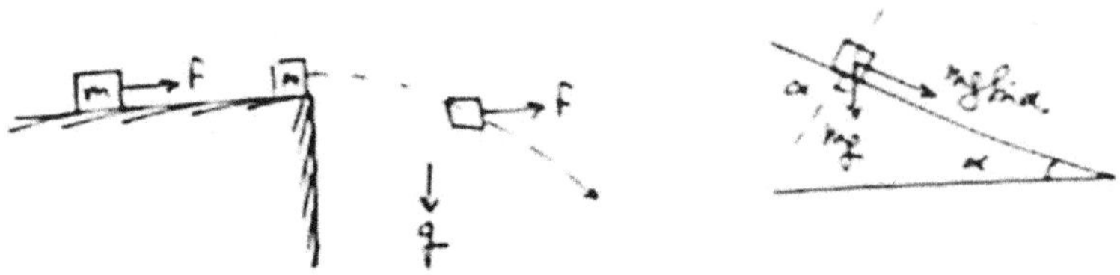


Scan for Video Explanation

Web Reference at www.physicsgalaxy.com

**Module-1: Constrained Motion**

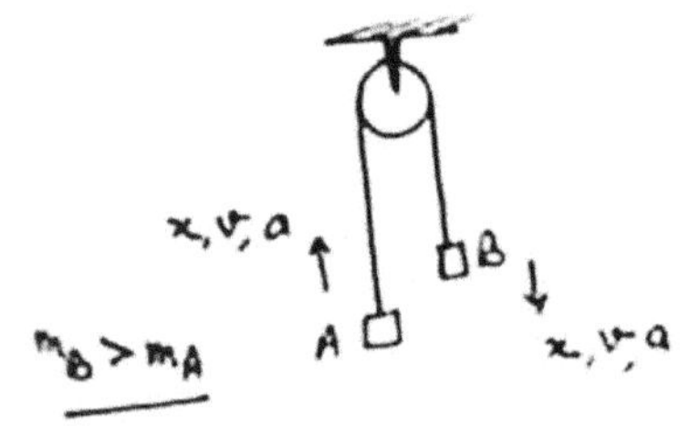


blocks A & B are Constrained to move with same speed & same accⁿ mag.

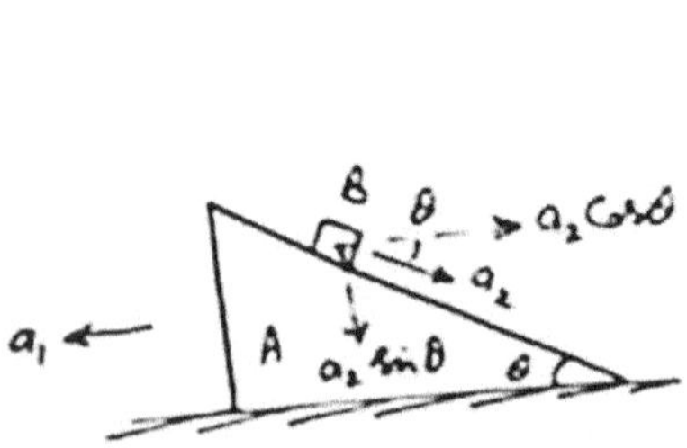


Block B is Constrained to move on Surface of A.

for B.

in x dir $a_x = a_2\cos\theta - a_1$

$a_y = a_2\sin\theta$.

net accⁿ of block B $a_B = \sqrt{a_x^2 + a_y^2} = ---$

Scan for Video Explanation

Web Reference at www.physicsgalaxy.com

**Module-1: Constrained Motion**

# Constrained Relations:

Mathematical relations by which motion of two or more bodies are related, are called Constrained Relations.

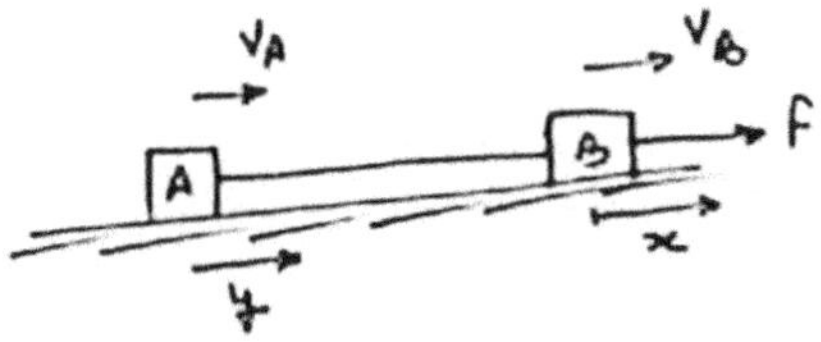


$$y = x$$

diff w.r.to t

$$\left.\begin{aligned} v_A &= v_B \\ a_A &= a_B \end{aligned}\right\} \text{Constrained Relations}$$

NOTE: To develop Constrained rel of vel & accⁿ try to relate displacement of bodies & diff w.r.to time.

Web Reference at www.physicsgalaxy.com | Scan for Video Explanation

**Module-2: Solved Example-1**

**Ex :** Find velocity of end A when rod is inclined at $\theta$ with horizontal.

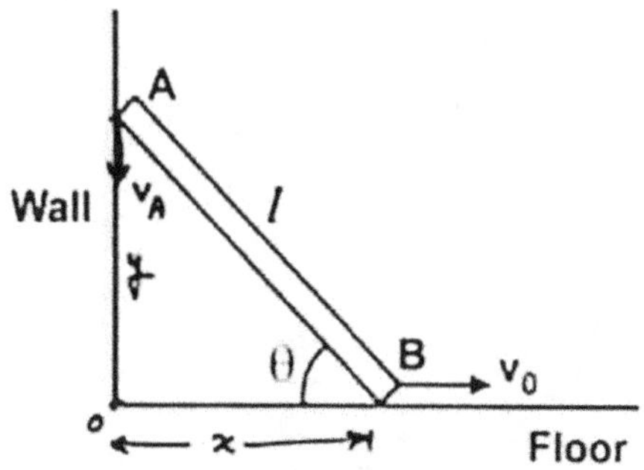


$$v_0 = \frac{dx}{dt} \qquad v_A = -\frac{dy}{dt}$$

Solⁿ:

rel in x in y $\quad x^2 + y^2 = l^2$

diff w.r.to t $\quad 2x\frac{dx}{dt} + 2y\frac{dy}{dt} = 0$

$$xv_0 - yv_A = 0 \Rightarrow v_A = \frac{x}{y}v_0$$

$$v_A = v_0 \cot\theta \quad \text{Ans.}$$

Web Reference at www.physicsgalaxy.com | Scan for Video Explanation

**Module-2: Solved Example-1**

## Alternative method :

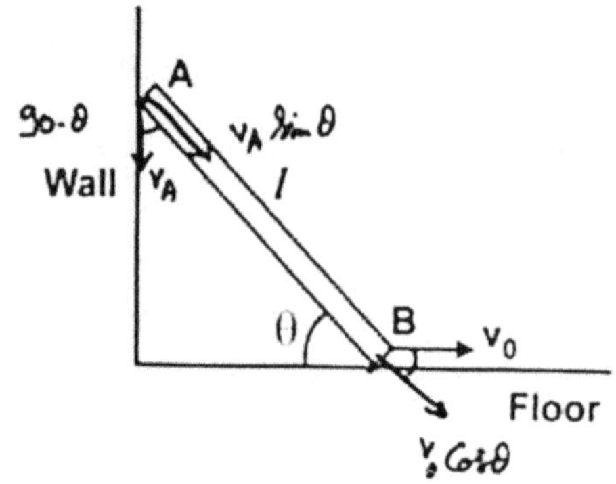


For $l$ to be constant

$$v_A \sin\theta = v_0 \cos\theta$$

$$v_A = v_0 \cot\theta \quad \text{Ans.}$$

v.imp. NOTE: For two pts, if their separation is constant then during motion their velocity components along the line joining these pts always remain equal.

Scan for Video Explanation

Web Reference at www.physicsgalaxy.com

**Module-3: Solved Example-2**

**Ex :** A toy boat in a swimming pool is pulled by a boy by a string as shown. If boy pulls the string at a constant speed u, find the speed with which boat will move when string makes an angle θ with horizontal.

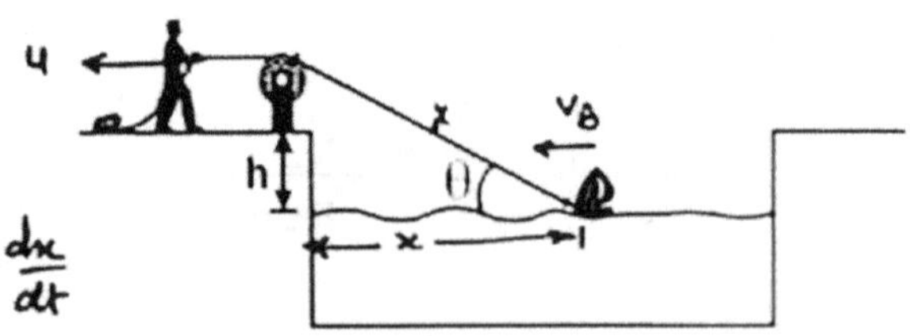


Sol$^n$:

$$v_B = -\frac{dx}{dt}$$

$$u = -\frac{dy}{dt}$$

Rel in x & y.

$$x^2 + h^2 = y^2$$

diff w.r.to t

$$2x\frac{dx}{dt} = 2y\frac{dy}{dt}$$

$$x\, v_B = y\, u$$

$$v_B = \frac{u}{x} \cdot y = u \sec\theta \quad \text{Ans.}$$

Scan for Video Explanation

Web Reference at www.physicsgalaxy.com

**Module-3: Solved Example-2**

## Alternative method :

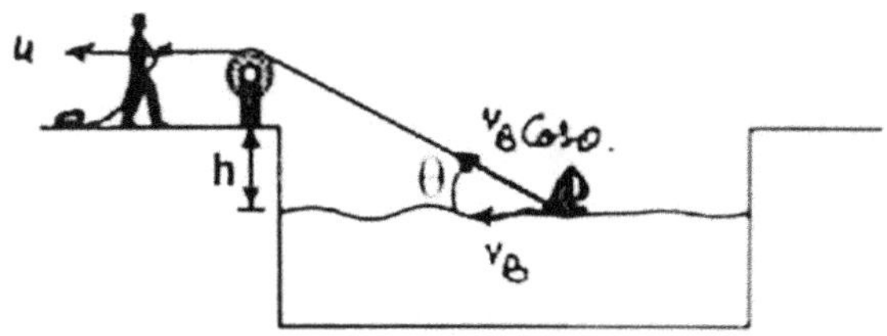


for string to remain tight $\quad u = v_B \cos\theta$

$$v_B = u \sec\theta \quad \text{Ans.}$$

NOTE: For an inextensible string always velocity components of all points on string along its length remain same (equal).



**Module-4: Solved Example-3**

Ex : In the system shown in figure block A is pulled toward left at speed $v_1$, find velocity of block B when string makes an angle θ with horizontal.

Alternative. $v_B = v_1 \cos\theta$.

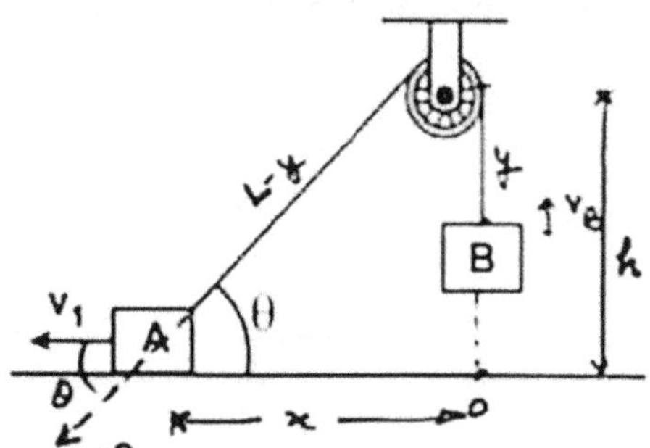


$v_1 = \frac{dx}{dt}$

$v_B = -\frac{dy}{dt}$

Sol<sup></sup>n: rel in x & y.

$$x^2 + h^2 = (L-y)^2$$

diff w.r.t t $\quad 2x\frac{dx}{dt} = 2(L-y)\left(-\frac{dy}{dt}\right)$

$$x v_1 = (L-y) v_B \;\longrightarrow\; v_B = \left(\frac{x}{L-y}\right)\cdot v_1 = v_1 \cos\theta \quad \text{Ans.}$$

**Module-5: Solved Example-4**

**Ex :** In the system shown in figure. Find velocity of block A, if B is pulled toward right speed $v_0$.

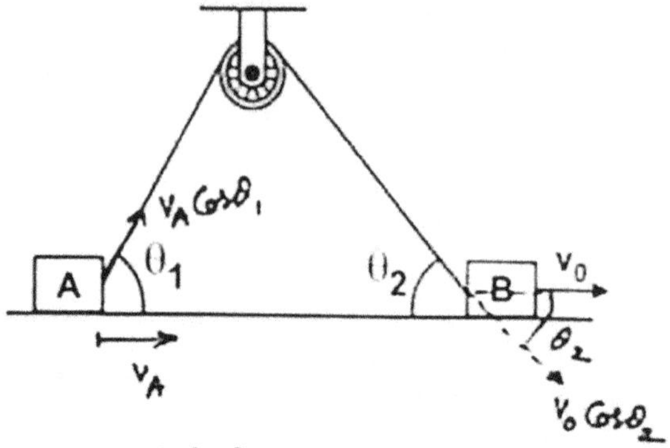


Sol.

$$v_A \cos\theta_1 = v_0 \cos\theta_2$$

$$v_A = \frac{v_0 \cos\theta_2}{\cos\theta_1} \quad \text{Ans.}$$

Scan for Video Explanation

Web Reference at www.physicsgalaxy.com

**Module-6: Simple Pulley Constrained Motion**

# Simple Pulley-Block Constrained Relations:

In pulley system these are developed by analyzing loosening & tightening of strings due to motion of blocks.

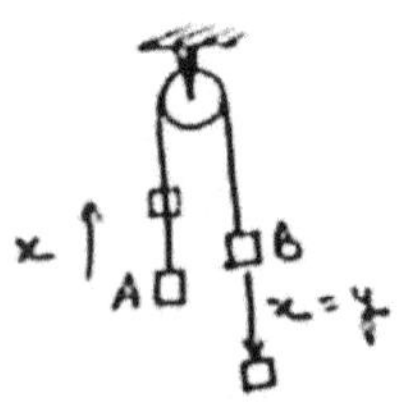


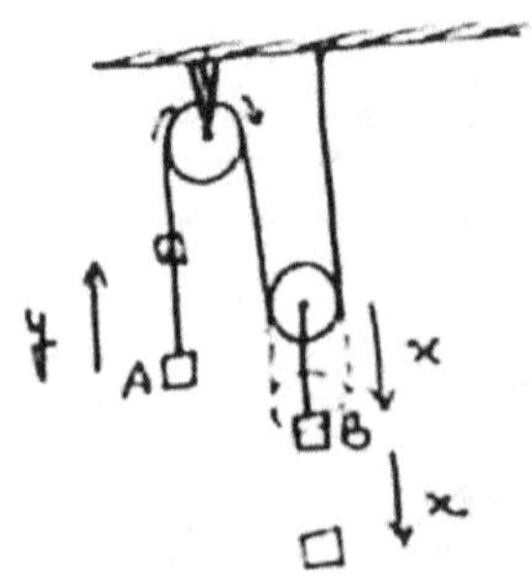


$y = 2x$
$v_A = 2v_B$
$a_A = 2a_B$

$y = x$
$v_B = v_A$
$a_B = a_A$

Scan for Video Explanation

Web Reference at www.physicsgalaxy.com

**Module-6: Simple Pulley Constrained Motion**

**Ex :** In the system shown in figure, if block A moves upward at 3 m/s, find speed of block B.

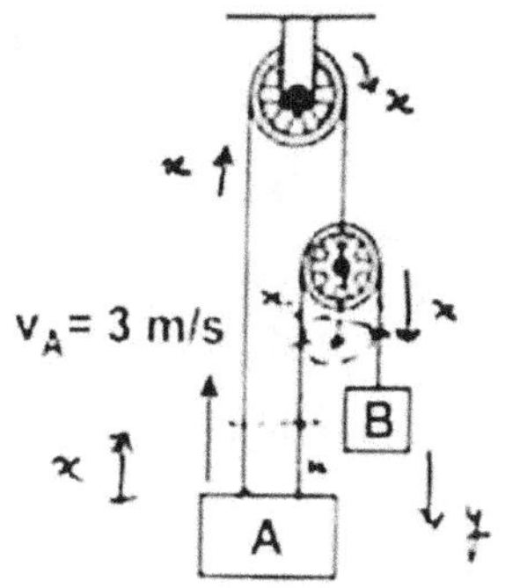


Sol:

$y = 3x$

$v_B = 3v_A$

$v_B = 3 \times 3 = 9$ m/s ↓ Ans.

Web Reference at www.physicsgalaxy.com

Scan for Video Explanation

**Module-7: Solved Example-5**

**Ex :** In the system shown in figure find speed of block C.

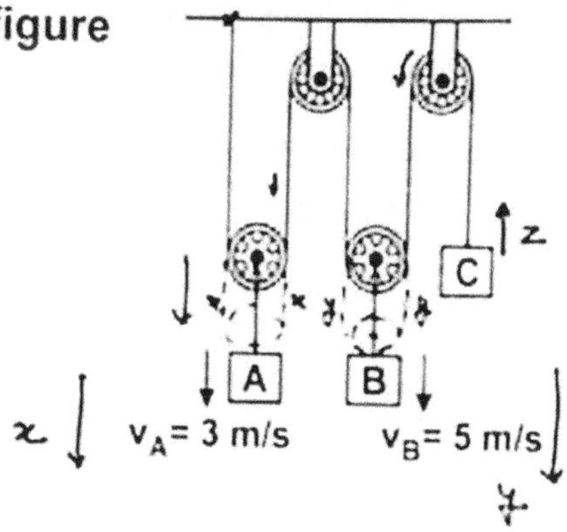


Sol:

$z = 2x + 2y$

$v_C = 2v_A + 2v_B$

$= 2 \times 3 + 2 \times 5$

$= 16$ m/s ↑ Ans.

Web Reference at www.physicsgalaxy.com

Scan for Video Explanation

**Module-8: Solved Example-6**

**Ex :** In the system shown in figure, the truck is moving toward left at 10 m/s and block P goes up at 3 m/s. Find speed of block Q.

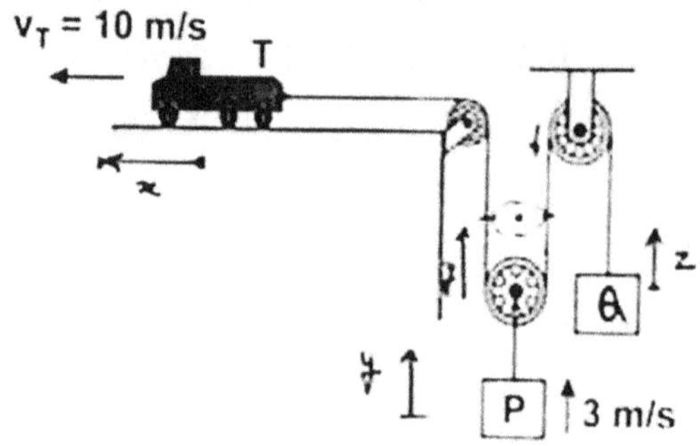


**Sol:**

$$x = 2y + z$$

$$v_T = 2v_P + v_Q$$

$$v_Q = v_T - 2v_P$$

$$= 10 - 6 = 4 \text{ m/s} \uparrow \quad \text{Ans.}$$

Scan for Video Explanation

Web Reference at www.physicsgalaxy.com

**Module-9: Solved Example-7**

**Ex :** Find acceleration of block B in the system shown in figure.

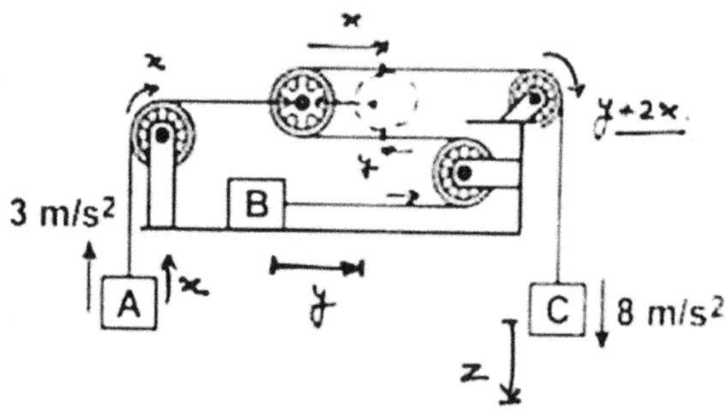


**Sol:**

$$z = 2x + y$$

$$a_C = 2a_A + a_B$$

$$a_B = a_C - 2a_A$$

$$a_B = 8 - 6 = 2 \text{ m/s}^2 \rightarrow \quad \text{Ans.}$$

Scan for Video Explanation

Web Reference at www.physicsgalaxy.com

**Module-10: Wedge Constrained Motion**

# Wedge Constrained Motion:

Whenever a body moves over another body under some constrained rules then such motion are called Wedge Constrained Motions.

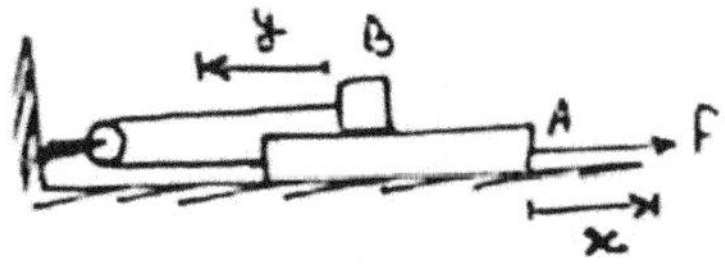


$$\left.\begin{aligned} y &= x \\ v_B &= v_A \\ a_B &= a_A \end{aligned}\right\}$$

**Web Reference at www.physicsgalaxy.com** | **Scan for Video Explanation**

**Module-10: Wedge Constrained Motion**

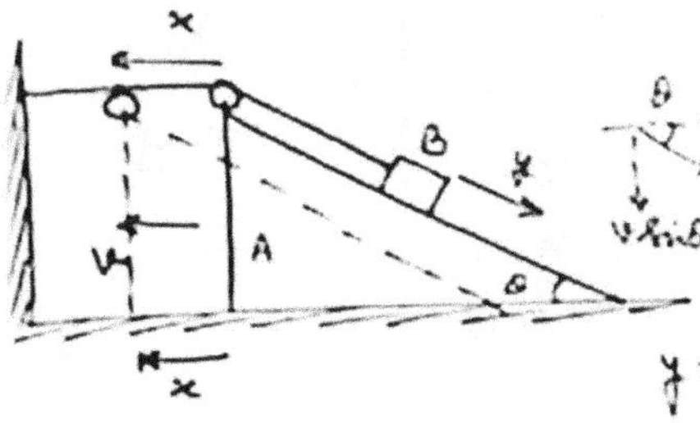


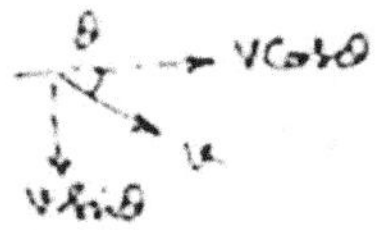


$y = x$

$v_B = v$ (relative to wedge A)

for B.

$$v_x = v - v\cos\theta$$

$$v_y = v\sin\theta$$

$$v_B = \sqrt{v_x^2 + v_y^2} = \sqrt{v^2(1-\cos\theta)^2 + v^2\sin^2\theta}$$

$$= \sqrt{v^2 + v^2 - 2v^2\cos\theta}$$

$$= \sqrt{v^2}\sqrt{2 - 2\cos\theta}$$

$$= 2v\sin\theta/2$$

**Web Reference at www.physicsgalaxy.com** | **Scan for Video Explanation**

**Module-11: Solved Example-8**

**Ex :** In the system shown in figure, the block A moves toward left at 2 m/s$^2$, find acceleration of rod B which is constrained to move vertically over the wedge.

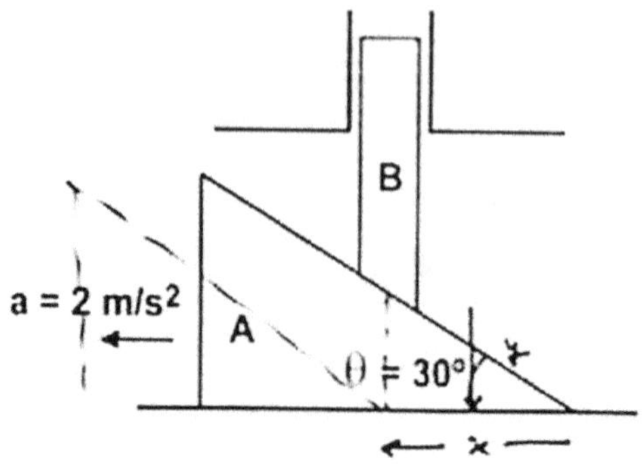


Solⁿ:

$$\tan 30^\circ = \frac{y}{x} \Rightarrow y = \frac{x}{\sqrt{3}}$$

$$a_B = \frac{a_A}{\sqrt{3}} = \frac{2}{\sqrt{3}} \text{ m/s}^2 \quad \text{Ans.}$$

Scan for Video Explanation

Web Reference at www.physicsgalaxy.com

**Module-12: Solved Example-9**

**Ex :** In the system shown in figure, the block A is pulled toward right at 5 m/s. Find the speed of block B.

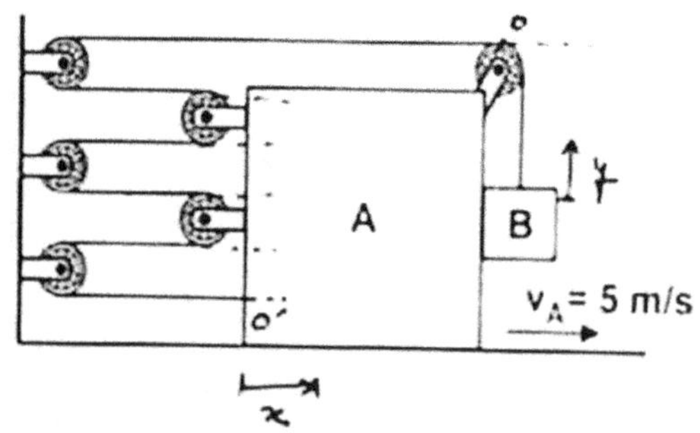


Solⁿ:

$$y = 6x$$

$$v_{By} = 6\,v_A = 6 \times 5 = 30 \text{ m/s}$$

$$v_{Bx} = 5 \text{ m/s}$$

$$v_B = \sqrt{v_{Bx}^2 + v_{By}^2} = \sqrt{25 + 900}$$

$$v_B = \sqrt{925} \text{ m/s} \quad \text{Ans.}$$

Scan for Video Explanation

Web Reference at www.physicsgalaxy.com

**Module-13: Solved Example-10**

**Ex :** In the system shown in figure, if block A is pulled toward right at acceleration a, find acceleration of block B.

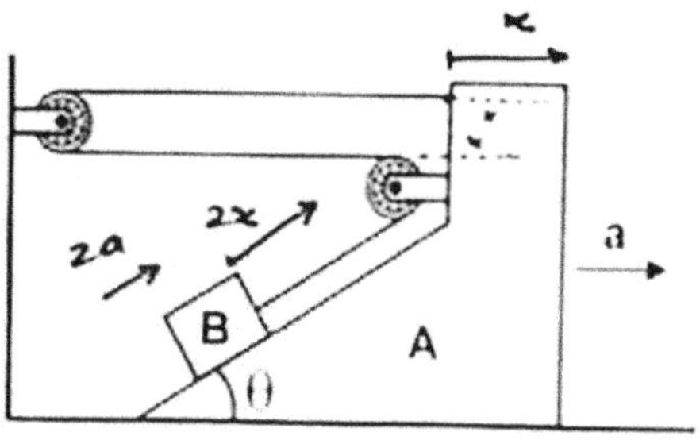


Sol$^n$: Acc$^n$ of block B.

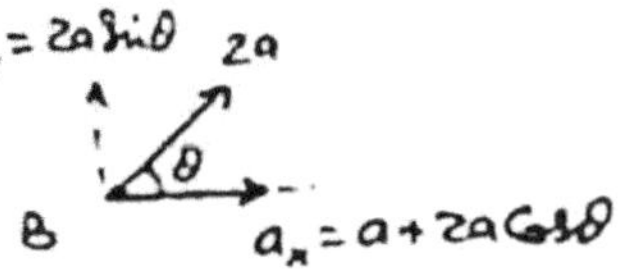


$$a_B = \sqrt{a_x^2 + a_y^2}$$

$$= \sqrt{a^2(1+2\cos\theta)^2 + (2a\sin\theta)^2}$$

$$= \sqrt{a^2 + 4a^2 + 4a^2\cos\theta}$$

$$a_B = a\sqrt{5 + 4\cos\theta} \quad \text{Ans.}$$

# Chapter 5
# Newton's Laws of Motion

## Lecture Notes Modules


**Module-1:** Forces of Nature
**Module-2:** Force can Change the State of Motion
**Module-3:** Balanced & Unbalanced Forces
**Module-4:** Newton's Laws of Motion
**Module-5:** Free Body Diagram
**Module-6:** Normal Reaction
**Module-7:** Tension in String
**Module-8:** Applications of Newton's Second Law
**Module-9:** Solved Example-1
**Module-10:** Solved Example-2
**Module-11:** Solved Example-3
**Module-12:** Solved Example-4
**Module-13:** Solved Example-5
**Module-14:** Concept of Weighing Machine
**Module-15:** Spring Force
**Module-16:** Concept of Spring Balance
**Module-17:** Concept of Pseudo Force
**Module-18:** Solved Example-6
**Module-19:** Solved Example-7
**Module-20:** Solved Example-8
**Module-21:** Solved Example-9
**Module-22:** Solved Example-10

**Module-1: Forces of Nature**

# Force:

A push or pull which can cause change in state of motion, is called a force.

state of motion → $\vec{v}$ → velocity

Force can change $\vec{v}$ → magnitude, → direction

All forces of Nature are classified in 4 fundamental Categories.

Scan for Video Explanation

Web Reference at www.physicsgalaxy.com

**Module-1: Forces of Nature**

# Fundamental Forces of Nature:

(1) Electromagnetic forces ] → large range forces
weakest ⇒ (2) Gravitational forces ]

Nuclear range forces [ (3) Strong forces ] → short range force (appear only at separation $< 10^{-12}$ m / $10^{-15}$ m)
Nuclear range forces [ (4) Weak forces ]

Scan for Video Explanation

Web Reference at www.physicsgalaxy.com

**Module-1: Forces of Nature**

# Relative Strength of Fundamental forces.

for a given pair of particles at short range

(1) Grav. force → reference.

(2) Weak forces → $10^{25}$ times stronger than grav forces

(3) Electromagnetic forces → $10^{36}$ times stronger than grav force

(4) Strong forces → $10^{38}$ times stronger than grav. forces.

Web Reference at www.physicsgalaxy.com | Scan for Video Explanation

**Module-2: Force can Change the State of Motion**

# Force can change State of Motion:

↓ $\vec{v}$

Force causes acceleration ⇒ change in $\vec{v}$.

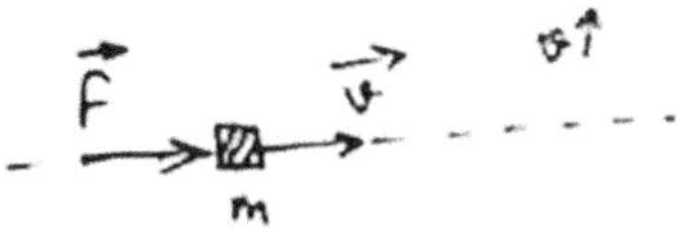

mag is increasing

dir of $\vec{v}$ is changing

a parallel force → causes change in mag of $\vec{v}$

a normal force → causes change in dir of $\vec{v}$

Web Reference at www.physicsgalaxy.com | Scan for Video Explanation

**Module-2: Force can Change the State of Motion**

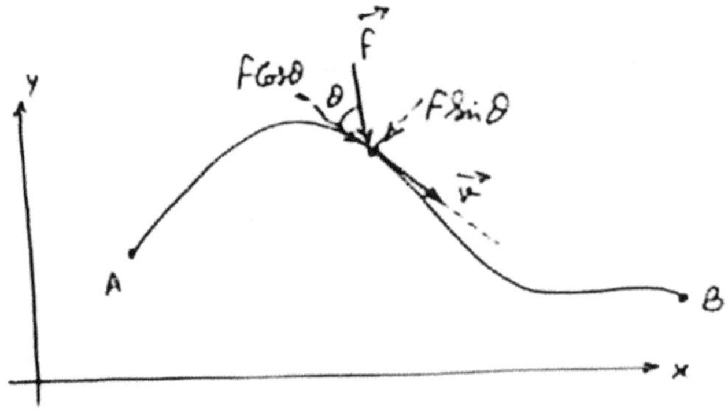


[ FCosθ → responsible for change in mag of $\vec{v}$
FSinθ → responsible for change in dir of $\vec{v}$ ]

Scan for Video Explanation



**Module-3: Balanced & Unbalanced Forces**

# Classification of Group of Forces:

When two or more forces act on a body, then based on its state of motion, these forces are classified in two categories.

(1) Balanced Forces: $\Sigma\vec{F}=0 \Rightarrow \vec{a}=0$, $\vec{v}$ = Constant < rest / uniform motion

↓ Equilibrium

(2) Unbalanced Forces: $\Sigma\vec{F}\neq 0$

accelerated motion $\vec{a}=\left(\frac{\Sigma\vec{F}}{m}\right)$

$\vec{v}$ changes.

Scan for Video Explanation

**Module-3: Balanced & Unbalanced Forces**

# Static Equilibrium:

When under the action of two or more forces a body is at rest then it is said to be in 'Static Equilibrium'

$$\Sigma \vec{F} = 0.$$

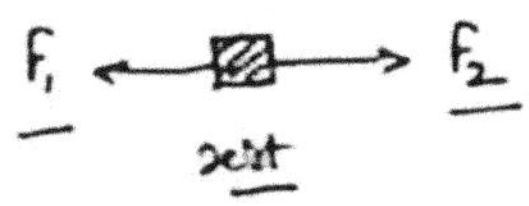


$\Rightarrow F_1 = F_2$

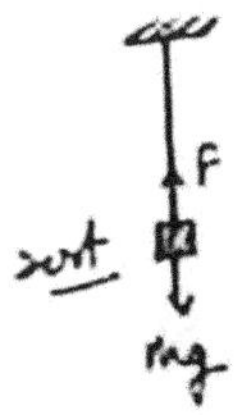


$F = mg.$

Web Reference at www.physicsgalaxy.com | Scan for Video Explanation

**Module-3: Balanced & Unbalanced Forces**

When a body is under the state of Equilibrium in the influence of 3 forces then these forces must successively represent the sides of a triangle.

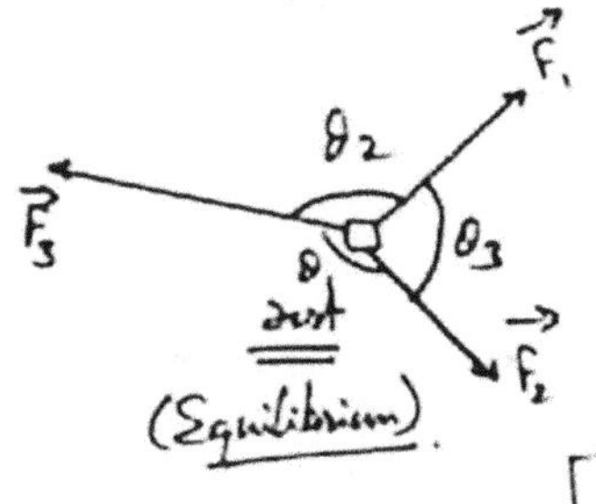


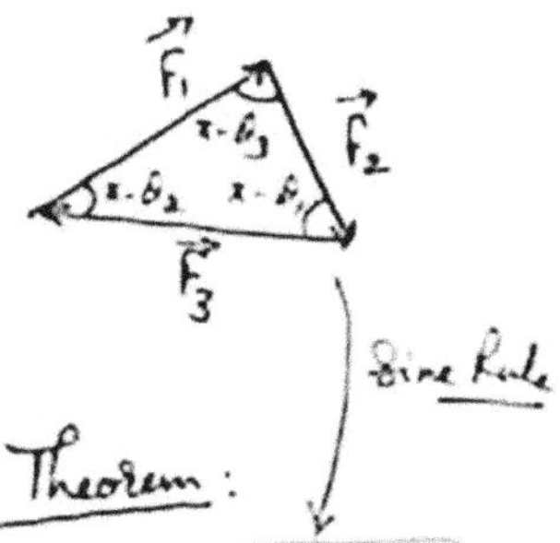


Lami's Theorem:

$$\frac{F_1}{\sin\theta_1} = \frac{F_2}{\sin\theta_2} = \frac{F_3}{\sin\theta_3}$$

Web Reference at www.physicsgalaxy.com | Scan for Video Explanation

**Module-4: Newton's Laws of Motion**

# Newton's Laws of Motion:

I law: Inertia law:

Inertia of a body $\propto$ mass

Inertia: property of a body to resist change in state of motion

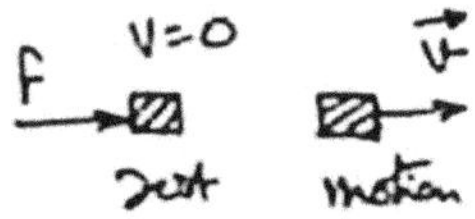


to change $\vec{v}$ we need $\vec{F}$

heavy body → large inertia ⇒ change in $\vec{v}$ will be less for a given $\vec{F}$

light body → low inertia ⇒ change in $\vec{v}$ will be large for a given $\vec{F}$.

Scan for Video Explanation

Web Reference at www.physicsgalaxy.com

**Module-4: Newton's Laws of Motion**

II law: law of unbalanced forces.

Force causes acceleration

Unbalanced forces on a body accelerates it & we use

$$\Sigma \vec{F} = m\vec{a}$$

$\vec{a} = \frac{d\vec{v}}{dt}$ $\quad$ $\Sigma \vec{F} = m\frac{d\vec{v}}{dt}$ $\quad$ $d\vec{p} = m d\vec{v}$

$$\Sigma \vec{F} = \frac{d\vec{p}}{dt}$$ ← rate of change of momentum.

Scan for Video Explanation

Web Reference at www.physicsgalaxy.com

**Module-4: Newton's Laws of Motion**

## III law: Law of Actions & Reaction

→ Force always act in pairs

⇓

Whenever a force is applied, always there exist an equal & opposite Reaction

$$\left[\begin{array}{c} |\vec{F}_1| = |\vec{F}_2| \\ \vec{F}_1 = -\vec{F}_2 \end{array}\right]$$

→ Action & Reaction are always equal & opposite

→ Action & Reaction both always act on different bodies.



Scan for Video Explanation

**Module-4: Newton's Laws of Motion**

## # Commonly used Forces in Mechanics:

In study of Rest & Motion most Commonly used forces are -

Electromagnetic in Nature ←
(1) Normal Reaction at Contacts (N)
(2) Tension in a String (T)

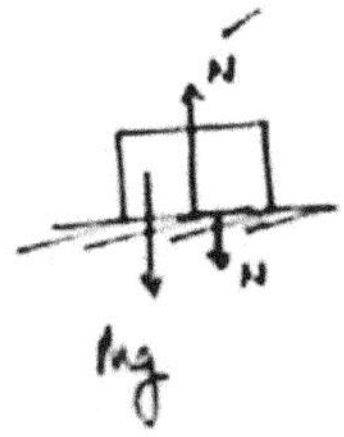


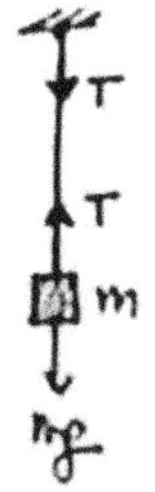




Scan for Video Explanation

## Module-5: Free Body Diagram

# Free Body Diagram: (FBD)

A diagram showing all forces acting on a body using arrows, considering body as a point mass, is called FBD of body.

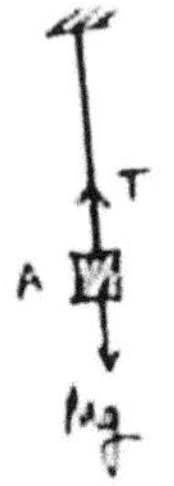


FBD of A

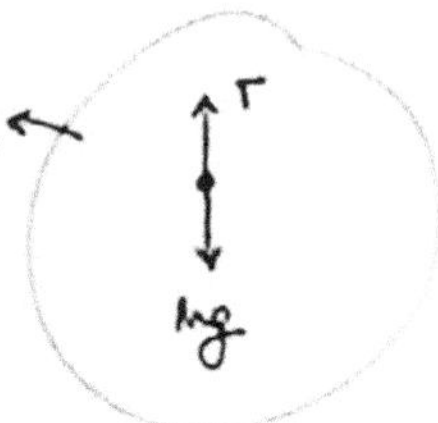


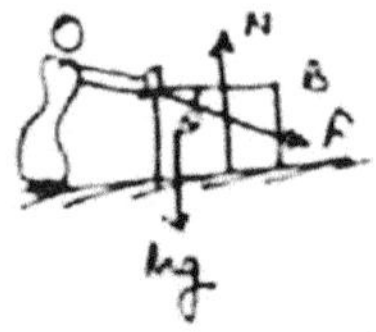


FBD of B

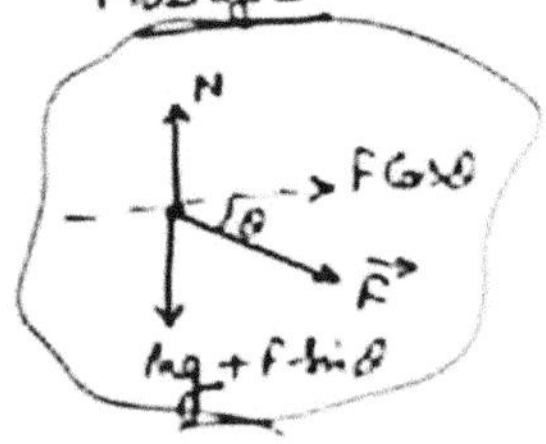


Scan for Video Explanation

Web Reference at www.physicsgalaxy.com

## Module-6: Normal Reaction

# Normal Reaction:

→ It is a force which exist when two bodies are in contact and one pushes the other.

→ It always act in the direction normal to the surfaces in contact.

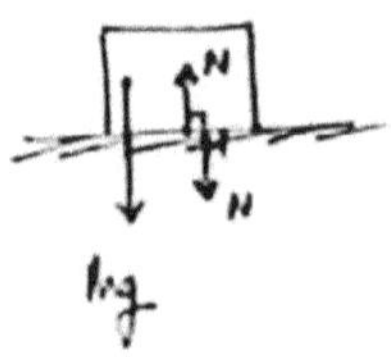


FBD of body

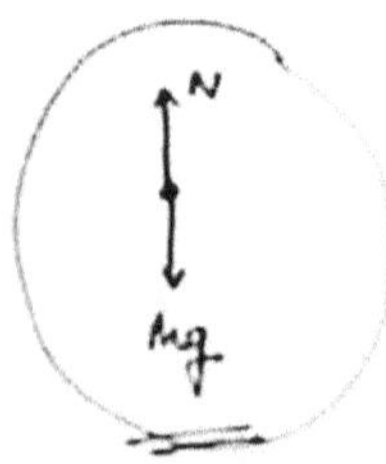


Scan for Video Explanation

Web Reference at www.physicsgalaxy.com

**Module-6: Normal Reaction**

**Ex :** Show all forces acting on the block & rod in their free body diagrams.

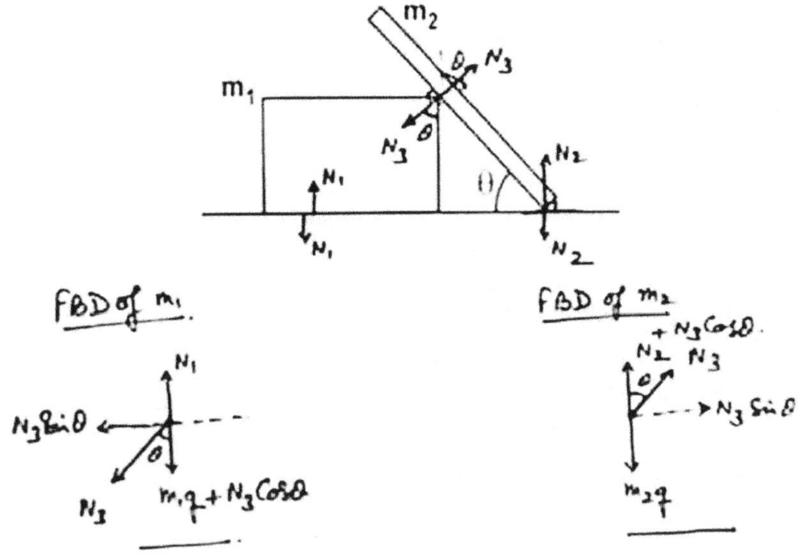




**Module-6: Normal Reaction**

NOTE: Incase of a circular, Spherical or cylindrical object, normal reaction always pass through the centre of object.

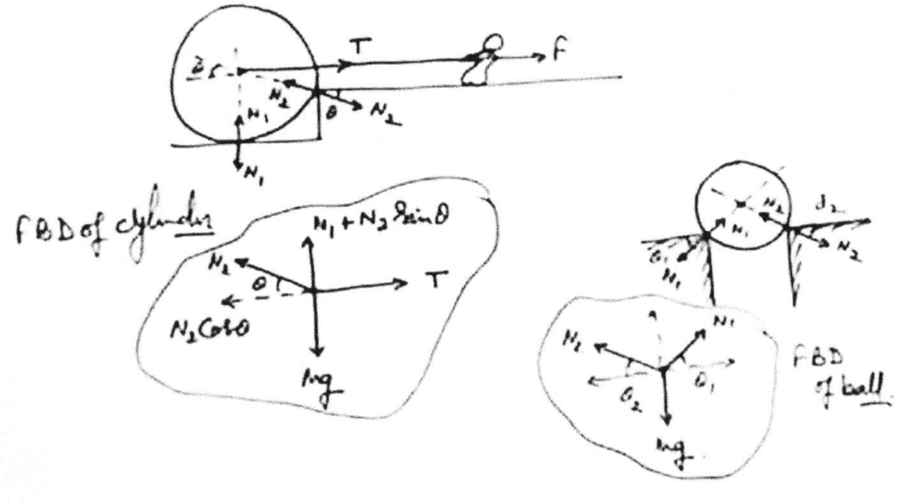

**Module-7: Tension in String**

# Tension in a String:

It is an internal molecular force which appear in a string when it is taut by some external force.

→ force of tension always act along the length of string

→ dir of tension is always away from tied ends or pts of contact of string

→ tension always has pulling tendency.

→ In an ideal string always, throughout its length the tension remains constant.

(ideal string ↓ massless & frictionless)

Scan for Video Explanation

Web Reference at www.physicsgalaxy.com

**Module-7: Tension in String**

Ex : Draw free body diagram of all blocks shown in figure.

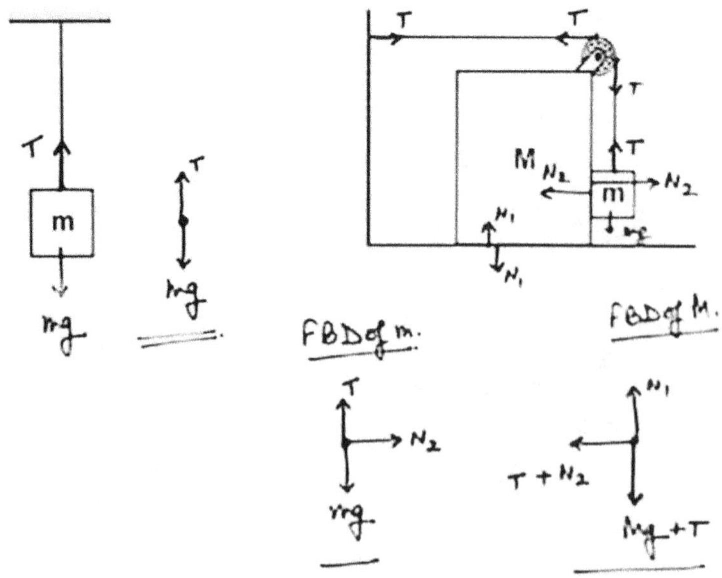


Scan for Video Explanation

Web Reference at www.physicsgalaxy.com

**Module-7: Tension in String**

**Ex :** Draw free body diagram of all blocks shown in figure.

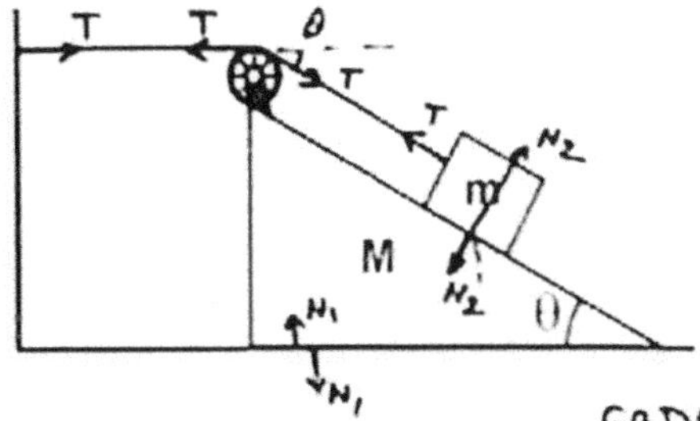


FBD of M

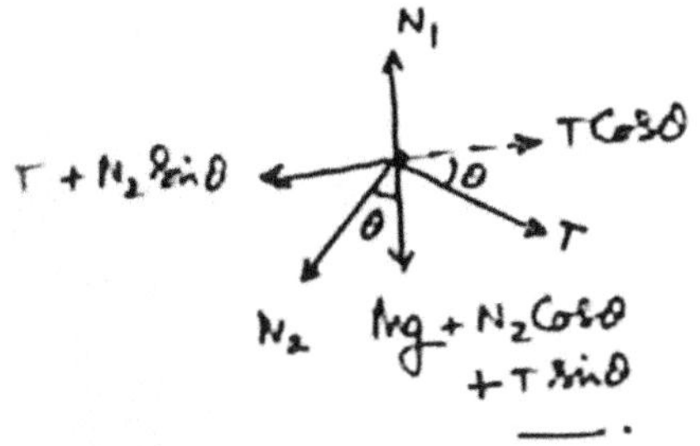


FBD of m

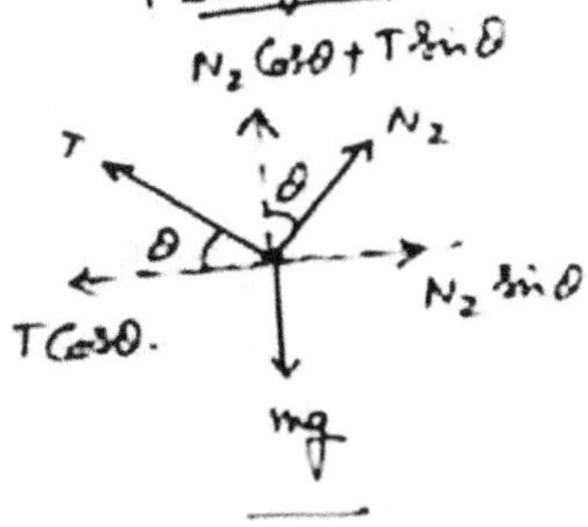


Web Reference at www.physicsgalaxy.com | Scan for Video Explanation

**Module-7: Tension in String**

**Ex :** Draw free body diagram of all blocks shown in figure.

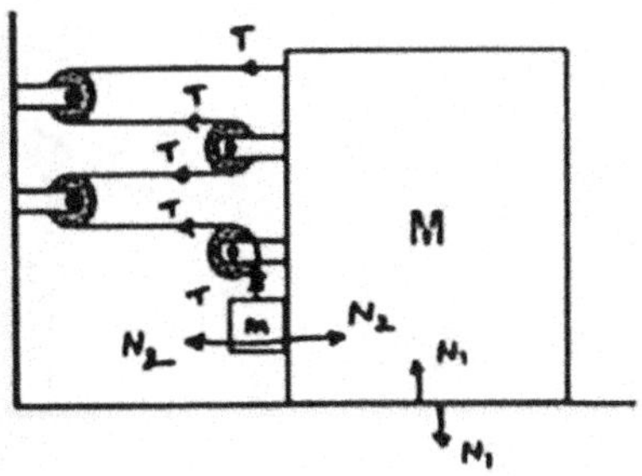


Sol:

FBD of M.

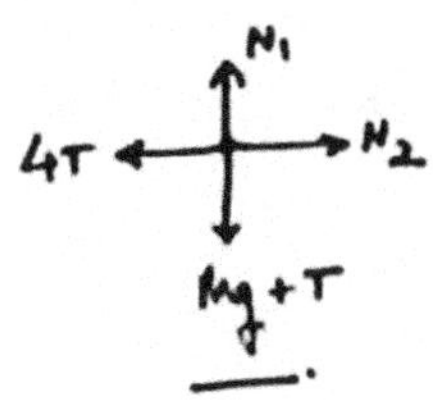


FBD of m

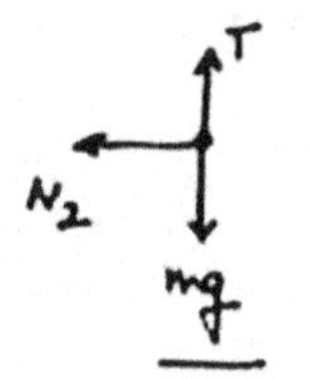


Web Reference at www.physicsgalaxy.com | Scan for Video Explanation

**Module-8: Applications of Newton's Second Law**

# Applications of Newton's Second Law:

$$\sum \vec{F} = m\vec{a}$$

**Working of Atwood's Machine:**

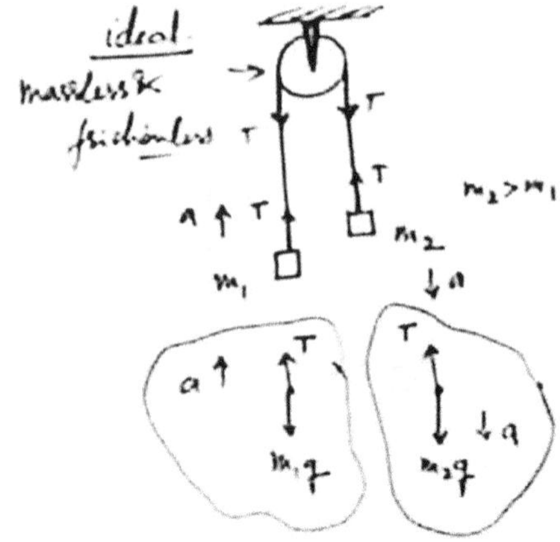


eqⁿ of motion of $m_1$ and $m_2$.

$$T - m_1 g = m_1 a \quad —(1)$$

$$m_2 g - T = m_2 a \quad —(2)$$

$$(m_2 - m_1) g = (m_1 + m_2) a$$

v. imp.: $$a = \left(\frac{m_2 - m_1}{m_1 + m_2}\right) g$$

from (1)

$$T - m_1 g = m_1 \left(\frac{m_2 - m_1}{m_1 + m_2}\right) g$$

$$T = \left(\frac{m_1 m_2 - m_1^2}{m_1 + m_2}\right) g + m_1 g$$

$$T = \frac{2 m_1 m_2}{m_1 + m_2} g$$

Scan for Video Explanation

Web Reference at www.physicsgalaxy.com

**Module-8: Applications of Newton's Second Law**

Ex : Find acceleration of blocks when released from rest.

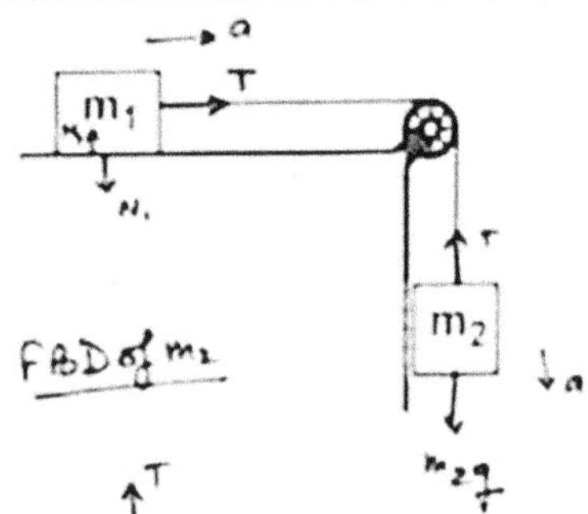


Sol: FBD of $m_1$.

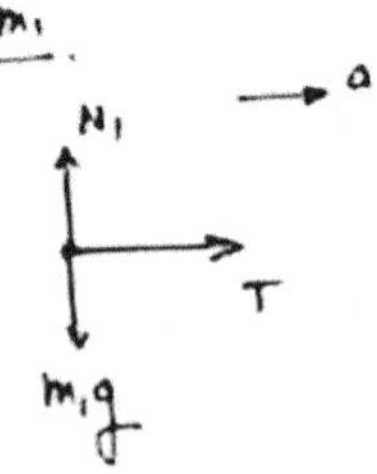


FBD of $m_2$

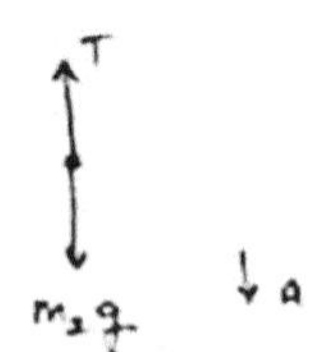


Eqⁿ of motions

$$T = m_1 a \quad —(1)$$

$$m_2 g - T = m_2 a \quad —(2)$$

(1) + (2) ⇒ $$m_2 g = (m_1 + m_2) a$$

$$a = \frac{m_2 g}{m_1 + m_2} \quad \text{Ans.}$$

Scan for Video Explanation

Web Reference at www.physicsgalaxy.com

**Module-8: Applications of Newton's Second Law**

Ex : Find acceleation of $m_1$ and $m_2$ in the system shown in figure.

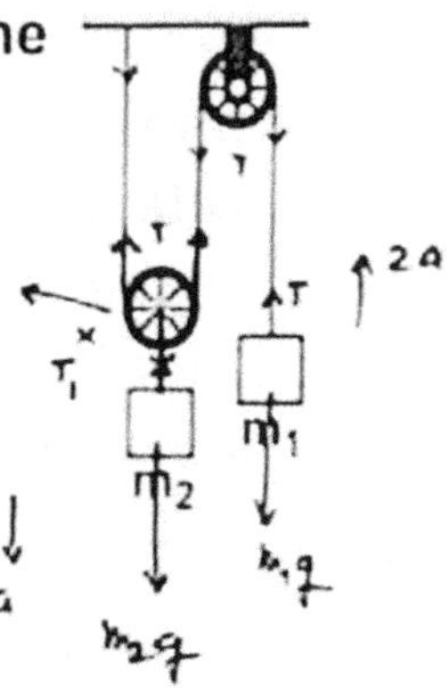


Sol$^n$:

NOTE: For a massless body, always Sum of all forces acting on it is zero.

$$\Sigma \vec{F} = m\vec{a} = 0$$

for pulley x we use $T_1 = 2T$

eq$^n$ of motions for $m_1$ & $m_2$ ⟶

$$T - m_1 g = m_1(2a) \times 2$$

$$m_2 g - 2T = m_2 a$$

$$(m_2 - 2m_1)g = (m_2 + 4m_1)a$$

$a_{m_2} = a \downarrow$

$a_{m_1} = 2a \uparrow$

$$a = \left(\frac{m_2 - 2m_1}{m_2 + 4m_1}\right) g \quad \text{Ans}$$

Web Reference at www.physicsgalaxy.com | Scan for Video Explanation

**Module-9: Solved Example-1**

Ex : Three blocks A, B and C of masses shown in figure are pulled by applying a force F = 60 N on 6 kg block. During motion find the tensions in string X and Y.

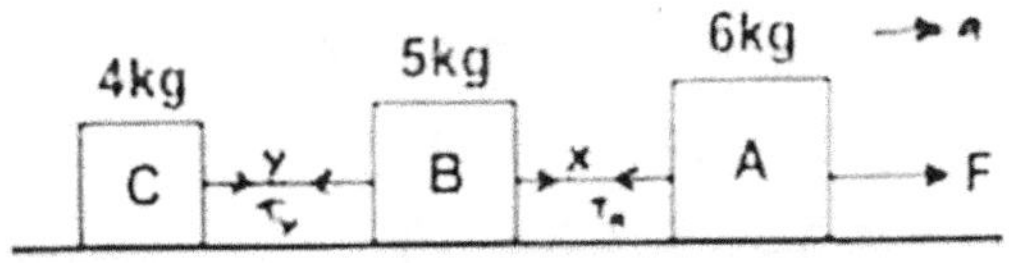


Sol$^n$: Common acc$^n$ of blocks $a = \frac{F}{M} = \frac{60}{15} = 4\ m/s^2$

FBD of A.

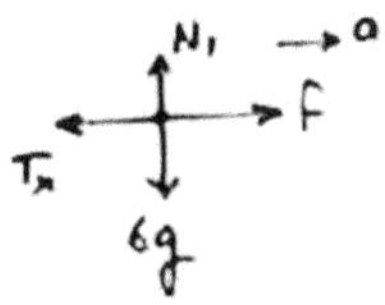


$$F - T_x = 6 \times 4$$

$$F - T_x = 24 \quad \text{Ans}$$

$$T_x = 60 - 24 = 36\ Nt.$$

FBD of C

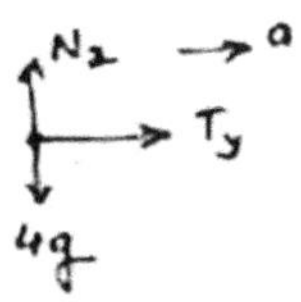


$$T_y = 4 \times 4 = 16\ Nt \quad \text{Ans.}$$

Web Reference at www.physicsgalaxy.com | Scan for Video Explanation

**Module-10: Solved Example-2**

**Ex :** When the system shown in figure is released, find the force applied by block A on B and tension in string.

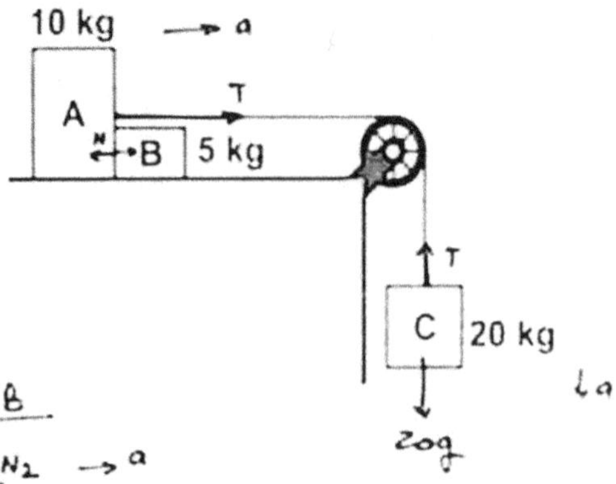


Sol$^n$: FBD for A   B   C

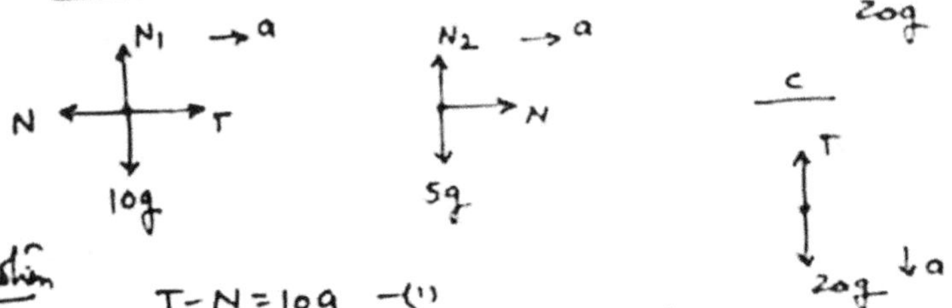


Eq$^n$ of Motion

$T - N = 10a \quad -(1)$

$N = 5a \quad -(2) \longrightarrow N = 5\left(\frac{4g}{7}\right)$

$20g - T = 20a \quad -(3) \qquad = \frac{200}{7}$ Nt Ans.

$20g = 35a$

$\left(a = \frac{4g}{7}\right)$

from (1) $T = 10a + N = \frac{400}{7} + \frac{200}{7} = \frac{600}{7}$ Nt Ans.

Scan for Video Explanation   Web Reference at www.physicsgalaxy.com

**Module-11: Solved Example-3**

**Ex :** Find acceleration of blocks A and B shown in figure when released from rest.

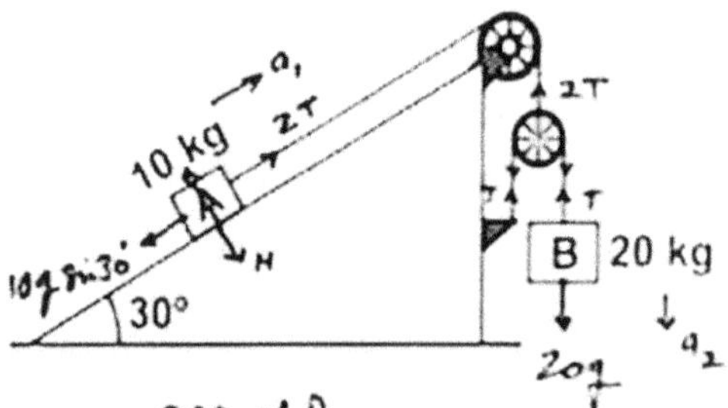


$a_2 = 2a_1$

Sol$^n$: FBD of A.   FBD of B

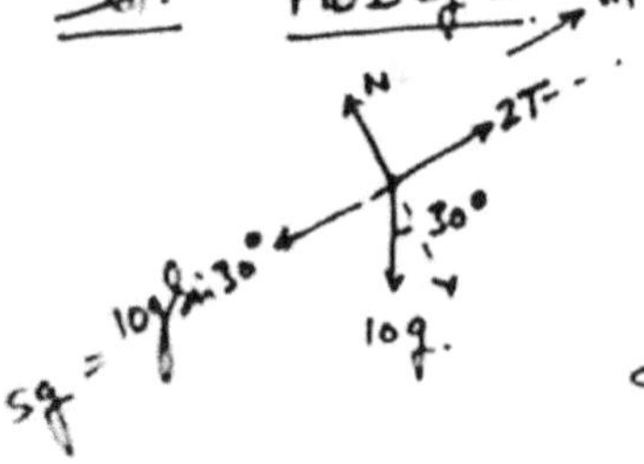


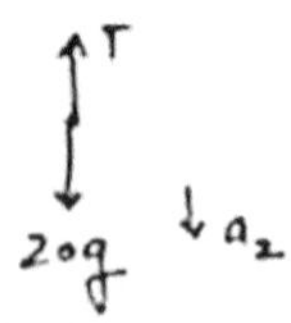


Eq$^n$ of Motion.

$2T - 5g = 10a_1 \quad -(1)$

$20g - T = 20(2a_1) \quad -(2)\times 2$

$35g = 90a_1$

$a_1 = \frac{35g}{90} = \frac{35}{9}$ m/s$^2$

$a_A = \frac{35}{9}$ m/s$^2$

$a_B = \frac{70}{9}$ m/s$^2$

Ans.

Scan for Video Explanation   Web Reference at www.physicsgalaxy.com

**Module-12: Solved Example-4**

**Ex :** In the system shown in figure if it is released find the acceleration of the rod which is free to move vertically. Given that mass of wedge A is 2 m and that of rod B is m.

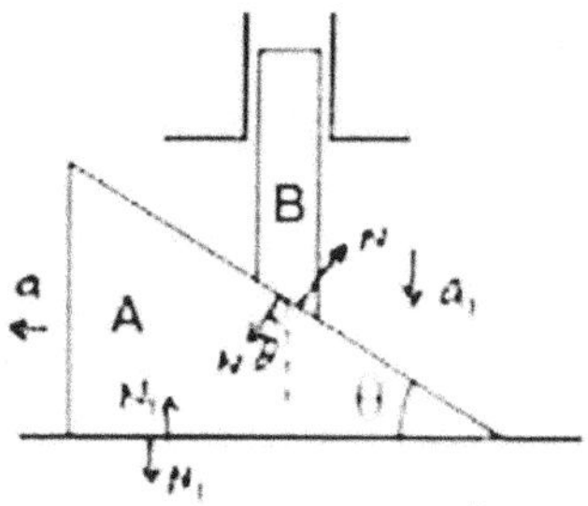


$a_1 = a \tan\theta$

**Sol$^n$:**

FBD of A

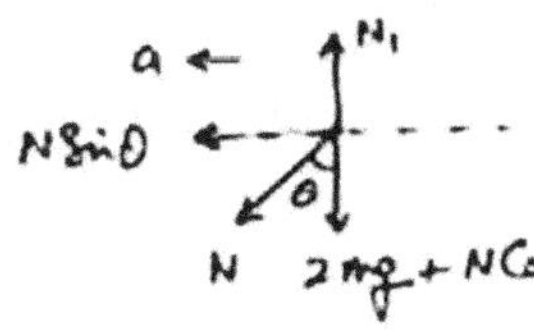


FBD of B

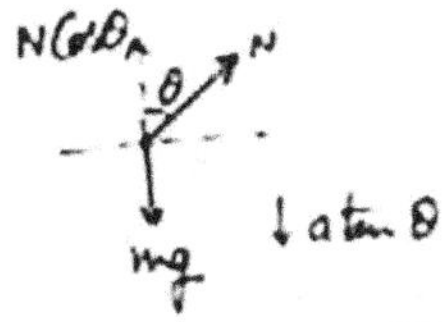


eq$^n$ of motions

$N\sin\theta = 2ma \longrightarrow N = \dfrac{2ma}{\sin\theta}$

$mg - N\cos\theta = ma\tan\theta$

$\cancel{m}g - 2\cancel{m}a\cot\theta = \cancel{m}a\tan\theta \Rightarrow$ $\boxed{a_A = \dfrac{g}{\tan\theta + 2\cot\theta}}$ Ans

$a_B = \dfrac{g\tan\theta}{\tan\theta + 2\cot\theta}$ Ans.

Web Reference at www.physicsgalaxy.com | Scan for Video Explanation

**Module-13: Solved Example-5**

**Ex :** Find acceleration of masses shown in the figure when this system is released from rest.

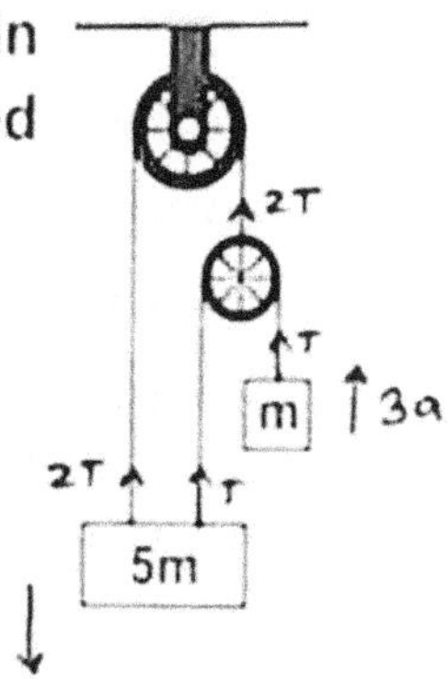


**Sol$^n$:**

FBD of 5m

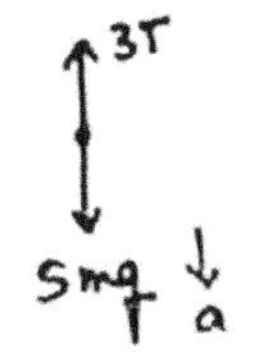


FBD of m

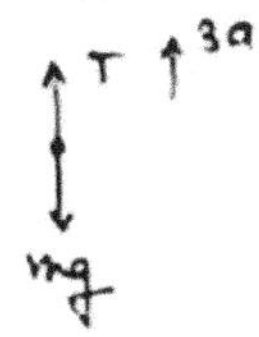


eq$^n$ of motions

$5mg - 3T = 5ma \quad -(1)$

$T - mg = m(3a) - (2) \times 3$

$2\cancel{m}g = 14\cancel{m}a$

$a = g/7$

$a_{5m} = g/7 \downarrow$

$a_m = 3g/7 \uparrow$ Ans.

Web Reference at www.physicsgalaxy.com | Scan for Video Explanation

**Module-14: Concept of Weighing Machine**

# Concept of Weighing Machine :

A weighing machine always measures the normal reaction acting on it in kgf.

$1\,kgf = g$ Newton.

If N is the normal reaction acting on a weighing machine then its reading will be

$$Reading = \left(\frac{N}{g}\right) kg$$

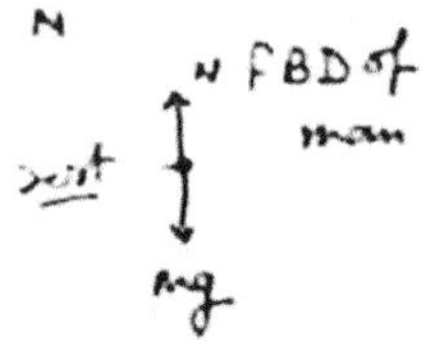


$N = mg$

$Reading = \frac{N}{g} = m$

Scan for Video Explanation — Web Reference at www.physicsgalaxy.com

**Module-14: Concept of Weighing Machine**

**Ex :** A man of mass 60 kg is standing on a weighing machine inside an elevator accelerating up at 2 m/s$^2$. Find the reading of weighing machine.

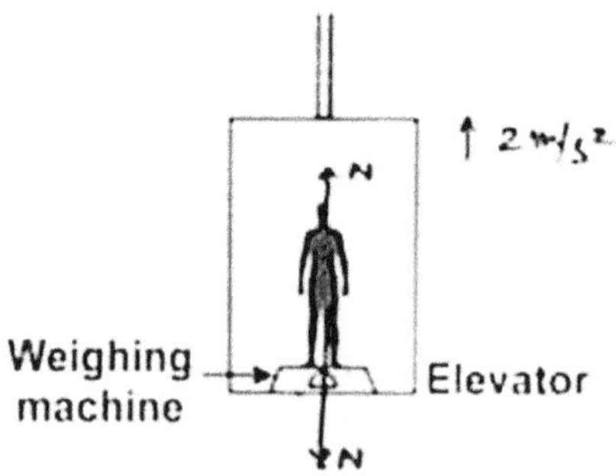


**Sol:**

FBD of man:

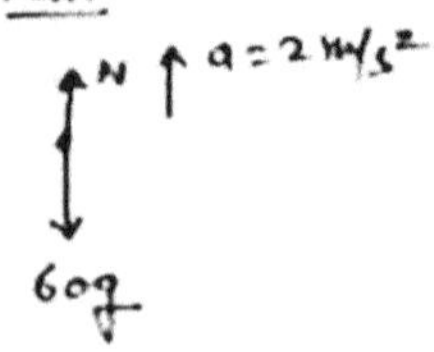


eq$^n$ of motion

$$N - 60g = 60a$$

$$N = 60(g + a) = 60 \times 12$$

$$= 720\,Nt.$$

$$Reading\ of\ machine = \frac{N}{g} = \frac{720}{10} = 72\,kg.$$

Scan for Video Explanation — Web Reference at www.physicsgalaxy.com

**Module-14: Concept of Weighing Machine**

# Cases of Motion of an Elevator:

$g_{eff} = g + a$

$N - mg = ma$

$N = m(g+a)$

$\text{Reading} = \dfrac{m(g+a)}{g}$ kg.

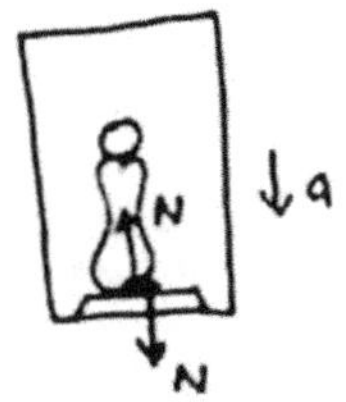

$g_{eff} = g - a$

$mg - N = ma$

$N = m(g-a)$

$\text{Reading} = \dfrac{m(g-a)}{g}$ kg.

$\vec{v}$ = Constant, $a = 0$

$g_{eff} = g$

$N = mg$

Reading = $m$ kg.

Web Reference at www.physicsgalaxy.com | Scan for Video Explanation

**Module-14: Concept of Weighing Machine**

Ex : In the figure a man of mass 80 kg is holding the cabin of mass 40 kg, in which he is standing on a weighing machine. Find the reading of weighing machine.

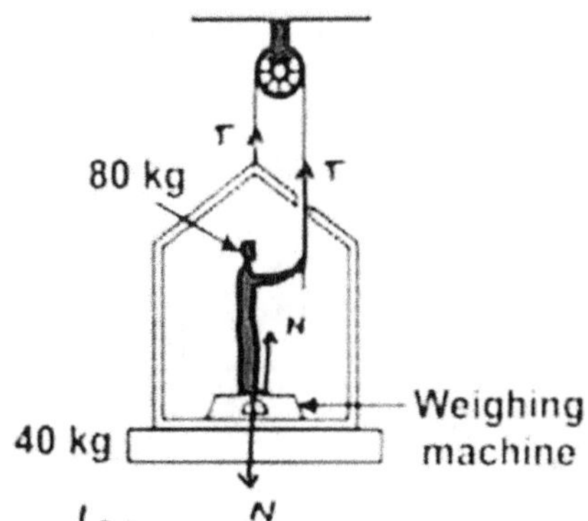


Sol: for cabin (rest): $T = 40g + N$

for man (rest): $T + N = 80g$

$T = 40g + N \longrightarrow N = T - 400 = 200$ Nt

$T + N = 80g$

$2T = 120g \rightarrow T = 600$ Nt.

$\text{Reading} = \dfrac{N}{g} = 20$ kg. Ans.

Web Reference at www.physicsgalaxy.com | Scan for Video Explanation

**Module-15: Spring Force**

# Spring Force:

ideal → massless.

→ Spring has a tendency to apply a restoring force opposite to the deformation of spring.

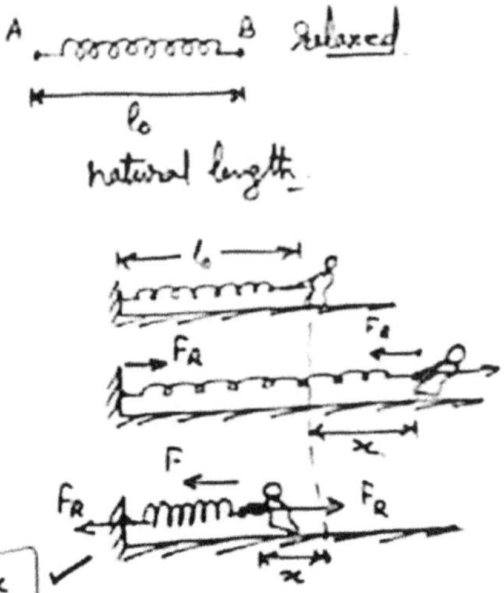


→ Restoring force is directly proportional to the deformation length.

$F_R \propto x$

$$\boxed{F_R = kx}$$ ✓

↓ force constant of Spring (Spring Constant)

→ Restoring force is always exerted on both ends of spring in opposite direction.

Scan for Video Explanation

Web Reference at www.physicsgalaxy.com

**Module-15: Spring Force**

Ex : In the situation shown in figure if acceleration 15 kg mass at an instant is 2 m/s$^2$, find acceleration of 10 kg mass at that instant.

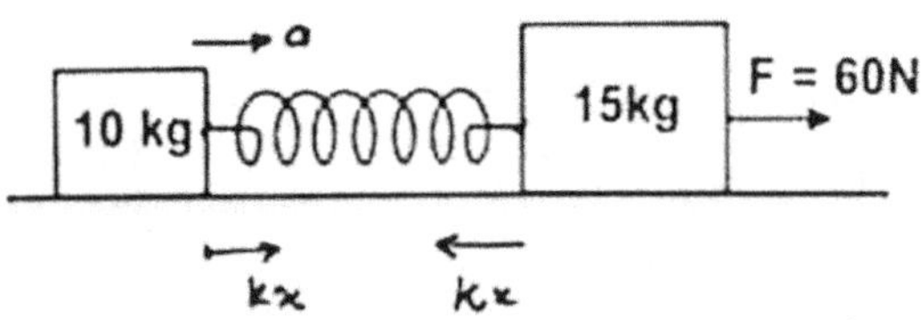


Sol:

$60 - kx = 15 \times 2$

$kx = 60 - 30$

$= 30$ Nt

$kx = 10a$

$a = \frac{30}{10} = 3$ m/s$^2$ Ans.

Scan for Video Explanation

Web Reference at www.physicsgalaxy.com

**Module-15: Spring Force**

**Ex :** In the situation shown in figure all surfaces are smooth. When the system is set in motion under gravity, find the extension in spring under the state of uniform acceleration of block.

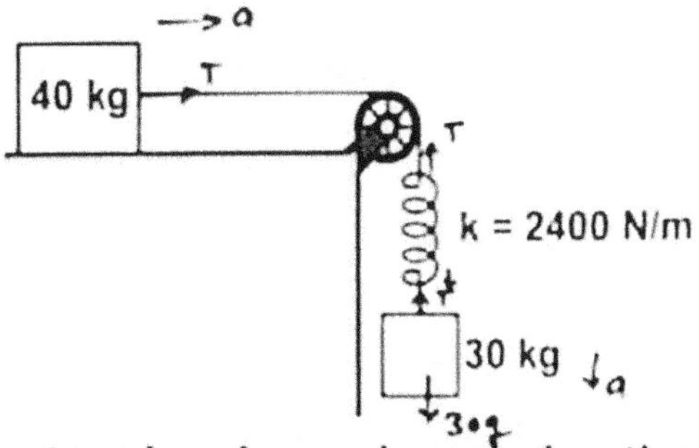


Sol: if $x$ is the extension in spring $\Rightarrow$ $T = kx$

$$30g - T = 30a$$
$$T = 40a \longrightarrow T = \frac{120g}{7} = \frac{1200}{7}\ \text{Nt}$$
$$30g = 70a \qquad kx = \frac{1200}{7}$$
$$a = \frac{3g}{7} \qquad 2400x = \frac{1200}{7} \Rightarrow x = \frac{1}{14}\ \text{m Ans}$$

Web Reference at www.physicsgalaxy.com | Scan for Video Explanation

**Module-16: Concept of Spring Balance**

# Concept of Spring Balance:

A spring balance always measures the tension in the string connected to it in units of kgf.

Reading of sp. balance $= \left(\frac{T}{g}\right)$ kg.

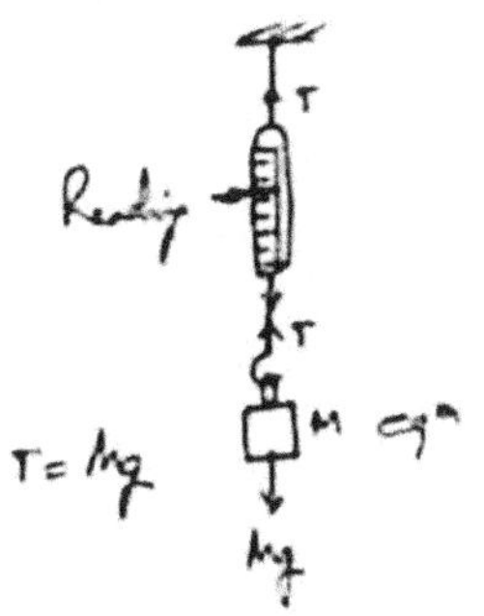


Ex:

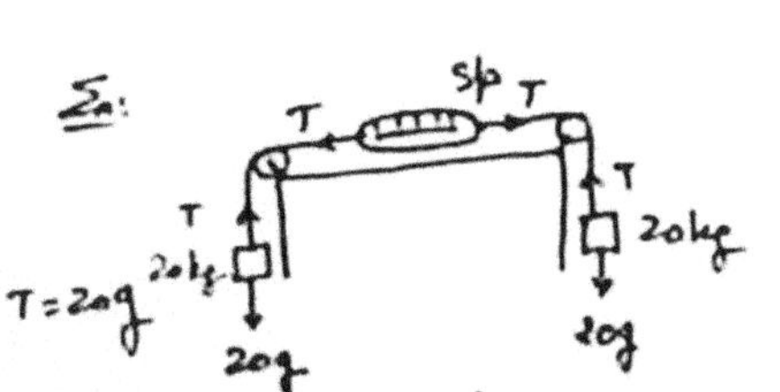


Reading $= \frac{T}{g} = 20$ kg.

Ex:

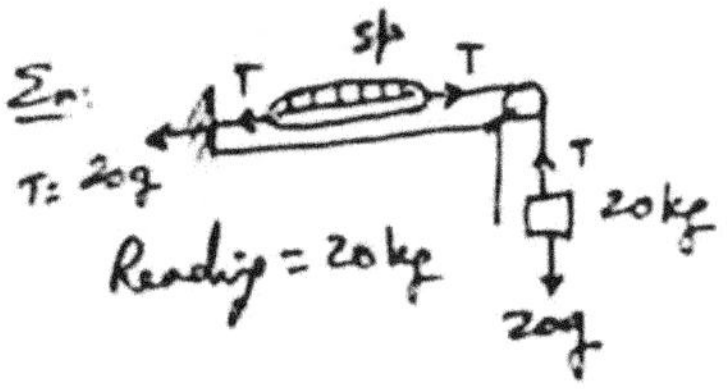


Reading = 20 kg

Web Reference at www.physicsgalaxy.com | Scan for Video Explanation

## Module-16: Concept of Spring Balance

**Ex :** In the system shown in figure. Find the reading of spring balance.

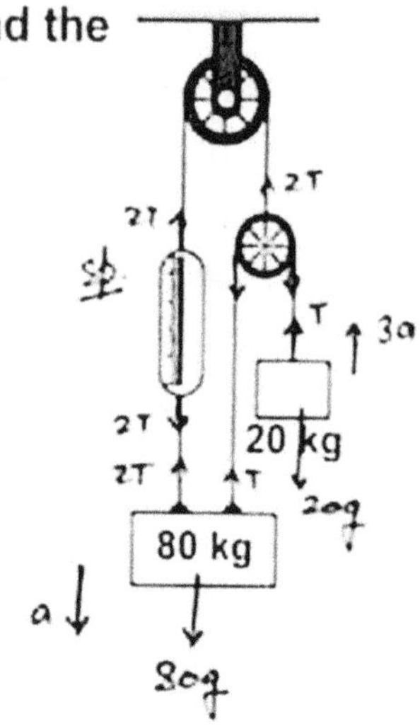


Sol: Eqⁿ of Motion

$$T - 20g = 20(3a) \quad -(1) \times 3$$

$$80g - 3T = 80a \quad -(2)$$

$$20g = 260a$$

$$a = \frac{2g}{26} = \frac{g}{13}$$

eqⁿ (1)

$$T = 20g + 60a$$

$$= 20g + 60(g/13) = \frac{320g}{13} = \frac{3200}{13}\ \text{Nt}$$

$$\text{Reading of Sb} = \frac{2T}{g} = \frac{6400}{13 \times 10} = \frac{640}{13}\ \text{kg. Ans.}$$

Scan for Video Explanation

Web Reference at www.physicsgalaxy.com

## Module-17: Concept of Pseudo Force

# Concept of Pseudo Force:

Pract. Illustration

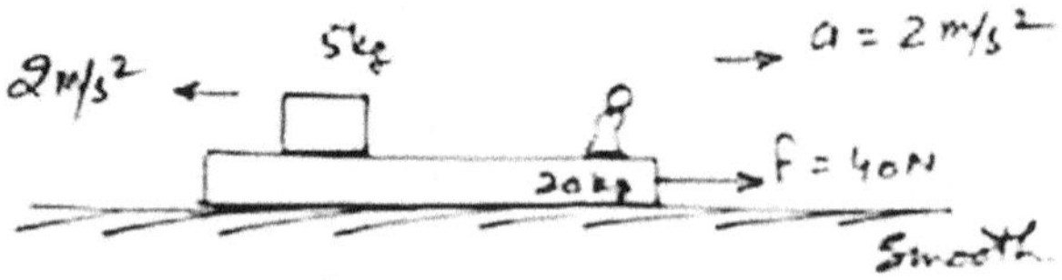


FBD of 5kg block w.r.t. to boy on plank

$2\ m/s^2 \leftarrow$

5kg

$F = 10N$

Pseudo force ← (not an actual force)

Scan for Video Explanation

Web Reference at www.physicsgalaxy.com

**Module-17: Concept of Pseudo Force**

NOTE: Pseudo force appear to act on a body when seen by an accelerating observer.

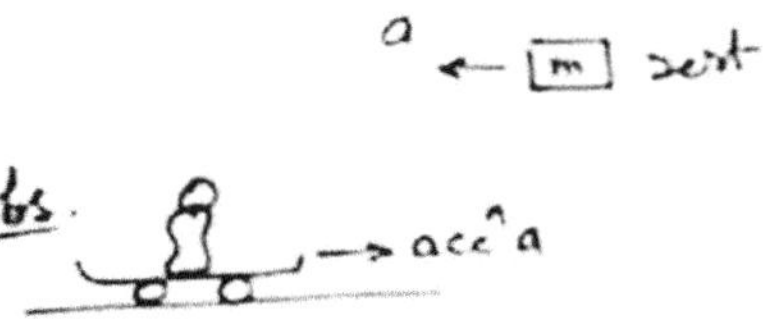


to this observer a pseudo force appear to act on mass m.

→ Pseudo force = mass of object × acc$^n$ of observer

→ Dir of Pseudo force is opp. to the dir of acc$^n$ of observer.

 Scan for Video Explanation

**Module-17: Concept of Pseudo Force**

# Pseudo Force does not obey Newton's laws:

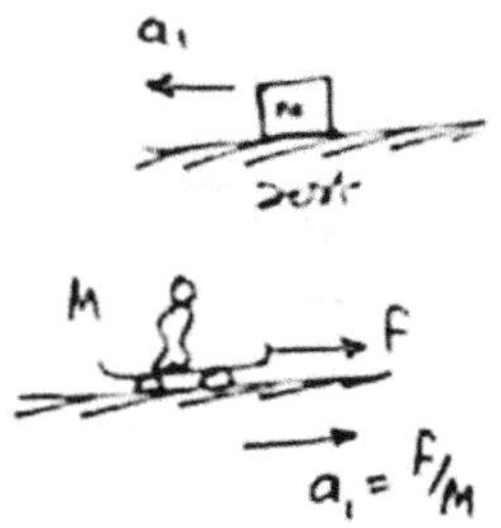


I law → × without force block started accelerating

II law → $\Sigma \vec{F} = m\vec{a}$ × with force mass is accelerating

III law → × No reaction exist for Pseudo force.

# NOTE: Based on acc$^n$ of observer, ref. frames are classified in two categories. These are -

(i) Inertial ref. frames → $a_{obs} = 0$ → uniform motion or Rest.

(ii) Non-inertial ref frames → $a_{obs} \neq 0$. (Pseudo forces)

 Scan for Video Explanation

**Module-17: Concept of Pseudo Force**

# Inertial & Non-inertial Reference frames:

Inertial ref frame are those which are at rest or in uniform motion. In Inertial ref. frames all Newton's laws are directly applicable.

The ref. frames which have some non zero acc$^n$ are called NI ref frames & in these frames NL's are not directly applicable. The observer in such frames sees a Pseudo force on all objects in dir opp to ref frame acc$^n$.

Pseudo force = mass of body × acc$^n$ of ref. frame

Scan for Video Explanation

Web Reference at www.physicsgalaxy.com

**Module-18: Solved Example-6**

**Ex :** Find the magnitude of F for which the blocks A and B will remain at rest w.r. to the bigger block.

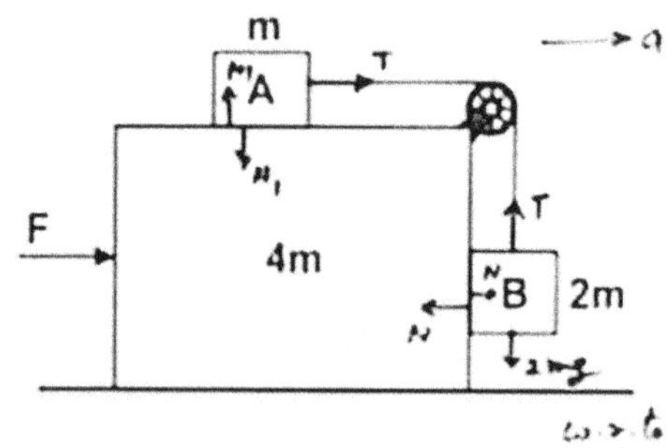


Sol$^n$

w.r.to 4m.

FBD of A.

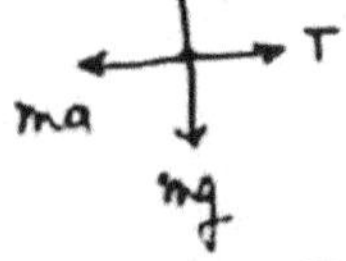


$$\begin{bmatrix} T = ma \\ N_1 = mg \end{bmatrix}$$

Common acc$^n$ of blocks $\quad a = \frac{F}{7m}$

w.r.to 4m

FBD of B.

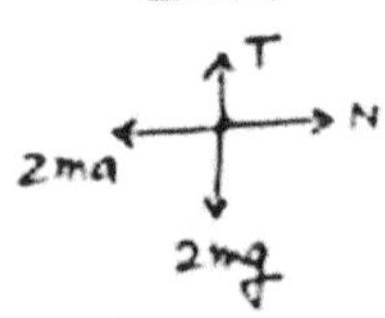


$$\begin{bmatrix} T = 2mg \\ N = 2ma \end{bmatrix}$$

$ma = 2mg$

$a = 2g$.

$F = 14mg$ Ans.

Scan for Video Explanation

Web Reference at www.physicsgalaxy.com

**Module-19: Solved Example-7**

**Ex :** When the system shown in figure is released from rest, find acceleration of 2m.

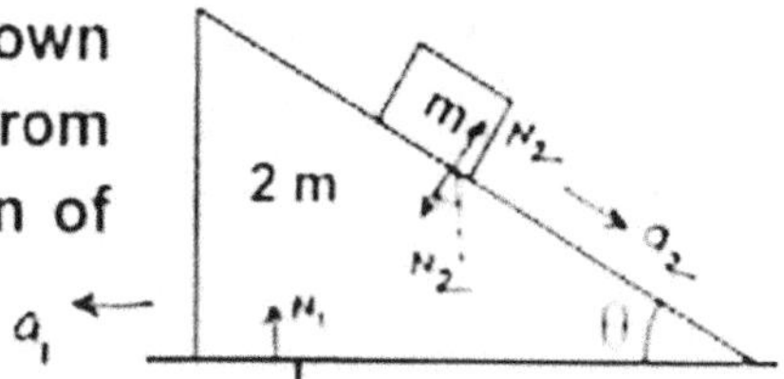


Sol^n FBD of 2m

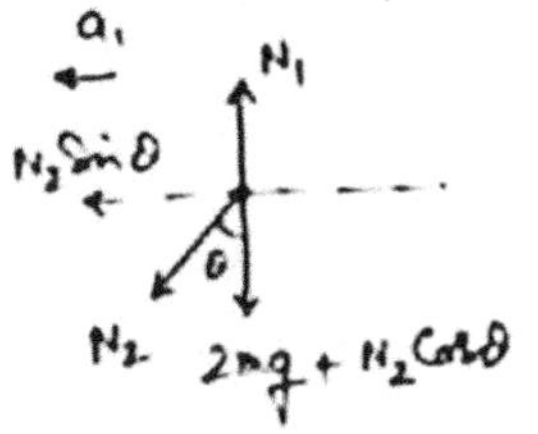


FBD of m w.r.t. wedge.

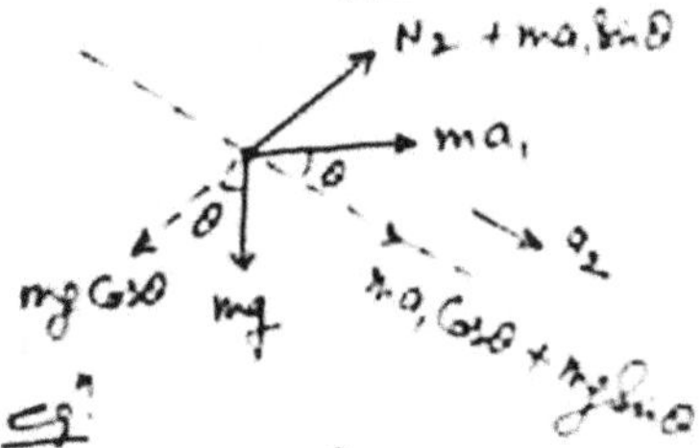


eq^n $N_2\sin\theta = 2ma_1$ —(1)

eq^n $ma_1\cos\theta + mg\sin\theta = ma_2$ —(2)

$N_2 + ma_1\sin\theta = mg\cos\theta$ —(3)

(1)&(3) ⇒ $\frac{2ma_1}{\sin\theta} + ma_1\sin\theta = mg\cos\theta$

$a_1 = \left(\frac{g\sin\theta\cos\theta}{2+\sin^2\theta}\right)$ Ans.

**Web Reference at www.physicsgalaxy.com** **Scan for Video Explanation**

**Module-20: Solved Example-8**

**Ex :** Find the value of F for which block of mass m will remain at rest over the wedge.

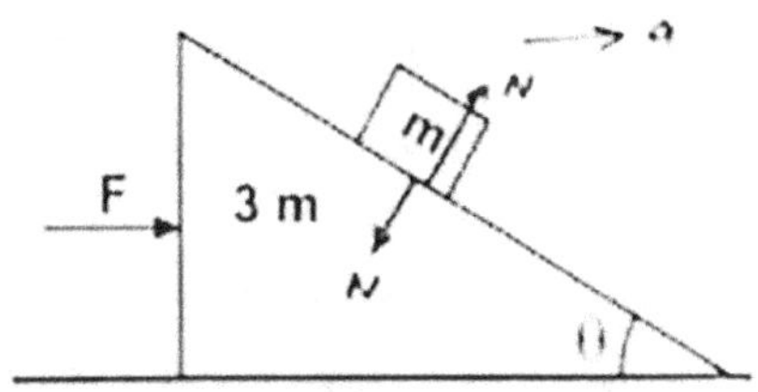


Sol^n

Common acc^n of system $a = \frac{F}{4m}$

FBD of m w.r.t. 3m.

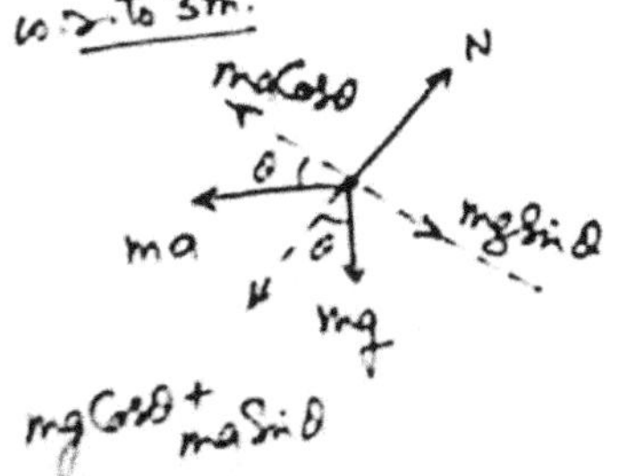


$mg\sin\theta = ma\cos\theta$

$a = g\tan\theta$ v.imp.

$F = 4mg\tan\theta$ Ans.

**Web Reference at www.physicsgalaxy.com** **Scan for Video Explanation**

**Module-21: Solved Example-9**

Ex : Find the magnitude of mass M so that 2 kg block will remain at rest on 5 kg block.

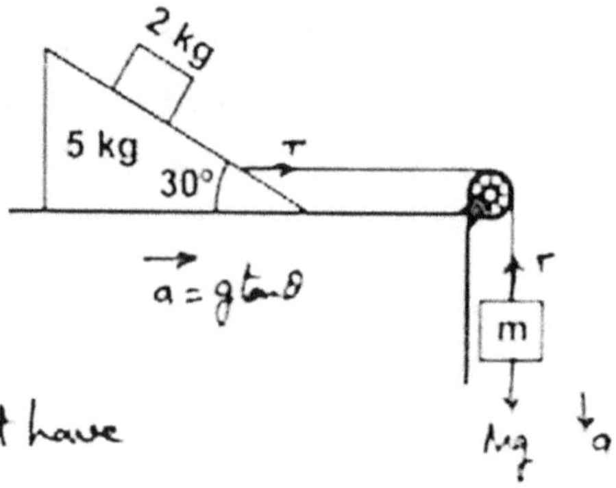


Sol: for 2kg block to be at rest on 5kg wedge we must have

$$a = g\tan 30° = g/\sqrt{3}$$

$$Mg - T = Ma$$
$$T = 7a$$
$$Mg = (M+7)a$$
$$a = \frac{Mg}{M+7}$$

$$\frac{g}{\sqrt{3}} = \frac{M}{M+7} \cdot g$$

$$M + 7 = \sqrt{3}M$$

$$M = \left(\frac{7}{\sqrt{3}-1}\right) \text{ kg Ans.}$$

Scan for Video Explanation

Web Reference at www.physicsgalaxy.com

**Module-22: Solved Example-10**

Ex : A pendulum bob of mass 5 kg is hanging from the ceiling of a 25 kg box car, which is moving under the effect of a 225 N force as shown. Find the angle made by the string of pendulum inside the moving car.

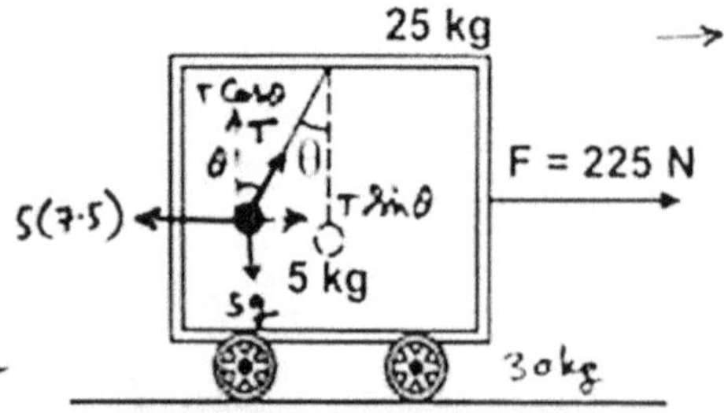


Sol$^n$:

$$\rightarrow a = \frac{225}{30} = 7.5 \text{ m/s}^2$$

As w.r.t. car bob is at rest

$$\left.\begin{aligned} T\sin\theta &= 5(7.5) \\ T\cos\theta &= 5(10) \end{aligned}\right\} \rightarrow \tan\theta = \frac{7.5}{10} = \frac{3}{4}$$

$$\theta = \tan^{-1}(3/4) = 37° \text{ Ans.}$$

Scan for Video Explanation

Web Reference at www.physicsgalaxy.com

# Chapter 6
# Friction

## Lecture Notes Modules

**Module-1:** Friction and its Classification
**Module-2:** Types of Friction-Static Friction
**Module-3:** Limiting & Kinetic Friction
**Module-4:** Solved Example-1
**Module-5:** Solved Example-2
**Module-6:** Solved Example-3
**Module-7:** Solved Example-4
**Module-8:** Angle of Friction & Angle of Repose
**Module-9:** Cases of Sliding on Rough Surface
**Module-10:** Solved Example-6
**Module-11:** Solved Example-7

**Module-1: Friction and its Classification**

# Friction: The force which opposes motion or tendency of motion is called friction. Based on opposition process, friction is divided in two categories –

(1) Dry Friction – due to irregularities of surfaces in contact.

(2) Wet friction – due to the continuous collisions of medium particles on body.
Ex: air friction

Scan for Video Explanation

Web Reference at www.physicsgalaxy.com

**Module-1: Friction and its Classification**

# Dry Friction:

Friction due to roughness of surfaces.

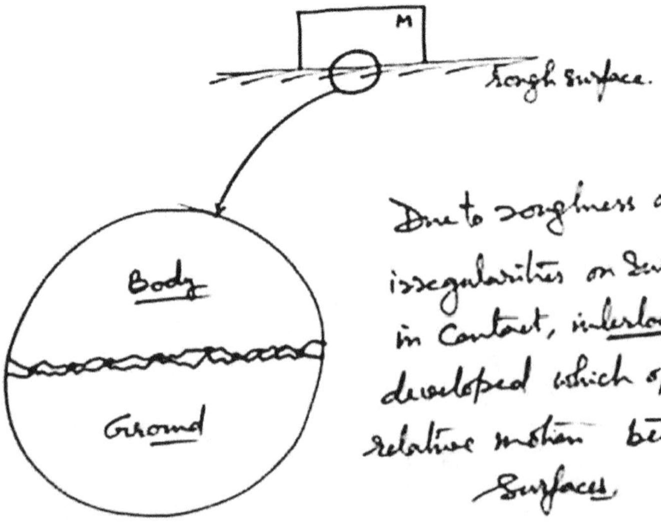


Due to roughness and irregularities on surfaces in contact, interlocks are developed which opposes relative motion between surfaces.

Scan for Video Explanation

Web Reference at www.physicsgalaxy.com

**Module-1: Friction and its Classification**

Due to these interlocks at various pts of Contact the Contact forces are distributed in various direction over the surface

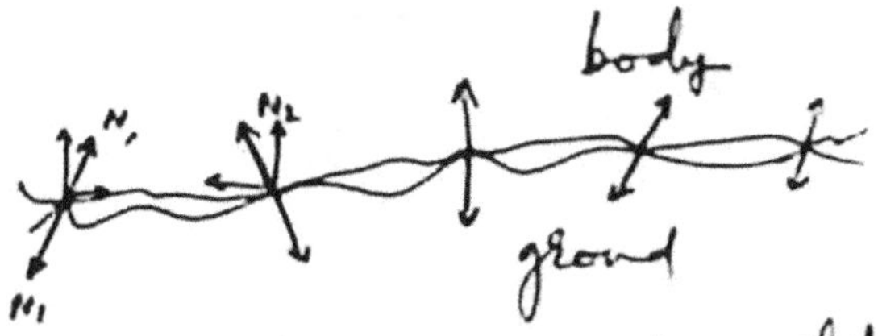


friction → Cumulative effect of horizontal Components of all Contact forces between the surface.

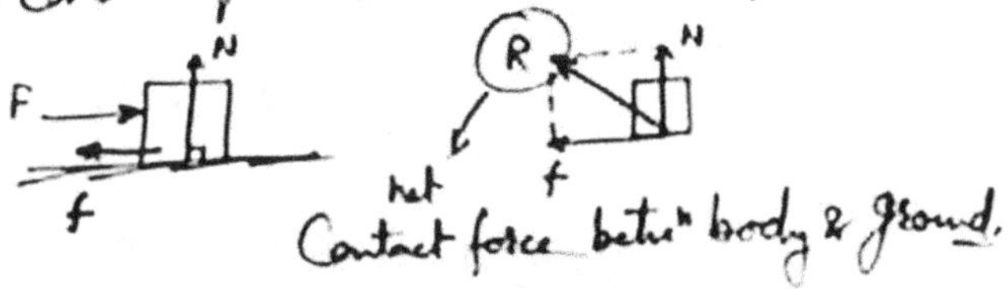


Web Reference at www.physicsgalaxy.com

Scan for Video Explanation

**Module-2: Types of Friction-Static Friction**

# Types of friction between two Surfaces in Contact:

When two Surfaces are in Contact the friction between these surfaces can be classified in 3 categories. –

(1) Static friction → acts when body is at rest

(2) Limiting friction → max limit of Static friction

(3) Kinetic friction → when body is in motion

Web Reference at www.physicsgalaxy.com

Scan for Video Explanation

**Module-2: Types of Friction-Static Friction**

# Static Friction: body is at rest.
& some external force is acting on it

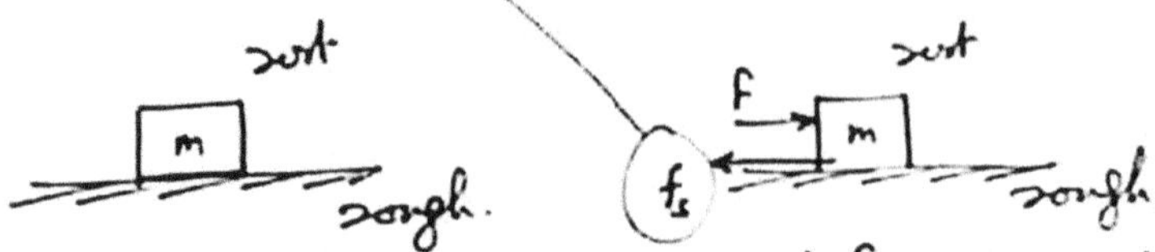


$f = 0.$
no friction

$f_s = F.$ for $F < f_L$
static friction.

max limit upto which the static friction balances the external force acting on body is termed as "Limiting Friction"

Limiting $f_s = f_L$

body will start moving when $F \geq f_L$

Scan for Video Explanation

Web Reference at www.physicsgalaxy.com

**Module-3: Limiting & Kinetic Friction**

# Magnitude of Limiting Friction:

magnitude depends on two factors

(1) Normal Reaction between Surfaces

(2) Nature of roughness.

$$f_L \propto N$$

$$\boxed{f_L = \mu_s N}$$

$\mu_s$ → Coefficient of Static friction (depends on roughness of Surfaces)

Scan for Video Explanation

Web Reference at www.physicsgalaxy.com

**Module-3: Limiting & Kinetic Friction**

# Practical Illustration:

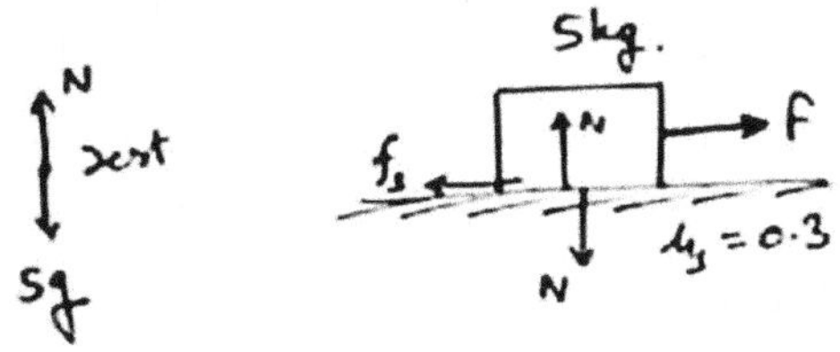


$N = 5g = 50$ Nt

limiting fr. $f_L = \mu_s N$
$= 0.3 \times 50$
$= 15$ Nt

| | | | | |
|---|---|---|---|---|
| $F =$ | 5N | 12N | 14.99N | 15N |
| $f_s =$ | 5N | 12N | 14.99N | $f_k$ → kinetic friction $f_k < f_L$ |
| State of motion → | rest | rest | rest | motion |

(rest, rest, rest) ↓ Static friction



**Module-3: Limiting & Kinetic Friction**

# Kinetic Friction: It acts when body is in motion

→ no external force is necessary for kinetic friction

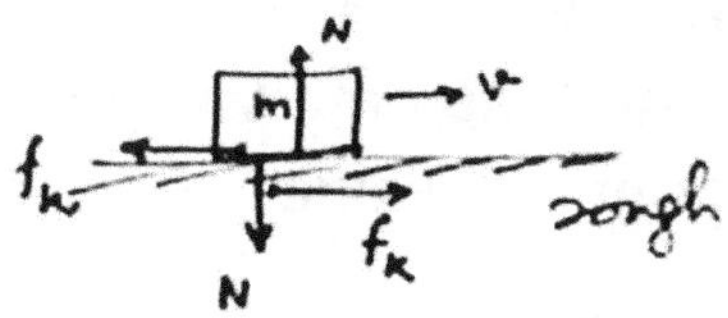


$f_k \propto N$

$$\boxed{f_k = \mu_k N}$$

↓ Coefficient of kinetic fr.

NOTE: When a body starts moving due to less interlocking between surfaces $f_k < f_L$

$\mu_k < \mu_s$.

**Module-4: Solved Example-1**

**Ex :** When the system shown in figure is released, it starts moving. Find the acceleration of blocks. The friction coefficient between $m_1$ and ground is μ.

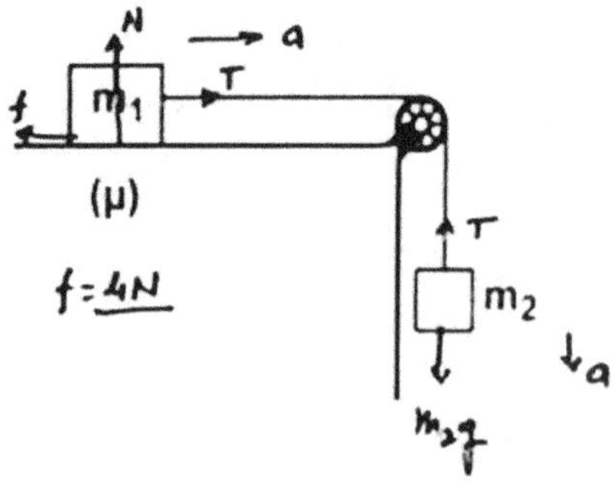


Sol$^n$: eq$^n$ of motion for $m_1$ & $m_2$

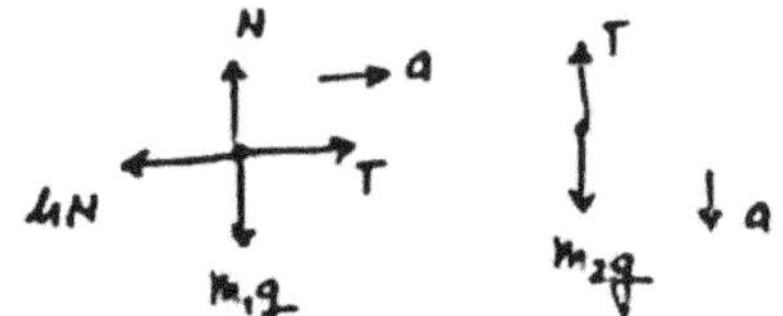


$N = m_1 g$.

$$m_2 g - T = m_2 a \quad —(1)$$
$$T - \mu m_1 g = m_1 a \quad —(2)$$
$$(1)+(2) \Rightarrow (m_2 - \mu m_1) g = (m_1 + m_2) a$$
$$a = \left(\frac{m_2 - \mu m_1}{m_1 + m_2}\right) g. \quad \text{Ans.}$$

Scan for Video Explanation

Web Reference at www.physicsgalaxy.com

**Module-5: Solved Example-2**

**Ex :** A block of mass m is placed and released from the top of an inclined plane of length *l* which has a friction coefficient μ for the block. If it slides down, find the time taken by the block to reach the bottom of incline.

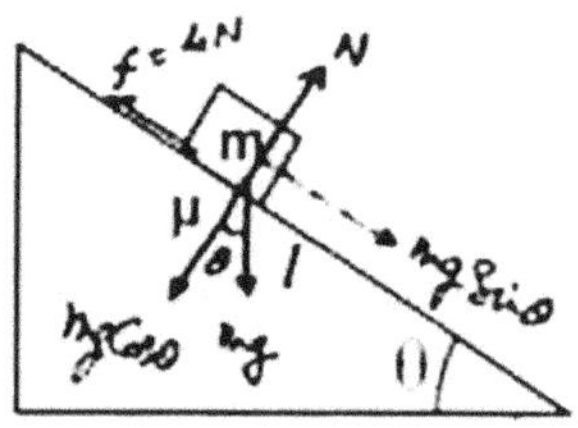


$N = mg\cos\theta$

Sol$^n$: If a is the acc$^n$ of block

$$mg\sin\theta - \mu mg\cos\theta = ma$$
$$a = g\sin\theta - \mu g\cos\theta$$

length covered

$$l = \frac{1}{2} a t^2$$
$$t = \sqrt{\frac{2l}{a}} = \sqrt{\frac{2l}{g(\sin\theta - \mu\cos\theta)}} \quad \text{Ans.}$$

Scan for Video Explanation

Web Reference at www.physicsgalaxy.com

**Module-6: Solved Example-3**

**Ex :** In the situation shown in figure a block of mass m is supported on the front surface of a bigger block of mass 3 m. The friction coefficient between vertical surface of 3 m and m is μ. Find a minimum magnitude of horizontal force F required to be applied so that m will remain at rest w.r.t. to 3 m.

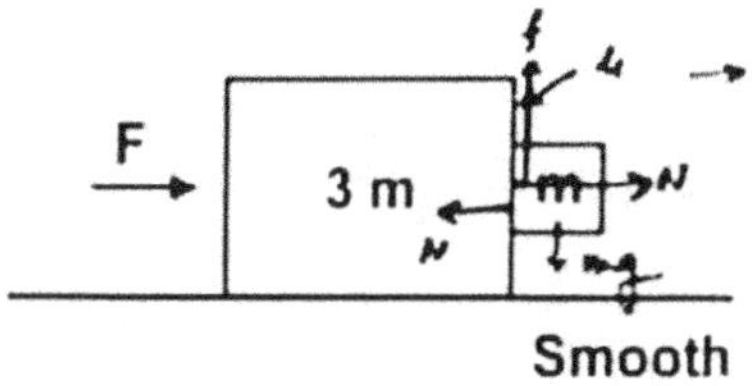


**Sol$^n$:** block will remain at rest

if $mg < \mu N$

$mg < \mu F/4$

or $\boxed{F > \frac{4mg}{\mu}}$ Ans.

Common acc$^n$ $a = \frac{F}{4m}$

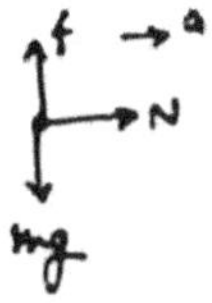

for m. $N = ma = m\left(\frac{F}{4m}\right)$

$N = \frac{F}{4}$

Web Reference at www.physicsgalaxy.com | Scan for Video Explanation

**Module-7: Solved Example-4**

**Ex :** A block of mass m is placed on a rough horizontal surface having friction coefficient μ. Find minimum force required to slide the block. Also find the direction θ at which this force is to be applied.

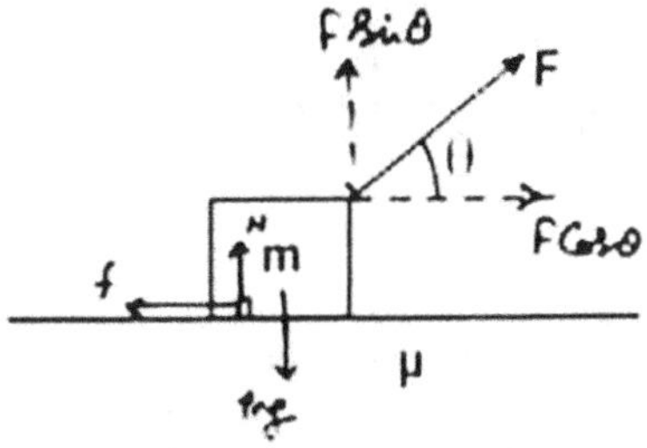


**Sol$^n$:** block will slide when $F\cos\theta \ge \mu N$

in vert dir.

$N + F\sin\theta = mg$

$N = mg - F\sin\theta$

$F\cos\theta = \mu(mg - F\sin\theta)$

$$F = \frac{\mu mg}{\cos\theta + \mu\sin\theta} \quad \text{—(1)}$$

F is min when

$\frac{dF}{d\theta} = 0 \Rightarrow -\sin\theta + \mu\cos\theta = 0$

$\tan\theta = \mu$

$\theta = \tan^{-1}\mu$ (Ans)

$$F_{min} = \frac{\mu mg}{\left(\frac{1}{\sqrt{1+\mu^2}}\right) + \mu\left(\frac{\mu}{\sqrt{1+\mu^2}}\right)}$$

$$F_{min} = \frac{\mu mg}{\sqrt{1+\mu^2}} \text{ Ans.}$$

Web Reference at www.physicsgalaxy.com | Scan for Video Explanation

**Module-8: Angle of Friction & Angle of Repose**

# Angle of Friction:

When a body is at the verge of sliding, the angle between the normal reaction and total Contact force is termed as angle of friction

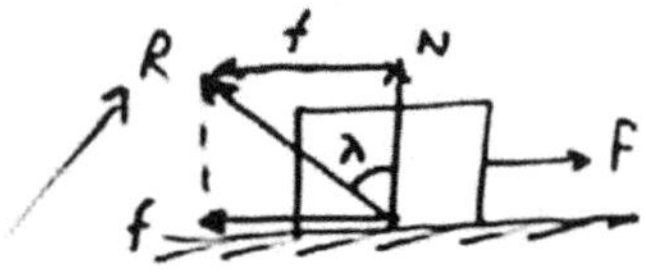


Total Contact force. $f = \mu N$.

$\tan\lambda = \frac{f}{N} = \frac{\mu N}{N}$

$\tan\lambda = \mu$

Angle of friction $\rightarrow \lambda = \tan^{-1}(\mu)$

Scan for Video Explanation

Web Reference at www.physicsgalaxy.com

**Module-8: Angle of Friction & Angle of Repose**

# Angle of Repose:

$f_L = \mu N = \mu mg\cos\theta$.

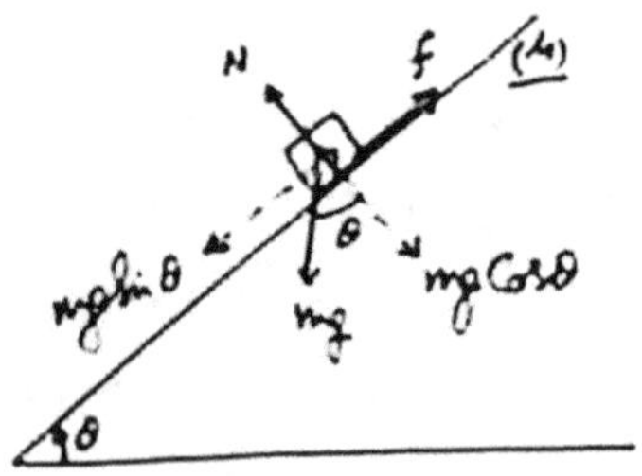


$N = mg\cos\theta$

if m is at rest

$\Rightarrow f = mg\sin\theta$.

block will start sliding down at an angle of inclination

$\theta = \theta_R$

when $mg\sin\theta_R = \mu mg\cos\theta_R$

$\tan\theta_R = \mu$

Angle of Repose. $\leftarrow \theta_R = \tan^{-1}(\mu)$

Scan for Video Explanation

Web Reference at www.physicsgalaxy.com

**Module-9: Cases of Sliding on Rough Surface**

# Cases of Sliding on Rough Surfaces:

On a rough surface a body at rest will slide only if by some external force, friction approaches to the value $\mu N$. (Limiting friction)

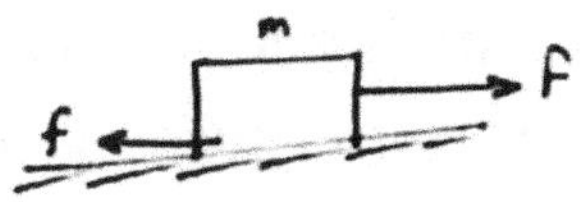


$f = F = \mu N \Rightarrow$ body will start sliding

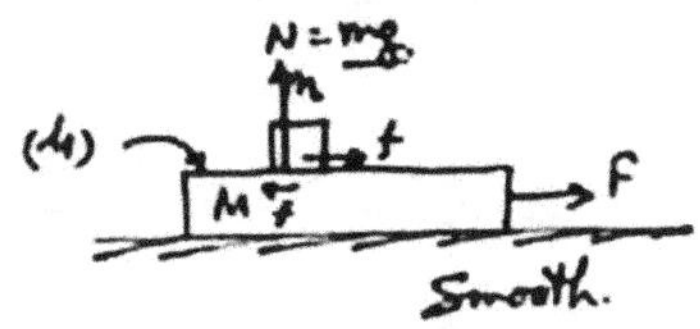


Common acc$^n$ $a = \frac{F}{M+m}$

for m $f = m\left(\frac{F}{M+m}\right)$

m will slide over M when

$F = \mu(M+m)g \quad \leftarrow \quad f = \frac{mF}{M+m} = \mu mg$



**Module-9: Cases of Sliding on Rough Surface**

**Ex :** Find mass M such that on releasing the system, block A will start sliding on B.

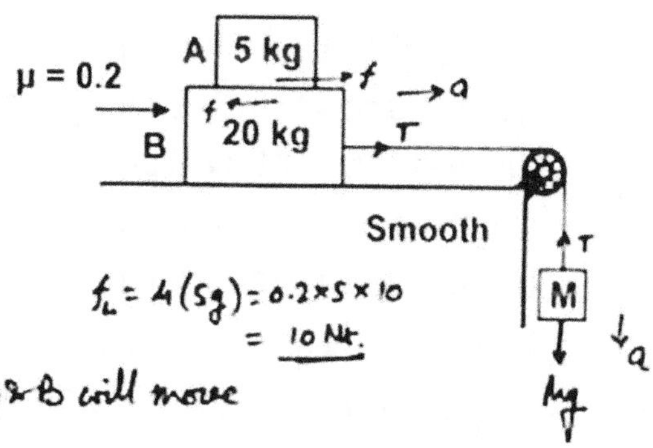


$f_L = \mu(5g) = 0.2 \times 5 \times 10 = 10$ Nt.

Sol$^n$: for low values of M, A & B will move together

$$\left.\begin{aligned} Mg - T &= Ma \\ T &= 25a \end{aligned}\right] \rightarrow Mg = (25+M)a$$

$$a = \left(\frac{M}{25+M}\right)g$$

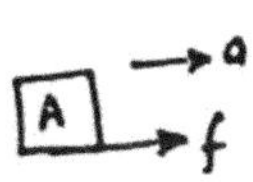


for block A.

$$f = 5\left(\frac{Mg}{M+25}\right) = 10$$

$$5M = M + 25$$

$$M = \frac{25}{4} = 6.25 \text{ kg. Ans.}$$

**Module-10: Solved Example-6**

**Ex :** In figure the friction coefficient between 4 kg block and ground is 0.2 and that between the two blocks is 0.3. A horizontal force F is applied on 4 kg block as shown, if F is gradually increased, find values of F at which sliding starts at bottom surface and then between the two blocks.

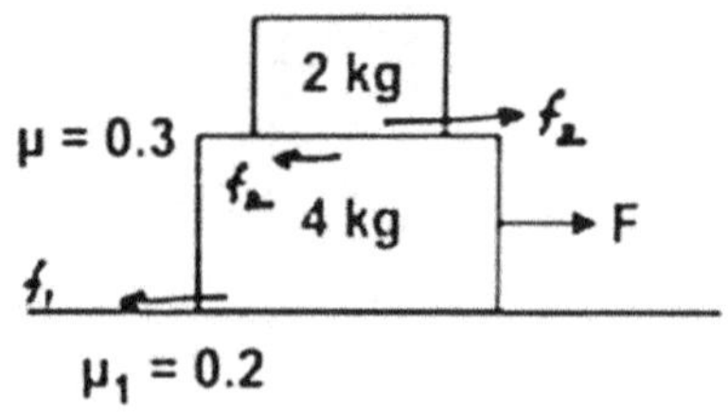


Sol<sup></sup>:

$f_{LB} = 0.2 \times 6 \times 10 = 12N$ ✓

$f_{LM} = 0.3 \times 2 \times 10 = 6N$

for $F > 12N$, both blocks start together with acc^n $a = \frac{F-12}{6}$ m/s²

for 2kg block. $f_2 = 2\left(\frac{F-12}{6}\right) = 6$ (Cond for sliding of 2kg block)

$F - 12 = 18$

$F = 30$ Nt Ans.

Scan for Video Explanation

Web Reference at www.physicsgalaxy.com

**Module-11: Solved Example-7**

**Ex :** In the situation shown in figure find the values of F at which sliding will start at different surfaces of blocks.

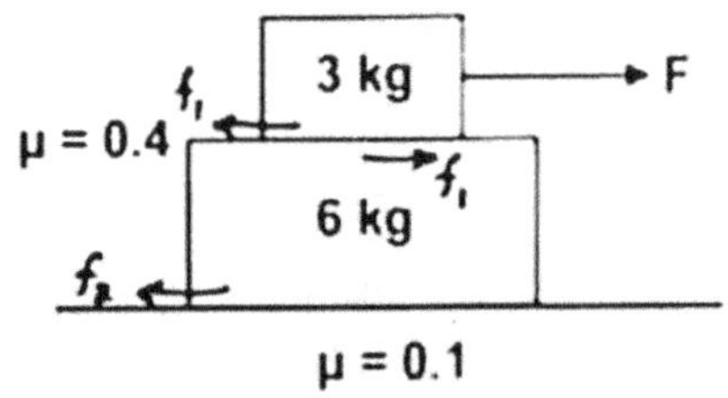


Sol:

$f_{LB} = 0.1 \times 9 \times 10 = 9N$

$f_{LM} = 0.4 \times 3 \times 10 = 12N.$

at $F = 9N$, Sliding at bottom surface will start.

for $F > 9$, Common acc^n of blocks $a = \frac{F-9}{9}$ m/s²

for 3kg block $F - f_1 = 3\left(\frac{F-9}{9}\right)$

3kg will start sliding when $f_1 \to 12N$

$F - 12 = \frac{F-9}{3}$

$3F - 36 = F - 9 \Rightarrow 2F = 27$

$F = 13.5$ Nt Ans.

Scan for Video Explanation

Web Reference at www.physicsgalaxy.com

# Chapter 7
# Work Energy Power

## Lecture Notes Modules


**Module-1:** Work & its Calculation
**Module-2:** Work Done by a Variable Force
**Module-3:** Solved Example-1
**Module-4:** Solved Example-2
**Module-5:** Energy & its Types
**Module-6:** Conservative & Non-Conservative Forces
**Module-7:** Potential Energy Stored in a Spring
**Module-8:** Work-Energy Theorem
**Module-9:** Relation in Force & Potential Energy
**Module-10:** Solved Example-3
**Module-11:** Solved Example-4
**Module-12:** Solved Example-5
**Module-13:** Work-Energy Equation for Rigid Bodies
**Module-14:** Solved Example-6
**Module-15:** Solved Example-7
**Module-16:** Solved Example-8
**Module-17:** Solved Example-9
**Module-18:** Concept of Equilibrium
**Module-19:** Power
**Module-20:** Solved Example-10
**Module-21:** Solved Example-11

**Module-1: Work & its Calculation**

# Work:

Work → Transfer of Energy.

→ Work is done when a force produces some displacement.

work = Force × disp of pt. of appⁿ of force in the dir of force.

$= F.S$ Joule.

Scan for Video Explanation

Web Reference at www.physicsgalaxy.com

**Module-1: Work & its Calculation**

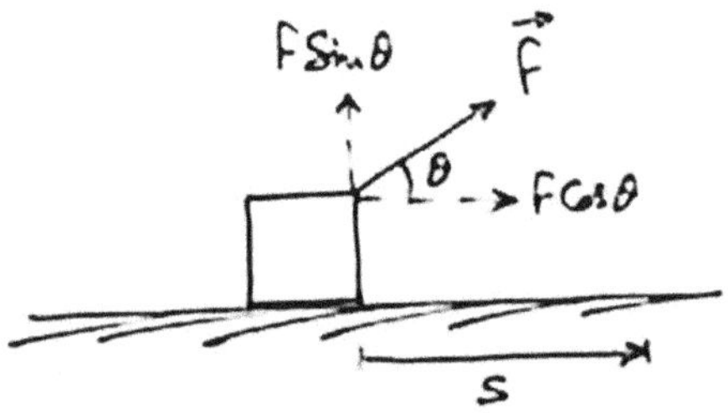


Work done by $\vec{F}$ $\quad W = F\cos\theta \times s = Fs\cos\theta$

vect $\quad W = \vec{F}.\vec{s}$

W → amount of energy transferred.

Scan for Video Explanation

Web Reference at www.physicsgalaxy.com

**Module-1: Work & its Calculation**

# Positive & Negative Work:

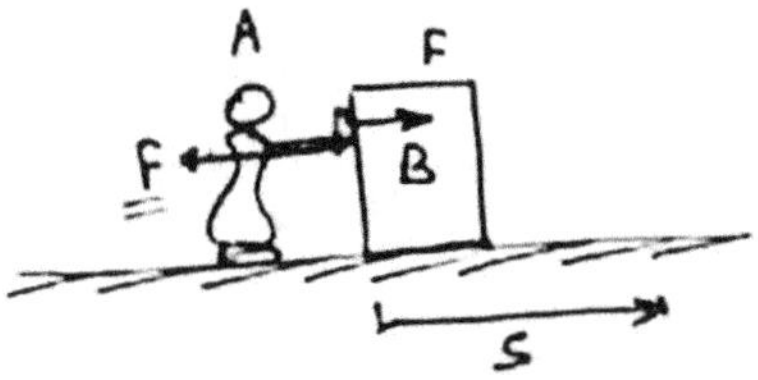


Work Done by boy $W = F.S$ (+ve)

W/D by block $W = -FS$ (−ve)

+ve work ⇒ Supply of Energy.

−ve work ⇒ Gain of Energy



**Module-2: Work Done by a Variable Force**

# Work Done by a variable Force:

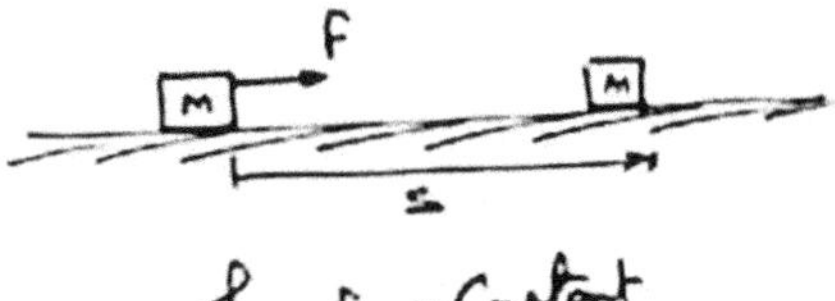


When F → Constant

Work = F.S

$W = \vec{F}.\vec{s}$

if $F = f(x)$

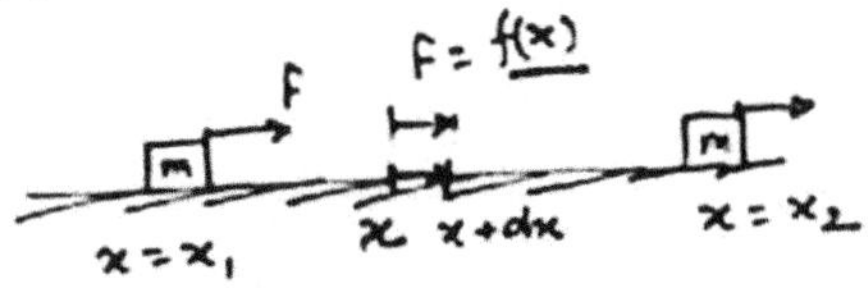


work done in elemental displ dx is $dW = \vec{F}.\vec{dx}$

Total work in displ from $x = x_1$ to $x = x_2$ $W = \int dW = \int_{x_1}^{x_2} \vec{F}.\vec{dx}$

**Module-2: Work Done by a Variable Force**

**Ex :** A block of mass 5 kg is pulled by an external force which depends on its position x as

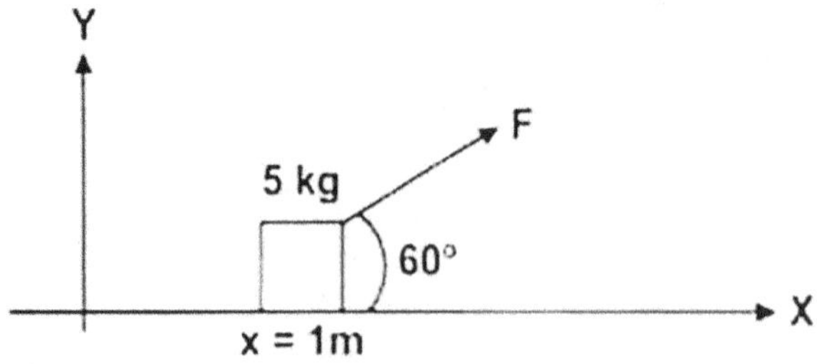


$$F = 4x^2 + 2x + 1 \text{ N}$$

Find work done by this force is displacing the block to the position x = 2m.

**Sol:**

$$W = \int \vec{F}\cdot\vec{dx} = \int F dx \cos 60^\circ$$

$$= \frac{1}{2}\int_1^2 (4x^2 + 2x + 1)\,dx$$

$$= \frac{1}{2}\left[\frac{4x^3}{3} + x^2 + x\right]_1^2 = \frac{1}{2}\left[\left(\frac{32}{3}+4+2\right) - \left(\frac{4}{3}+1+1\right)\right]$$

$$= \frac{20}{3} \text{ J} \quad \text{Ans.}$$

Scan for Video Explanation

Web Reference at www.physicsgalaxy.com

**Module-2: Work Done by a Variable Force**

# Calculation of Work using Graphs:

As we know $\text{Work} = \int_{x_1}^{x_2} \vec{F}\cdot\vec{dx}$

↑ area under F-x graph.

$F = f(x)$

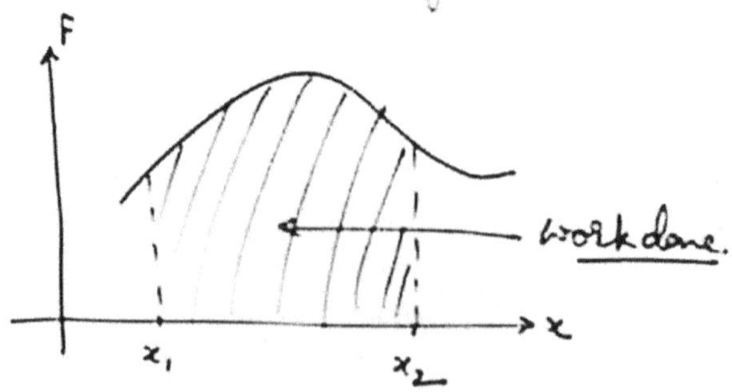


W = area under F-x curve.

Scan for Video Explanation

Web Reference at www.physicsgalaxy.com

**Module-3: Solved Example-1**

**Ex :** A body is constrained to move in Y direction. A force $\vec{F} = (-2\hat{i} + 15\hat{j} + 6\hat{k})$N is applied on it. What is the work done by the force in moving the body by 10 m in +Y direction.

**Sol$^n$:**

Displacement vect $\vec{s} = 10\hat{j}$ m.

force $\vec{F} = -2\hat{i} + 15\hat{j} + 6\hat{k}$ N

Work Done by force $W = \vec{F}\cdot\vec{s}$

$= 150$ J Ans.

Web Reference at www.physicsgalaxy.com | Scan for Video Explanation

**Module-4: Solved Example-2**

**Ex :** Find work done by a variable force in displacing the body from x = 0 to x = 35 m.

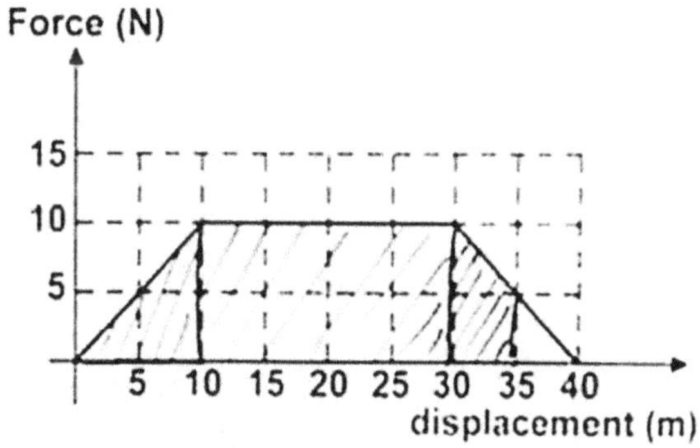


The force varies with displacement according to the graph shown.

**Sol$^n$:**

Work = Area under the curve from x = 0 to x = 35 m.

$W = \frac{1}{2} \times 10 \times 10 + 20 \times 10 + \frac{1}{2} \times 15 \times 5$

$= 50 + 200 + 37.5 = 287.5$ J Ans.

Web Reference at www.physicsgalaxy.com | Scan for Video Explanation

**Module-5: Energy & its Types**

# Energy & its Types:

Energy → ability to do **work**.

↑ required to develop a force.

Types of Energies → There are two ways in which energies can be classified, these are -

(i) Kinetic Energy: due to motion in a body.

$$KE = \frac{1}{2}mv^2$$

(ii) Potential Energy: Stored form of Energy which can be used in doing work.

Scan for Video Explanation

Web Reference at www.physicsgalaxy.com

**Module-5: Energy & its Types**

# Dissipated Energy:

Work Done ⇒ Transfer of Energy.

→ Kinetic Energy

→ Potential Energy

→ lost to surroundings in form of heat, Sound etc. → Dissipated Energy. → Can not be returned back to the system.

Work done by friction

$$W = -\mu mgS$$

↓ lost to surroundings (Dissipated).

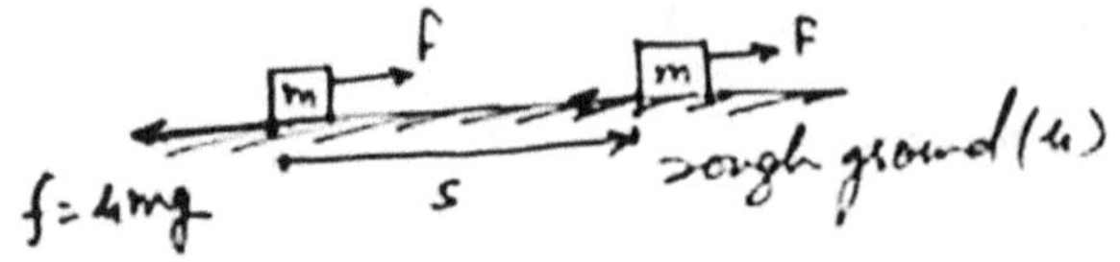


Scan for Video Explanation

Web Reference at www.physicsgalaxy.com

**Module-6: Conservative & Non-Conservative Forces**

# Classification of forces on the basis of Energy Transfer:

Based on the energy Transfer during W/D force can be classified in two categories.

→ (i) Conservative Forces

→ (ii) Non-Conservative Forces.

# Conservative Forces:

→ These are the forces which never dissipates energy.

→ Whenever these forces do work always this energy changes the potential energy of system.

+ve work ⇒ dec. in PE of System

−ve work ⇒ increase in PE of system.





**Module-6: Conservative & Non-Conservative Forces**

→ The region/field in which a body experiences Conservative forces, are called Conservative fields.

→ In Conservative fields always W/D depend on initial & final position of body & is independent from the path along which work is done.

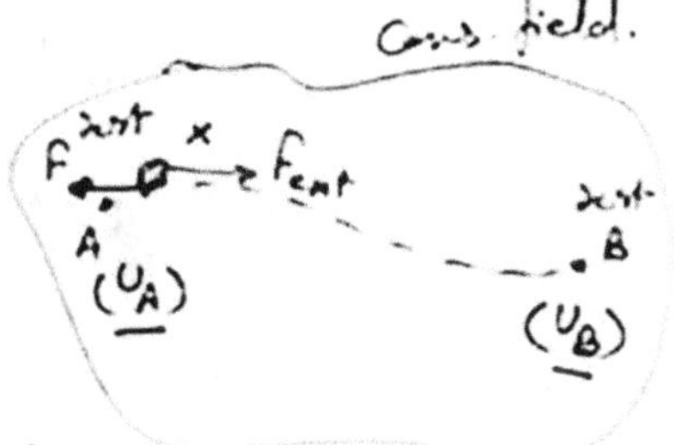


W/D by the field forces.

$$W = U_A - U_B$$

W/D by ext agent

$$W = U_B - U_A$$

Web Reference at www.physicsgalaxy.com

Scan for Video Explanation

**Module-6: Conservative & Non-Conservative Forces**

# Non-Conservative Forces:

These are forces of which W/D is always dissipated and the system energy always decreases

In non conservative field we never talk about Potential Energy.

Ex: friction.

Scan for Video Explanation

Web Reference at www.physicsgalaxy.com

**Module-7: Potential Energy Stored in a Spring**

# Potential Energy Stored in a Spring:

Force of an ideal Spring → Conservative Force.

Work done on a spring → Stored in form of Elastic Potential Energy

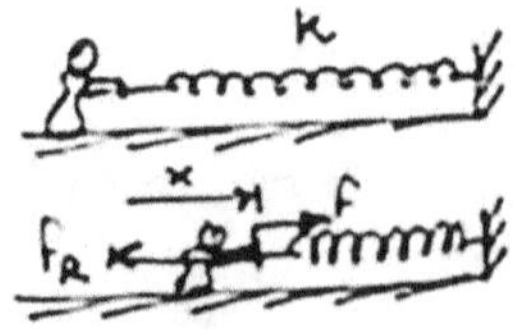


As spring is massless

$F = kx$

W/D in Compression of spring by x

$$W = \int_0^x kx \cdot dx$$

$$= \frac{1}{2}kx^2 \leftarrow \text{PE stored in spring}$$

Scan for Video Explanation

Web Reference at www.physicsgalaxy.com

## Module-7: Potential Energy Stored in a Spring

If a spring is elongated then also it

Stores energy $\longrightarrow \frac{1}{2}kx^2$

$\rightarrow$ under deformed state (Compressed / Elongated)

$\downarrow$

Spring has some PE stored ($\frac{1}{2}kx^2$)

Initially

$PE = \frac{1}{2}kx^2$

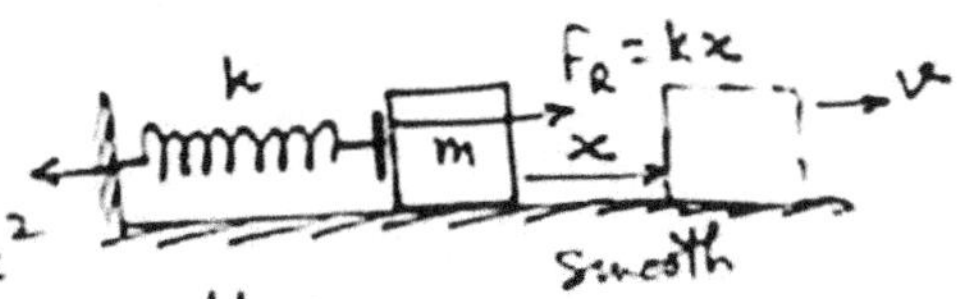


def = x.

Final PE = 0

spring is doing work on block

$\downarrow$

block gains KE

$\frac{1}{2}mv^2 = \frac{1}{2}kx^2 \Rightarrow v = \sqrt{\frac{k}{m}}\,x$

Web Reference at www.physicsgalaxy.com

Scan for Video Explanation

## Module-8: Work-Energy Theorem

\# Work- Energy Theorem:

Whenever work is done on an object or a system of particles, its energy increases and the inc in energy is equal to WD on system/object.

In case of rigid body. $\longrightarrow$ $\Delta W = \Delta K$ (inc in KE)

In Conservative fields $\longrightarrow$ $\Delta W = \Delta U$ (inc in PE)

For system of bodies $\longrightarrow$ $\Delta W = \Delta U + \Delta K$

Web Reference at www.physicsgalaxy.com

Scan for Video Explanation

**Module-8: Work-Energy Theorem**

Case-I.

A rigid body

Work = Fs

$\Delta W = \Delta K$

$Fs = \frac{1}{2}mv_2^2 - \frac{1}{2}mv_1^2$

Case-II

For Conservative forces

$\Delta W = \Delta U$

$F_R = kx$

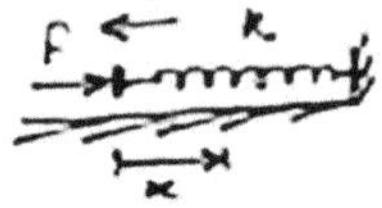


$U_i = 0$

$U_f = \frac{1}{2}kx^2$

$W_F = \frac{1}{2}kx^2$

Scan for Video Explanation

Web Reference at www.physicsgalaxy.com

**Module-8: Work-Energy Theorem**

Case-III

System of bodies (mixed)

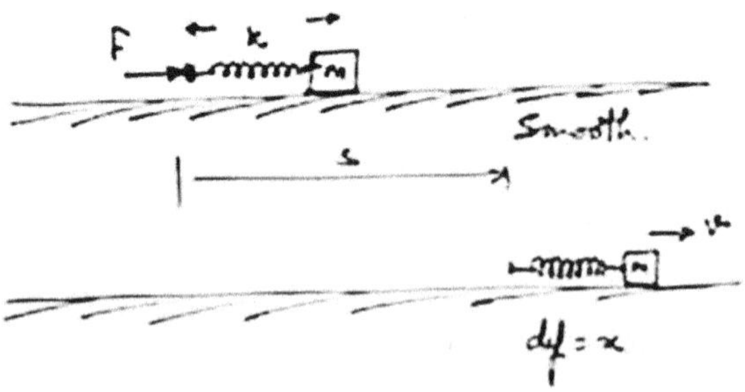


W/D by external force

$W_F = \frac{1}{2}mv^2 + \frac{1}{2}kx^2$

$$\boxed{W_F = \Delta K + \Delta U}$$

Scan for Video Explanation

Web Reference at www.physicsgalaxy.com

**Module-8: Work-Energy Theorem**

**Ex :** In a region potential energy of a body is given as

$$U = 4x^2 + 3x - 5 \text{ J}$$

Find work done by an external agent in slowly moving the body from point (3, 2) to (–1, 4).

Sol$^n$:

$$U_i = 4(3)^2 + 3(3) - 5 = 40 \text{ J}$$

$$U_f = 4(-1)^2 + 3(-1) - 5 = -4 \text{ J}$$

W/D by ext. agent $W = U_f - U_i$

$$= (-4) - (40) = -44 \text{ J} \quad \text{Ans. } (-ve)$$

$\Rightarrow$ In this process external agent will gain energy

Web Reference at www.physicsgalaxy.com | Scan for Video Explanation

**Module-9: Relation in Force & Potential Energy**

# Relation in Potential Energy & Forces:

Potential Energy $\Rightarrow$ Conservative Forces

$$\Delta W = \Delta U$$

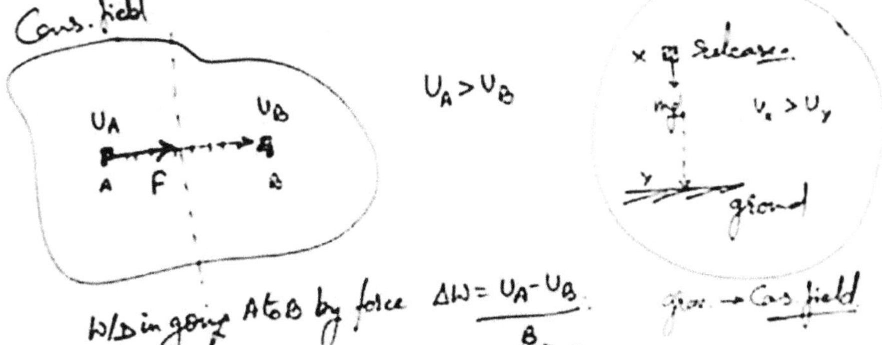


W/D in going A to B by force $\Delta W = U_A - U_B$ grav. → Cons. field

$$\Delta W = \int_A^B \vec{F} \cdot d\vec{x} = -\Delta U$$

Web Reference at www.physicsgalaxy.com | Scan for Video Explanation

**Module-9: Relation in Force & Potential Energy**

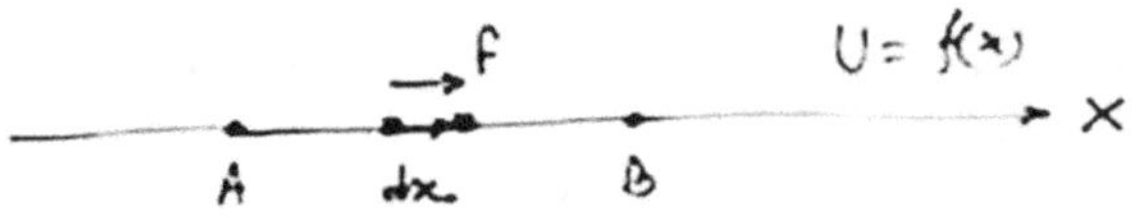


In displacing the body by dx, let dU be the decrease in PE of system then we can use

$$-dU = F\,dx$$

$$dU = -F\,dx$$

$$\text{or} \quad F = -\frac{dU}{dx}$$

vectorially $\boxed{\vec{F} = -\frac{dU}{dx}\hat{i}}$ for unidirectional variation in PE.

Scan for Video Explanation

Web Reference at www.physicsgalaxy.com

**Module-9: Relation in Force & Potential Energy**

In free space or 3D variation of PE we use

$$\boxed{\vec{F} = -\nabla U = -\text{gradient of } U}$$

$\nabla$ → gradient operator.

$$\nabla \equiv \left(\frac{\partial}{\partial x}\hat{i} + \frac{\partial}{\partial y}\hat{j} + \frac{\partial}{\partial z}\hat{k}\right)$$

$\partial$ → partial derivative operator

$$\vec{F} = -\frac{dU}{dx}\hat{i}$$

$$\vec{F} = -\left(\frac{\partial U}{\partial x}\hat{i} + \frac{\partial U}{\partial y}\hat{j} + \frac{\partial U}{\partial z}\hat{k}\right)$$

**Ex:** If in a region $U = 4x^3 + 3y^2 z$ find $\vec{F}$ in this region

**Sol:** $\vec{F} = -(12x^2)\hat{i} - (6yz)\hat{j} - (3y^2)\hat{k} = -12x^2\hat{i} - 6yz\hat{j} - 3y^2\hat{k}$ Ans.

Scan for Video Explanation

Web Reference at www.physicsgalaxy.com

**Module-10: Solved Example-3**

**Ex :** **In a region, potential energy of a particle of mass 4 kg depends on its x-coordinate as**

$$U = 5x^3 + 4x + 7 \text{ Joule}$$

**Find the acceleration of particle at position (2, 4).**

Sol$^n$: As we know, force on a particle

$$F = \left|\frac{dU}{dx}\right| = 15x^2 + 4 \ \text{N}$$

at $x = 2$, force on particle is

$$F = 15(2)^2 + 4 = 64 \text{ N}$$

$$a = \frac{F}{m} = \frac{64}{4} = 16 \text{ m/s}^2 \quad \text{Ans.}$$

**Web Reference at www.physicsgalaxy.com** | **Scan for Video Explanation**

**Module-11: Solved Example-4**

**Ex :** In a region potential energy of a particle of mass 2 kg is given by

$$U = 3x^2 + 4y^2 \text{ J}$$

Find the acceleration of particle when it is released from (1, 1)

Sol$^n$: force on particle $\vec{F} = -\nabla U = -\left(\frac{\partial U}{\partial x}\hat{i} + \frac{\partial U}{\partial y}\hat{j}\right)$

$$\vec{F} = -6x\hat{i} - 8y\hat{j}$$

for (1, 1)

$$\vec{F} = -6\hat{i} - 8\hat{j}$$

$$\vec{a} = \frac{\vec{F}}{m} = -\frac{(6\hat{i} + 8\hat{j})}{2} = -3\hat{i} - 4\hat{j} \text{ m/s}^2$$

$$a = \sqrt{3^2 + 4^2} = 5 \text{ m/s}^2$$

at an angle 53° with $-x$ dir (in III quad) Ans.

$\theta = \tan^{-1}(4/3) = 53°$

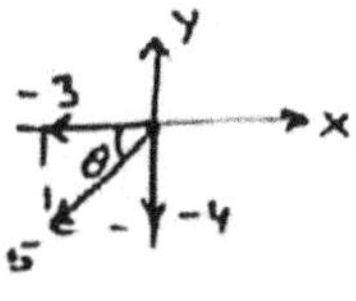


**Web Reference at www.physicsgalaxy.com** | **Scan for Video Explanation**

**Module-12: Solved Example-5**

**Ex :** In a conservative force field, work done in carrying a body from point P to Q is 5J and that from Q to R is 2J. Find work done in carrying the body from P to R.

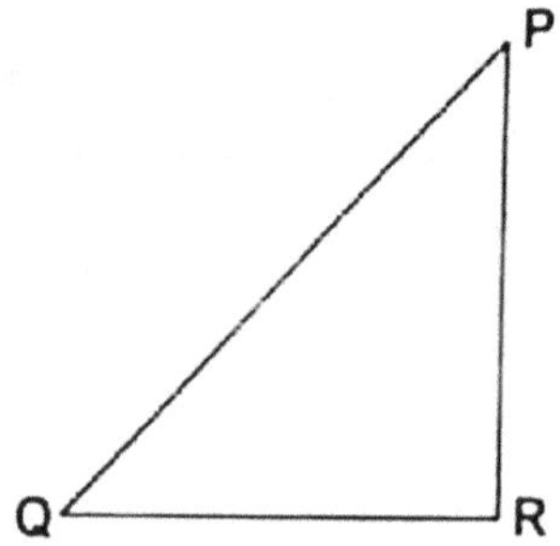


**Sol$^n$:**

$$W = U_f - U_i$$

$$W_{PQ} = U_Q - U_P = 5J \quad -(1)$$

$$W_{QR} = U_R - U_Q = 2J \quad -(2)$$

$$W_{PR} = U_R - U_P = 7J. \text{ Ans.}$$

Scan for Video Explanation

Web Reference at www.physicsgalaxy.com

**Module-13: Work-Energy Equation for Rigid Bodies**

# Work-Energy Equation for rigid bodies:

For rigid bodies → KE

$$\Delta W = \Delta K$$

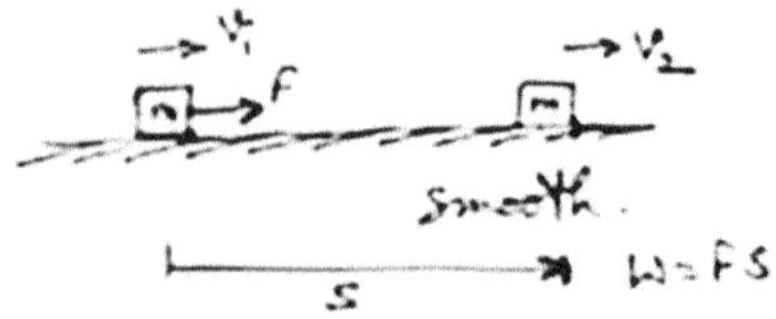


$$\underset{K_i}{\tfrac{1}{2}mv_1^2} + FS = \underset{K_f}{\tfrac{1}{2}mv_2^2}$$

[Initial KE of body (rigid)] + [W/D by forces acting in the dir of disp] − [W/D by forces acting opp. to the dir of disp] = [Final KE of rigid body]

W-E Eq. for motion of a rigid body.

Scan for Video Explanation

Web Reference at www.physicsgalaxy.com

**Module-13: Work-Energy Equation for Rigid Bodies**

Ex:

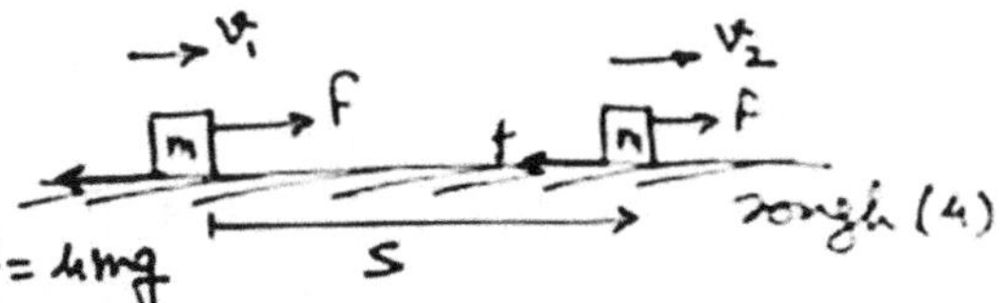


W-E eqⁿ for this motion

$$\overset{K_i}{\frac{1}{2}mv_1^2} + Fs - \mu mgs = \overset{K_f}{\frac{1}{2}mv_2^2}$$

Ex: Find the speed when body hits the ground.

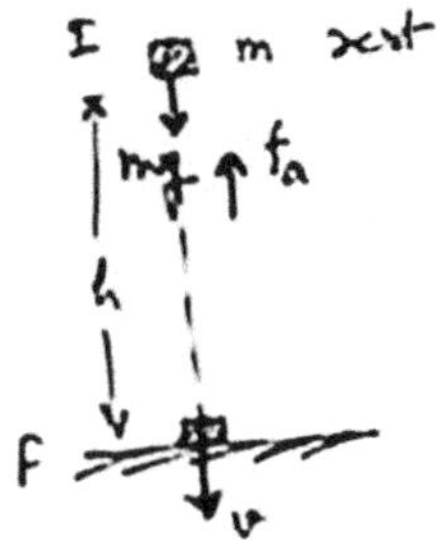


Solⁿ: W-E eqⁿ between I & F pos.

$$\overset{K_i}{0} + mgh - f_a \cdot h = \overset{K_f}{\frac{1}{2}mv^2}$$

$$v = \sqrt{\frac{2(mgh - f_a h)}{m}}$$



**Module-13: Work-Energy Equation for Rigid Bodies**

**Ex :** A rigid block of mass m moves at speed v toward an ideal spring of force constant k as shown, find the maximum distance to which the spring will be compressed by the block.

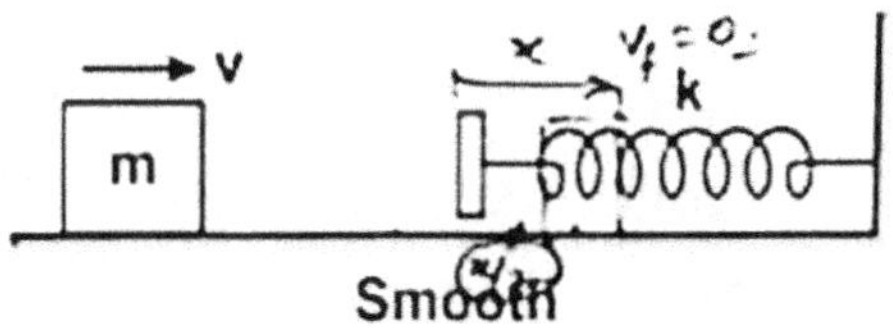


Solⁿ: Let finally spring is compressed by x.

using W-E eqⁿ we have

$$\overset{K_i}{\frac{1}{2}mv^2} - \frac{1}{2}kx^2 = \overset{K_f}{0}$$

$$\frac{1}{2}mv^2 = \frac{1}{2}kx^2$$

$$x = \sqrt{\frac{m}{k}}\, v \quad \text{Ans.}$$

if we wish to find vel of block when compression in spring is x/2 then we use →

$$\frac{1}{2}mv^2 - \frac{1}{2}k\left(\frac{x}{2}\right)^2 = \frac{1}{2}mv_1^2 \qquad v_1 = \ldots$$

**Module-14: Solved Example-6**

Ex : A block of mass m is released from rest from a height h above the light pan attached to a vertical spring of force constant k, mounted on floor as shown. Find the maximum compression in the spring by the block.

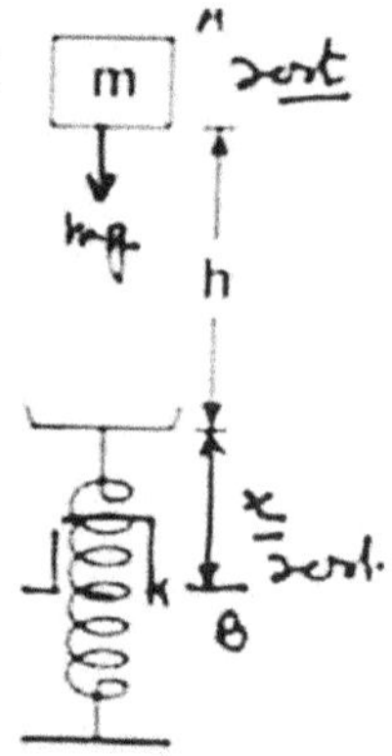


Sol: using WE eqⁿ from A to B.

$$\underbrace{0}_{K_i} + mg(h+x) - \frac{1}{2}kx^2 = \underbrace{0}_{K_f}$$

$$kx^2 - 2mgx - 2mgh = 0$$

$$x = \left(\frac{2mg + \sqrt{4m^2g^2 + 8mghk}}{2k}\right) \text{ Ans.}$$

Scan for Video Explanation

Web Reference at www.physicsgalaxy.com

**Module-15: Solved Example-7**

Ex : In the figure the length of incline is $2\sqrt{3}$ m and a block is projected up the incline at speed 10 m/s. Find the speed with which block will move off the top of incline plane.

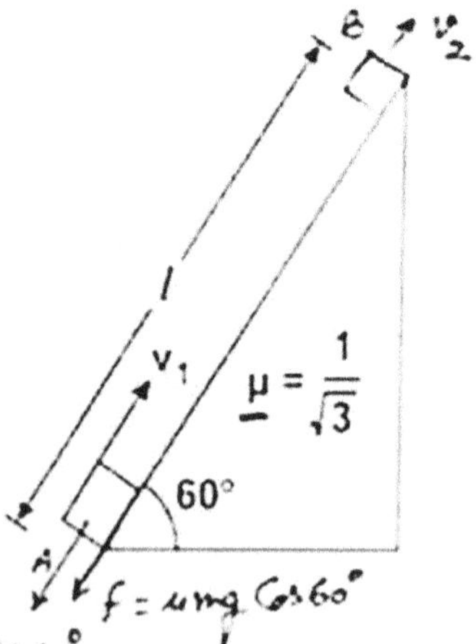


Sol: W-E eqⁿ from A to B.

$$\underbrace{\frac{1}{2}mv_1^2}_{K_i} - mg\sin 60° \times l - \mu mg \cos 60° \times l = \underbrace{\frac{1}{2}mv_2^2}_{K_f}$$

$$\frac{(10)^2}{2} - 10 \times \frac{\sqrt{3}}{2} \times 2\sqrt{3} - \frac{1}{\sqrt{3}} \times 10 \times \frac{1}{2} \times 2\sqrt{3} = \frac{1}{2}v_2^2$$

$$50 - 30 - 10 = \frac{1}{2}v_2^2 \Rightarrow v_2 = \sqrt{20} \text{ m/s. Ans.}$$

Scan for Video Explanation

Web Reference at www.physicsgalaxy.com

**Module-16: Solved Example-8**

**Ex :** A pendulum bob is projected horizontally as shown, find the speed of pendulum bob when string is at an angle $\theta$ with vertical. Also find the maximum angle to which string will be deflected.

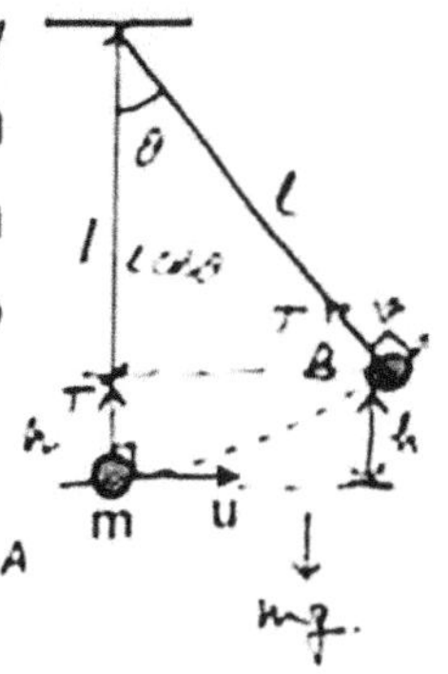


Sol$^n$: using W-E eq$^n$ betw$^n$ pts A and B.

$$\overset{K_A}{\frac{1}{2}mu^2} - mg\,l(1-\cos\theta) = \overset{K_B}{\frac{1}{2}mv^2}$$

$$h = l - l\cos\theta = l(1-\cos\theta)$$

$$v = \sqrt{u^2 - 2gl(1-\cos\theta)} \quad \text{Ans.}$$

If $\theta = \phi$ is the max angle to which string is deflected

$\Rightarrow$ at $\theta = \phi$, $v = 0$.

$$0 = u^2 - 2gl(1-\cos\phi)$$

$$\cos\phi = \frac{2gl - u^2}{2gl} \Rightarrow \phi = \cos^{-1}\left(\frac{2gl-u^2}{2gl}\right) \quad \text{Ans.}$$

Web Reference at www.physicsgalaxy.com Scan for Video Explanation

**Module-17: Solved Example-9**

**Ex :** A block B of mass 1 kg is attached to a spring of natural length 7 cm and force constant 2000 N/m of which other end is attached to a fixed point A. Initially the block B is placed at a position where the distance AB = 10 cm. If block B is released, find the maximum speed attained by the block during its motion.

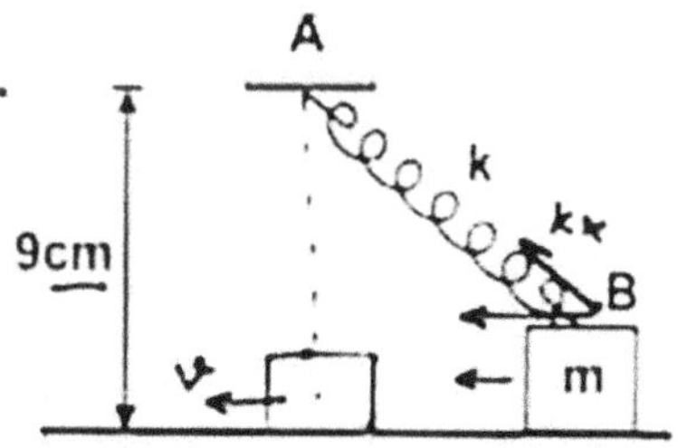


Sol$^n$: $U_{si} = \frac{1}{2}k(0.03)^2 \qquad U_{sf} = \frac{1}{2}k(0.02)^2$

WE eq$^n$

$$\overset{K_i}{0} + \left[\frac{1}{2}k(0.03)^2 - \frac{1}{2}k(0.02)^2\right] = \overset{K_f}{\frac{1}{2}mv^2}$$

$$v = \sqrt{\frac{k}{m}(5\times10^{-4})} = \sqrt{\frac{2000}{1}\times5\times10^{-4}} = 1\ \text{m/s} \quad \text{Ans.}$$

Web Reference at www.physicsgalaxy.com Scan for Video Explanation

**Module-18: Concept of Equilibrium**

# Concept of Equilibrium:

Equilibrium $\Rightarrow$ $\Sigma \vec{F} = 0$

In Conservative field.

$$\vec{F} = -\frac{dU}{dx}\hat{i}$$

When a body is under the state of eqn in Cons. field

$$F = \left|\frac{dU}{dx}\right| = 0$$

(1) If $U =$ Constant (2) If $U \to$ maximum (3) If $U \to$ minimum

Scan for Video Explanation

Web Reference at www.physicsgalaxy.com

**Module-18: Concept of Equilibrium**

Case-I: $U \to$ maximum $U = f(x)$

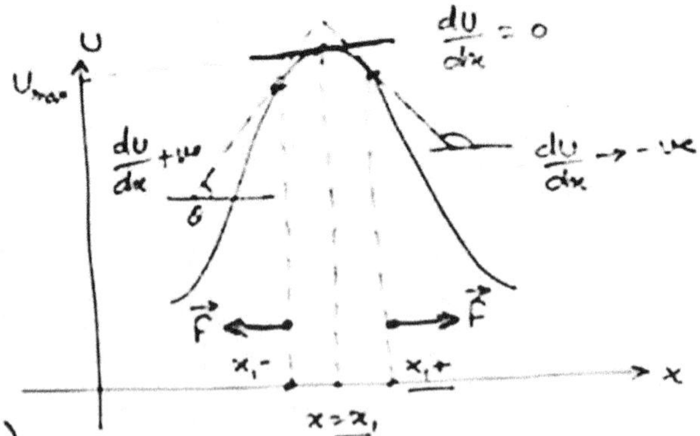


at $x = x_1$

$$F = \left|\frac{dU}{dx}\right| = 0$$

at $x = x_1+$ $\vec{F} = -\frac{dU}{dx}\hat{i}$ (+ve)

at $x = x_1-$

$\vec{F} = -\frac{dU}{dx}\hat{i}$ (-ve)

at $x = x_1 \longrightarrow$ body is said to be in "Unstable Equilibrium."

Scan for Video Explanation

Web Reference at www.physicsgalaxy.com

**Module-18: Concept of Equilibrium**

Case-II: $U \rightarrow$ minimum

$U = f(x)$

at $x = x_1$

$\vec{F} = -\frac{dU}{dx}\hat{i} = 0.$

at $x = x_1+$

$\vec{F} = -\frac{dU}{dx}\hat{i}$ (-ve)

at $x = x_1-$

$\vec{F} = -\frac{dU}{dx}\hat{i}$ (+ve)

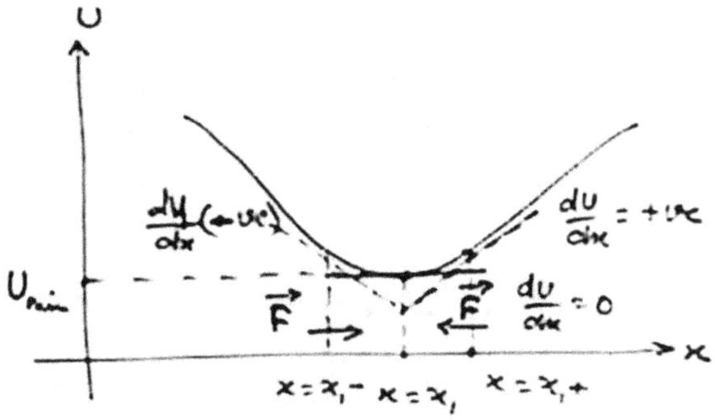


Stable Equilibrium

↓

minimum PE.

Web Reference at www.physicsgalaxy.com

Scan for Video Explanation

**Module-18: Concept of Equilibrium**

Case-III: $U \rightarrow$ Constant

everywhere

$U = $ Const

$F = \left|\frac{dU}{dx}\right| = 0$

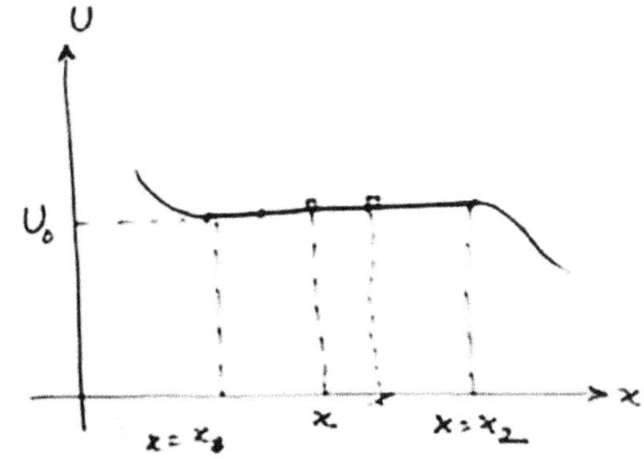


Neutral Equilibrium

for all positions $x_1 \leq x \leq x_2$

Web Reference at www.physicsgalaxy.com

Scan for Video Explanation

**Module-19: Power**

# Power:

Rate of Consumption of Energy

OR

Rate of Doing work

→ Power is a measure of how fast one can do work.

unit of power → J/sec or watt

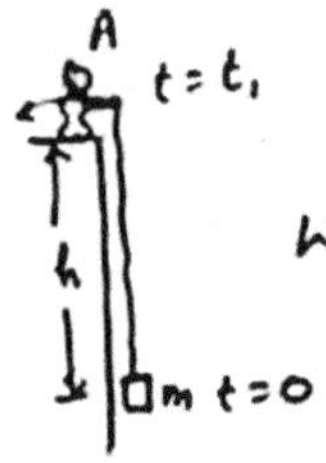



$$W_t = \frac{\text{Work}}{\text{time}} = \underline{P}$$

$$P_A = \frac{mgh}{t_1}$$

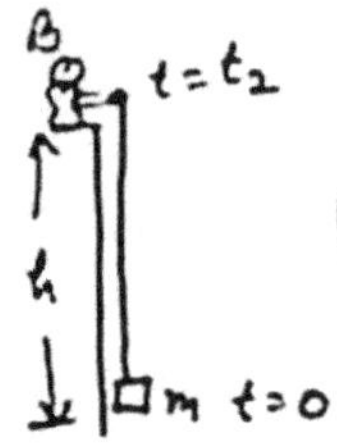



$t_2 < t_1$

$$P_B = \frac{mgh}{t_2}$$

$$\underline{P_B > P_A}$$

Scan for Video Explanation

Web Reference at www.physicsgalaxy.com

**Module-19: Power**

For a process we can define power in two ways -

(1) Average Power

(2) Instantaneous Power.

# Average Power:

$$\langle P \rangle_{avg} = \frac{\text{Total work done}}{\text{time taken}} = \frac{\Delta W}{\Delta t}$$

For a rigid body always $\Delta W = \underline{\Delta K}$ $\quad \langle P_{avg} \rangle = \frac{\Delta K}{\Delta t}$

★ Power associated with a force can be +ve or −ve.

+ Power ⇒ Supply of Energy/sec

− Power ⇒ Gain of Energy/sec.

Scan for Video Explanation

Web Reference at www.physicsgalaxy.com

**Module-19: Power**

**Ex :** Find average power of gravity in motion of a projectile of mass m from starting points O to the highest point of trajectory A as shown in figure.

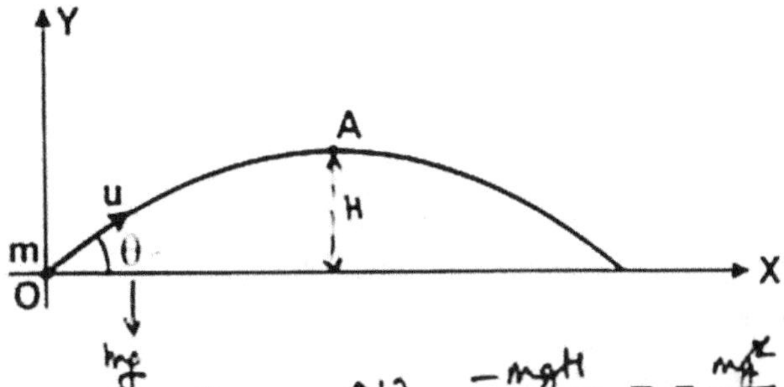


Sol:

$$P_{\underset{O\to A}{avg}} = \frac{\Delta W}{\Delta t} = \frac{-mgH}{(u\sin\theta)/g} = -\frac{mg^2}{u\sin\theta}\times\frac{u^2\sin^2\theta}{2g}$$

$$P_{avg} = -\frac{1}{2}mgu\sin\theta \quad \text{Ans.}$$



**Module-19: Power**

# Instantaneous Power:

In a process of doing work if for an elemental duration $dt$, W/D is $dW$ then for this instant, instantaneous power is defined as

$$P = \frac{dW}{dt}$$

if $\vec{F}$ is the force associated $\Rightarrow dW = \vec{F}\cdot\vec{dx}$

$$P = \frac{\vec{F}\cdot\vec{dx}}{dt} = \vec{F}\cdot\vec{v} = Fv\cos\theta$$

$$\boxed{P = \vec{F}\cdot\vec{v}}$$

[ for supporting forces P is +ve
for opposing forces P is −ve. ]

**Module-19: Power**

**Ex :** A body of mass m is thrown at an angle θ with the horizontal with initial velocity u. Find the instantaneous power developed by gravity as function of time.

Sol$^n$:

$\vec{F} = -mg\hat{j}$

vel vector in the motion of proj at time t is

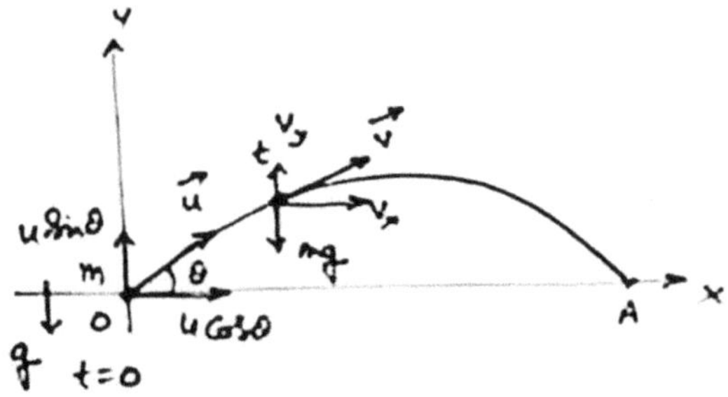


$\vec{v} = (u\cos\theta)\hat{i} + (u\sin\theta - gt)\hat{j}$

Instantaneous Power $P = \vec{F}\cdot\vec{v}$

$P = -mgu\sin\theta + mg^2t$.

Ans.

Scan for Video Explanation

Web Reference at www.physicsgalaxy.com

**Module-20: Solved Example-10**

**Ex :** A man lifts 100 stones to a height of 6 m in 2 minutes. If mass of each stone is 1 kg what is the average power of man. (Take g = 10 m/s$^2$)

Sol$^n$:

Total W/D by man in 120 sec is

$W = 100 \times mgh$

$= 100 \times 1 \times 10 \times 6 = 6000\,J$

Avg power of man $\langle P\rangle_{avg} = \frac{\Delta W}{\Delta t} = \frac{6000}{120} = 50\,\text{Watt}$

Ans.

Scan for Video Explanation

Web Reference at www.physicsgalaxy.com

**Module-21: Solved Example-11**

**Ex :** A block of mass m is placed at rest on a rough horizontal surface with fiction coefficient μ. A constant horizontal force is applied on it so that it starts accelerating with acceleration a. Find the power of the applied force at time t.

**Sol$^n$:**

for motion of block

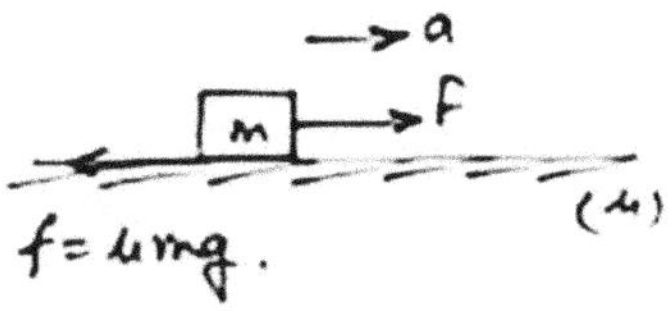


$$F - \mu mg = ma$$

$$F = ma + \mu mg$$

vel of block at time t is $v = at$.

Instantaneous Power of force F at time t is $P = Fv$

$$= m(a + \mu g)at$$

Ans.

# Chapter 8
# Circular Motion

## Lecture Notes Modules


**Module-1:** Circular Motion and Angular Properties
**Module-2:** Relation in Linear and Angular Velocity
**Module-3:** Angular Acceleration
**Module-4:** Solved Example-1
**Module-5:** Solved Example-2
**Module-6:** Solved Example-3
**Module-7:** Solved Example-4
**Module-8:** Analysis of Acceleration in 2D Motion
**Module-9:** Tangential and Normal Acceleration
**Module-10:** Solved Example-5
**Module-11:** Solved Example-6
**Module-12:** Solved Example-7
**Module-13:** Solved Example-8
**Module-14:** Horizontal Circular Motion of a Conical Pendulum
**Module-15:** Concept of Centrifugal Force
**Module-16:** Turning of an Automobile on a Road
**Module-17:** Solved Example-9
**Module-18:** Solved Example-10
**Module-19:** Solved Example-11
**Module-20:** Solved Example-12
**Module-21:** Vertical Circular Motion of a Pendulum Bob
**Module-22:** Cases of Projection of a Pendulum Bob
**Module-23:** Vertical Circular Motion of a Pendulum with a Rod
**Module-24:** Circular Motion inside a Vertical Circular Track
**Module-25:** Vertical Circular Motion inside a Tube
**Module-26:** Vertical Circular Motion outside a Track
**Module-27:** Solved Example-13
**Module-28:** Solved Example-14
**Module-29:** Solved Example-15
**Module-30:** Solved Example-16
**Module-31:** Solved Example-17

**Module-1: Circular Motion and Angular Properties**

# Circular Motion :

Motion of a particle along a circular path.

→ It is an accelerated motion.

In general CM is defined in two categories –

(1) Uniform Circular Motion (UCM) → $|\vec{v}|$ = Constant

(2) Non-Uniform Cir Motion (NUCM) → $|\vec{v}| \neq$ Constant

Scan for Video Explanation

Web Reference at www.physicsgalaxy.com

**Module-1: Circular Motion and Angular Properties**

# Standard Properties of Angular Motion :

(1) Angular Displacement : (in rad)
Angle traversed by p.v. of particle w.r.to a ref point (Centre of Circle)

(2) Angular velocity : rate of change of angular displacement.
(in rad/sec)

(3) Angular Acceleration : rate of change of angular velocity.
(in rad/sec$^2$)

NOTE : Dir of angular vectors are given by Right-Hand Thumb Rule.

Scan for Video Explanation

Web Reference at www.physicsgalaxy.com

**Module-1: Circular Motion and Angular Properties**

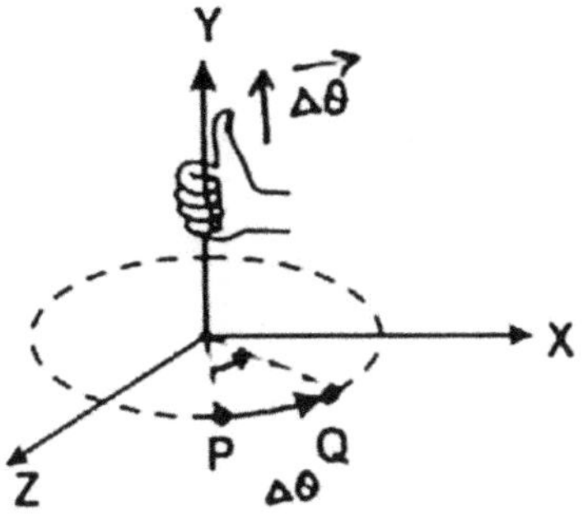


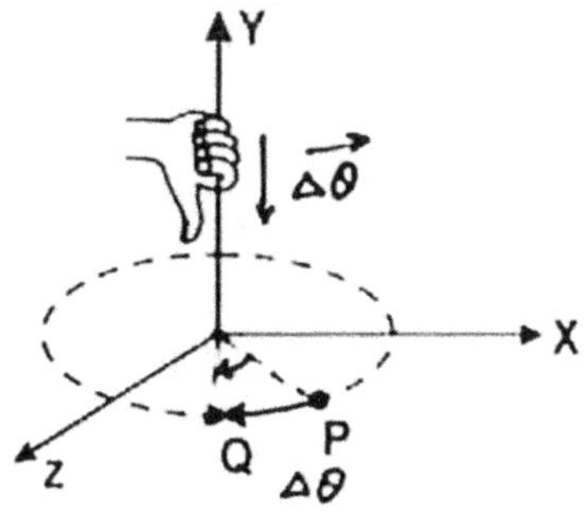


# Angular Displacement: → $\vec{\Delta\theta}$

# Angular Velocity → $\vec{\omega} = \frac{\vec{\Delta\theta}}{\Delta t}$ rad/sec

↑ Avg ang. velocity

Instantaneous ang vel $\omega = \frac{d\theta}{dt}$

Web Reference at www.physicsgalaxy.com | Scan for Video Explanation

**Module-1: Circular Motion and Angular Properties**

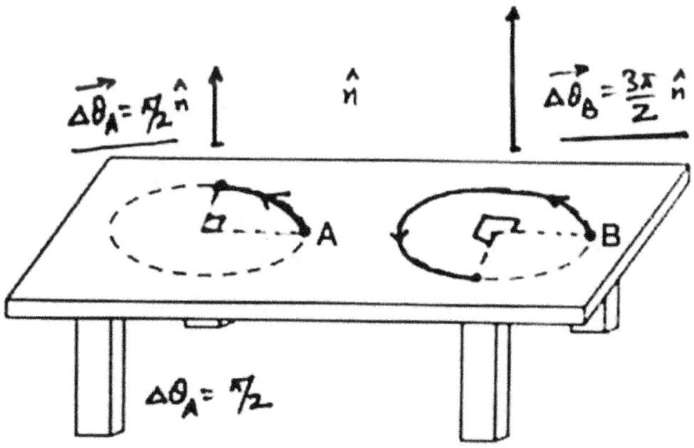


$\Delta\theta_B = \frac{3\pi}{2}$.

Actual dir of Angular disp → ckwise ↓
OR
ackwise ↑

Web Reference at www.physicsgalaxy.com | Scan for Video Explanation

**Module-2: Relation in Linear and Angular Velocity**

## # Relation in Angular & Linear Velocity:

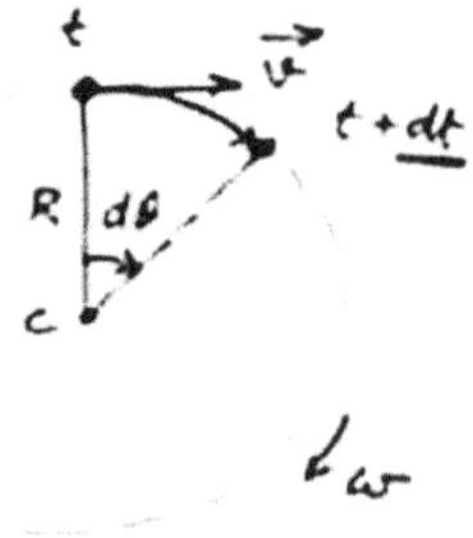


distance covered in time dt is

$$dS = v\,dt$$

Angular disp in time dt

$$d\theta = \frac{dS}{R} = \frac{v\,dt}{R}$$

$$\frac{d\theta}{dt} = \frac{v}{R}$$

Ang. vel. $\boxed{\omega = v/R}$ or $\boxed{v = R\omega}$

$\vec{\omega} \rightarrow$ Dir is given by RHTR.

Scan for Video Explanation

Web Reference at www.physicsgalaxy.com

**Module-2: Relation in Linear and Angular Velocity**

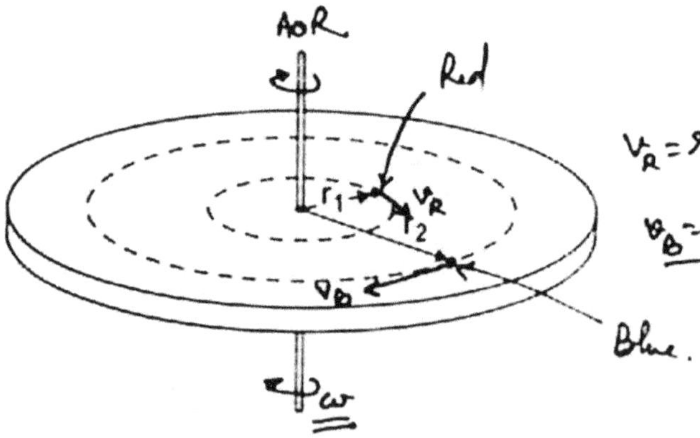


$v_R = r_1\omega$

$v_B = r_2\omega$.

NOTE: When a body rotates at an angular speed ω then all of its particles revolve in cir. motion at same ang. speed ω.

Scan for Video Explanation

Web Reference at www.physicsgalaxy.com

**Module-3: Angular Acceleration**

# Angular Acceleration: (denoted by 'α')

Acceleration → rate of change of velocity

Angular Acceleration → rate of change of ang. vel.

$$a = \frac{dv}{dt} \quad \text{OR} \quad a = v\frac{dv}{dx} \longrightarrow \alpha = \frac{d\omega}{dt} \times \frac{d\theta}{d\theta} = \frac{d^2\theta}{dt^2} \quad \text{OR} \quad \alpha = \omega\frac{d\omega}{d\theta}$$

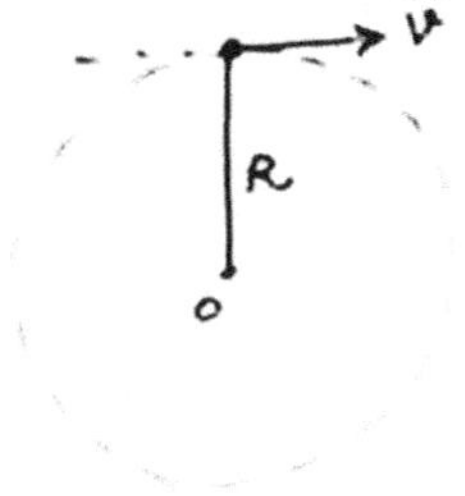


As we know $v = R\omega$

diff w.r.to t $\frac{dv}{dt} = R.\frac{d\omega}{dt}$

tangential acceleration $\boxed{a_t = R\alpha}$

Web Reference at www.physicsgalaxy.com

Scan for Video Explanation

**Module-3: Angular Acceleration**

# Speed Equations for Constant Angular Acceleration:

for linear motion with Constant acc$^n$ we use -

$$v = u + at$$
$$s = ut + \tfrac{1}{2}at^2$$
$$v^2 = u^2 + 2as$$
$$\Delta s_n = u + \frac{a}{2}(2n-1)$$

Similarly for uniform ang. acc$^n$ in circular motion we use angular speed equations, given as -

$$\omega = \omega_0 + \alpha t$$
$$\theta = \omega_0 t + \frac{1}{2}\alpha t^2$$
$$\omega^2 = \omega_0^2 + 2\alpha\theta$$
$$\Delta\theta_n = \omega_0 + \frac{\alpha}{2}(2n-1)$$

$\omega_0$ → Initial ang. vel

$\omega$ → Final ang. vel

$\alpha$ → Ang. acc$^n$

$\theta$ → ang. disp

$\Delta\theta_n$ → ang disp in $n^{th}$ sec

$t$ → time

Web Reference at www.physicsgalaxy.com

Scan for Video Explanation

**Module-3: Angular Acceleration**

If in a Case ang acc$^n$ is variable then we use

$$\left[\alpha = \frac{d\omega}{dt} \quad \text{or} \quad \alpha = \omega\frac{d\omega}{d\theta} \qquad \omega = \frac{d\theta}{dt}\right]$$

$$a = \frac{dv}{dt} \text{ or } a = \frac{v\,dv}{dx} \qquad v = \frac{dx}{dt}$$

Ex: If a body rotates such that its ang. acc$^n$ varies with time as-

$$\alpha = 4t^3 + 2t \text{ rad/s}^2$$

Find ang. speed of body at $t = 2$ sec if it starts from rest

Sol$^n$:

$$\frac{d\omega}{dt} = 4t^3 + 2t$$

$$\int_0^\omega d\omega = \int_0^2 (4t^3 + 2t)\,dt \Rightarrow \omega = \left[t^4 + t^2\right]_0^2$$

$$\omega = 16 + 4 = 20 \text{ rad/sec} \quad \text{Ans.}$$

Scan for Video Explanation

Web Reference at www.physicsgalaxy.com

**Module-4: Solved Example-1**

**Ex :** A solid body rotates about a stationary axis with an angular retardation $\alpha = k\sqrt{\omega}$, where $\omega$ is the angular velocity of body. Find the time after which body will come to rest if at t = 0, angular velocity of body was $\omega_0$.

Sol$^n$: We use

$$\alpha = -k\sqrt{\omega}$$

$$\frac{d\omega}{dt} = -k\sqrt{\omega}$$

$$\int_{\omega_0}^{\omega=0} \frac{d\omega}{\sqrt{\omega}} = -\int_0^t k\,dt$$

$$\left[2\sqrt{\omega}\right]_{\omega_0}^{0} = -kt$$

$$2\sqrt{\omega_0} = kt \Rightarrow t = \frac{2\sqrt{\omega_0}}{k} \quad \text{Ans.}$$

Scan for Video Explanation

Web Reference at www.physicsgalaxy.com

**Module-5: Solved Example-2**

**Ex :** A wheel is subjected to uniform angular acceleration about its axis. Initially it was at rest. In first 2 seconds it rotates through angle $\theta_1$. In next 2 seconds, it rotates through angle $\theta_2$. Find the ratio of $\theta_2/\theta_1$.

**Sol$^n$:** As initially wheel was at rest then at time t after starts its ang. disp is -

$$\theta = \omega_0 t + \frac{1}{2}\alpha t^2 \qquad [As\ \omega_0 = 0]$$

$$\theta = \frac{1}{2}\alpha t^2$$

at $t = 2s$ $\quad \theta_1 = \frac{1}{2}\alpha(2)^2 = 2\alpha$

at $t = 4s$ $\quad \theta_1 + \theta_2 = \frac{1}{2}\alpha(4)^2 = 8\alpha$.

$\theta_2 = 6\alpha$

$$\frac{\theta_2}{\theta_1} = 3 \quad \text{Ans.}$$

Web Reference at www.physicsgalaxy.com | Scan for Video Explanation

**Module-6: Solved Example-3**

**Ex :** The length of second's hand in a watch is 1 cm. Find the magnitude of change in velocity of its tip in 15 second. Also find the magnitude of average acceleration during this interval.

**Sol$^n$:** Time period of second's hand $T = 60$ sec

$\Rightarrow$ Ang vel $\quad \omega = \frac{2\pi}{T} = \frac{\pi}{30}$ rad/s

vel of tip of second's hand is $\quad v = r\omega = \frac{\pi}{30}$ cm/s

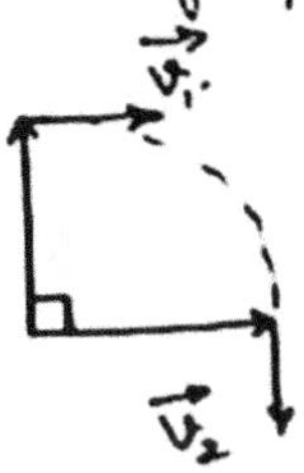


change in vel in 15 sec $\quad \vec{\Delta v} = \vec{v}_2 - \vec{v}_1$

$= \sqrt{2}\cdot\frac{\pi}{30}$ cm/s Ans

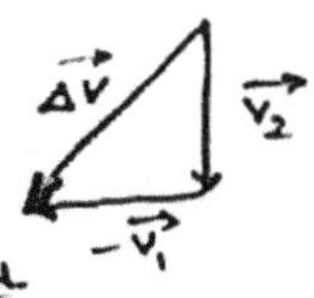


avg acc$^n$ $\quad \vec{a}_{avg} = \frac{\vec{\Delta v}}{\Delta t} = \frac{\sqrt{2}\pi/30}{15} = \frac{\sqrt{2}\pi}{450}$ cm/s$^2$ Ans.

Web Reference at www.physicsgalaxy.com | Scan for Video Explanation

**Module-7: Solved Example-4**

**Ex :** A particle starts moving in a circular path of radius 2 m with initial speed 2 m/s at constant angular acceleration. After two complete revolutions, its speed becomes 8 m/s. Find the angular acceleration of the particle.

**Sol$^n$:**

Initial ang speed $\omega_0 = \frac{v_i}{R} = \frac{2}{2} = 1$ rad/s

Final ang speed $\omega = \frac{v_f}{R} = \frac{8}{2} = 4$ rad/s.

We use $\omega^2 = \omega_0^2 + 2\alpha\theta$

$$\alpha = \frac{\omega^2 - \omega_0^2}{2\theta} = \frac{(4)^2 - (1)^2}{2(4\pi)} = \frac{15}{8\pi} \text{ rad/s}^2$$

Ans.

Scan for Video Explanation

Web Reference at www.physicsgalaxy.com

**Module-8: Analysis of Acceleration in 2D Motion**

# Analysis of Acceleration of a particle in 2D motion:

$\vec{a} \rightarrow$ change in $\vec{v}$ → magnitude, Direction

$\vec{a} = \frac{\vec{F}}{m}$

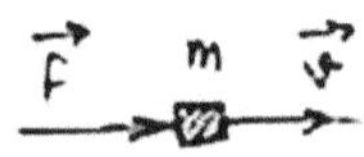


mag of $\vec{v}$ increases

$\vec{a} = \frac{\vec{F}}{m}$

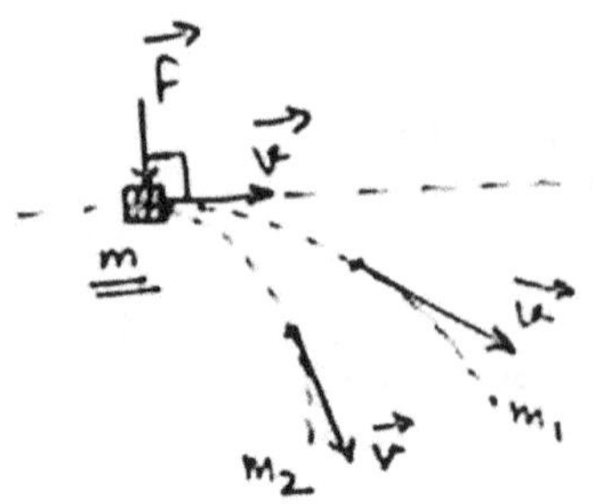


dir of $\vec{v}$ changes.

mag → Constant

$m_2 < m_1$

$a_2 > a_1$

Scan for Video Explanation

Web Reference at www.physicsgalaxy.com

**Module-8: Analysis of Acceleration in 2D Motion**

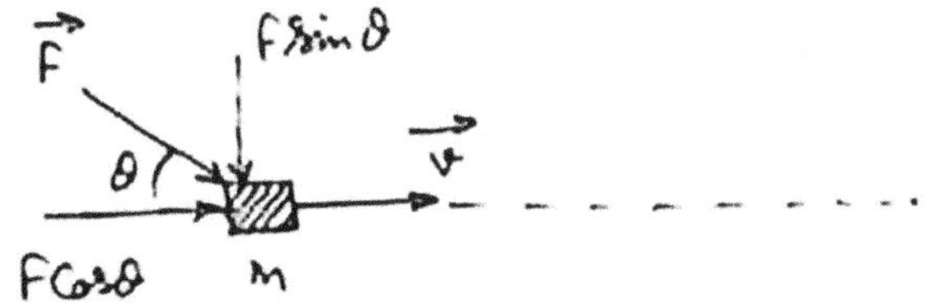


acceleration along $\vec{v}$ $\quad a_{\parallel} = \dfrac{F\cos\theta}{m}$ ← responsible for change in magnitude of velocity.

acceleration normal to $\vec{v}$ $\quad a_{\perp} = \dfrac{F\sin\theta}{m}$ ← responsible for change in direction of velocity.

net accⁿ $\quad \vec{a}_T = \vec{a}_{\parallel} + \vec{a}_{\perp}$

Web Reference at www.physicsgalaxy.com | Scan for Video Explanation

**Module-8: Analysis of Acceleration in 2D Motion**

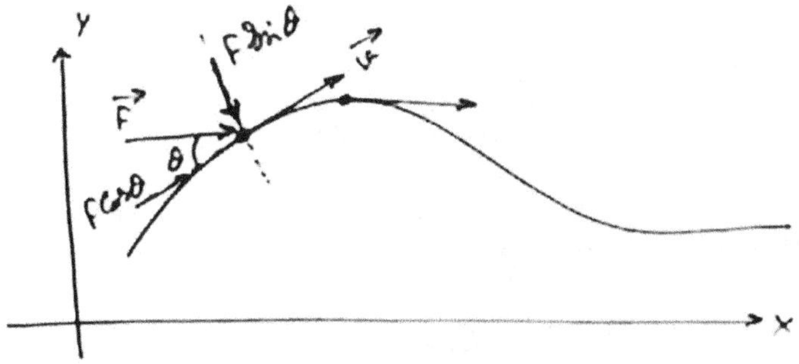


✓ Tangential Accⁿ of Particle $\quad a_t = \dfrac{F\cos\theta}{m}$

✓ Normal Accⁿ of Particle $\quad a_N = \dfrac{F\sin\theta}{m}$

Total accⁿ of Particle $\quad \boxed{a_T = \sqrt{a_t^2 + a_N^2}}$

Web Reference at www.physicsgalaxy.com | Scan for Video Explanation

**Module-9: Tangential and Normal Acceleration**

# Tangential Acceleration:

for Circular Motion

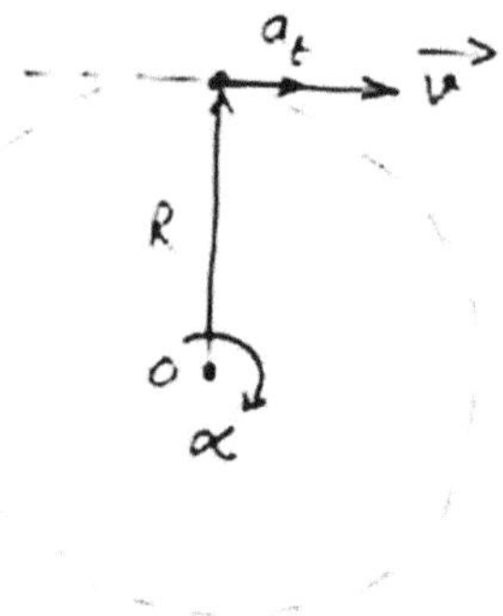


for UCM $\rightarrow$ $|\vec{v}| = \text{Constant}$

$a_t = 0$

for NUCM $\rightarrow$ $a_t \neq 0$

$$\left[\begin{array}{l} a_t = \dfrac{dv}{dt} = \dfrac{d}{dt}|\vec{v}| \\ a_t = v\dfrac{dv}{ds} \end{array}\right]$$

$$\boxed{a_t = R\alpha}$$

Scan for Video Explanation

Web Reference at www.physicsgalaxy.com

**Module-9: Tangential and Normal Acceleration**

# Normal Acceleration:

for Circular Motion

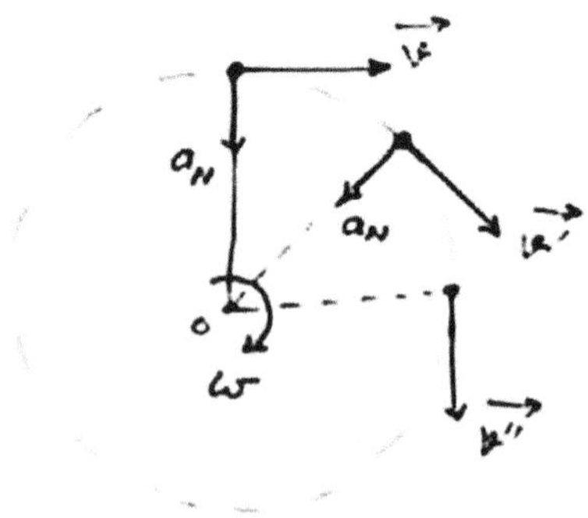


The normal accⁿ for a particle in circular motion can be given as-

$$\boxed{a_N = \frac{v^2}{R}}$$

$v = R\omega$

$$a_N = \frac{(R\omega)^2}{R} = \omega^2 R$$

Scan for Video Explanation

Web Reference at www.physicsgalaxy.com

**Module-9: Tangential and Normal Acceleration**

$$a_N = \frac{v^2}{R} = \omega^2 R.$$

Vel vect of particle at time t is

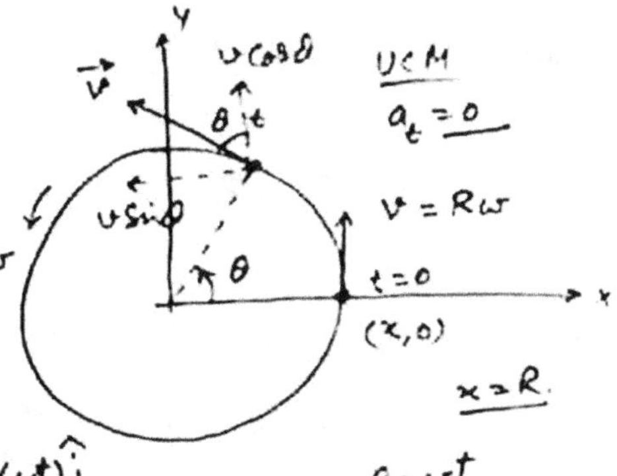


UCM

$a_t = 0$

$v = R\omega$

$x = R$

$\theta = \omega t$

$\omega = \frac{v}{R}$

$$\vec{v} = -v\sin\theta\,\hat{i} + v\cos\theta\,\hat{j}$$

$$\vec{v} = -v\sin(\omega t)\,\hat{i} + v\cos(\omega t)\,\hat{j}$$

$$\vec{a} = \frac{d\vec{v}}{dt} = -v\omega\cos(\omega t)\,\hat{i} - v\omega\sin(\omega t)\,\hat{j}$$

$$a_N = \sqrt{[v\omega\cos(\omega t)]^2 + [v\omega\sin(\omega t)]^2} = v\omega = \frac{v^2}{R} = \omega^2 R.$$

Web Reference at www.physicsgalaxy.com | Scan for Video Explanation

**Module-9: Tangential and Normal Acceleration**

# Normal Acceleration for a particle in 2D motion:

rad. of curv. $r = \dfrac{\left[1+\left(\frac{dy}{dx}\right)^2\right]^{3/2}}{\left|d^2y/dx^2\right|}$

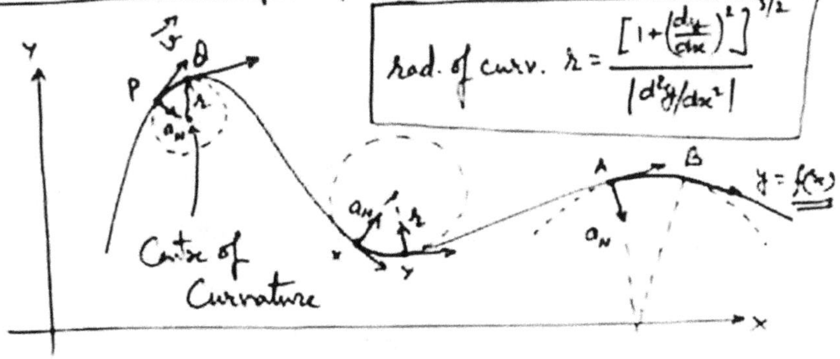


$r$ – Radius of curvature

If at a position on curve where rad. of curvature is $r$ & a particle is moving at speed $v$ then its normal accⁿ at this pos. can be given as

$$a_N = \frac{v^2}{r} \leftarrow \text{Centripetal acc}^n$$

Web Reference at www.physicsgalaxy.com | Scan for Video Explanation

**Module-9: Tangential and Normal Acceleration**

# Centripetal Force in Circular Motion :

$$a_N = \frac{v^2}{R}$$

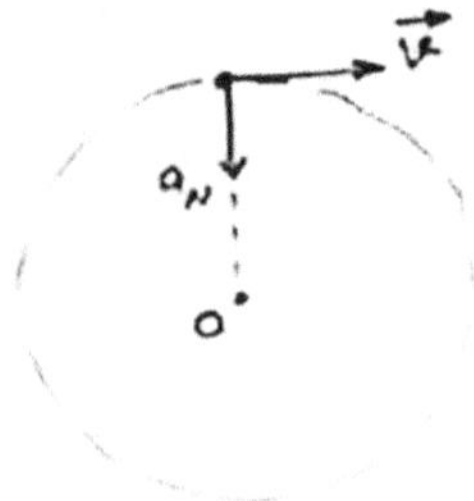


The force responsible for normal acc$^n$ in a cir. motion is called Centripetal force.

C.P. force always act toward centre of circle.

$$F_{cp} = m a_N = \frac{mv^2}{R} \text{ (toward centre of circle)}$$

NOTE: For circular motion of a particle Centripetal force is necessary.

Scan for Video Explanation

Web Reference at www.physicsgalaxy.com

**Module-9: Tangential and Normal Acceleration**

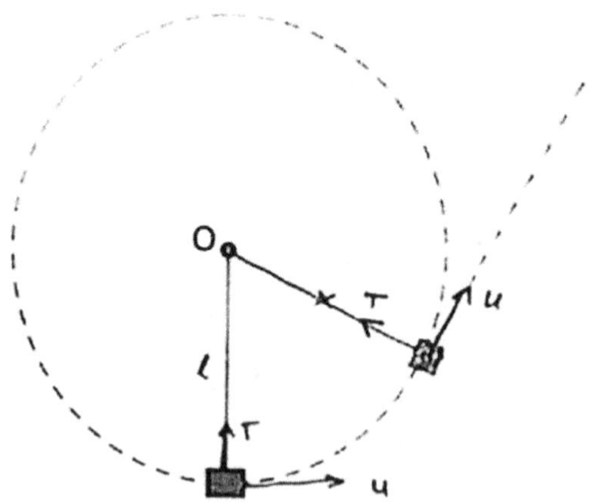


Here $T = \frac{mu^2}{l}$.

Scan for Video Explanation

Web Reference at www.physicsgalaxy.com

**Module-9: Tangential and Normal Acceleration**

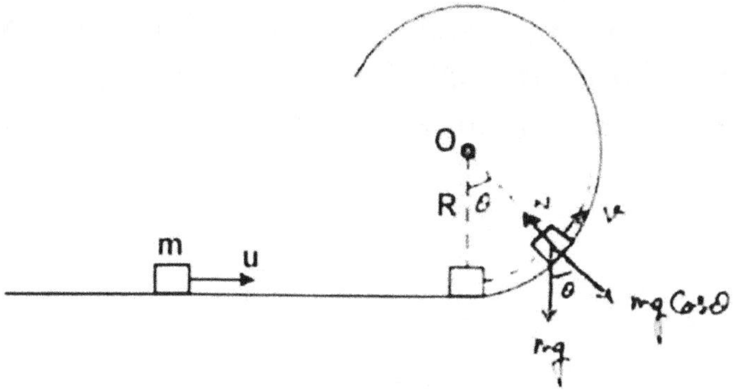


Here we use

$$N - mg\cos\theta = \frac{mv^2}{R}$$

in a CM net radially inward force $= \dfrac{mv^2}{R}$

↑ Centripetal force.

Web Reference at www.physicsgalaxy.com | Scan for Video Explanation

**Module-10: Solved Example-5**

**Ex :** A particle moves in a circular path of radius 3m at a speed given as $v = 3t^2$ m/s. Find at t = 2 sec.

(a) Tangential acceleration of particle.

(b) Normal acceleration of particle.

(c) Total acceleration of particle.

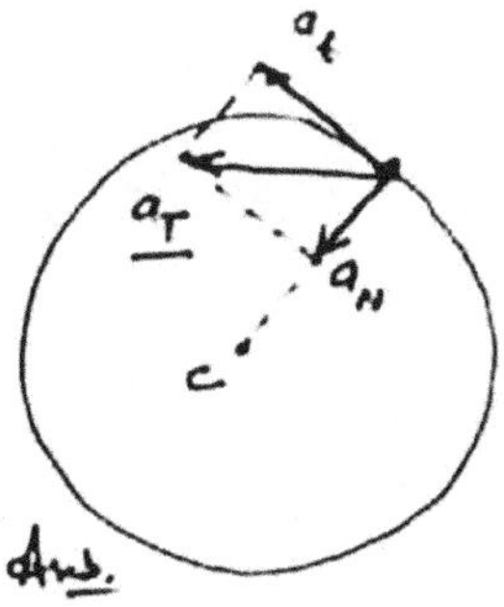


Soln:

$$v = 3t^2 \text{ m/s}$$

(a) $a_t = \dfrac{dv}{dt} = 6t = 6 \times 2 = 12 \text{ m/s}^2$ Ans.

(b) $a_N = \dfrac{v^2}{R} = \dfrac{(3t^2)^2}{R} = \dfrac{9t^4}{3}\Big|_{t=2} = 3 \times 16 = 48 \text{ m/s}^2$ Ans.

(c) $a_T = \sqrt{a_t^2 + a_N^2} = \sqrt{(12)^2 + (48)^2} = 12\sqrt{17} \text{ m/s}^2$ Ans.

Web Reference at www.physicsgalaxy.com | Scan for Video Explanation

**Module-11: Solved Example-6**

**Ex :** A particle is projected at a speed u at an angle θ with the horizontal. Find the radius of curvature at the highest point of the trajectory of projectile.

Sol$^n$:

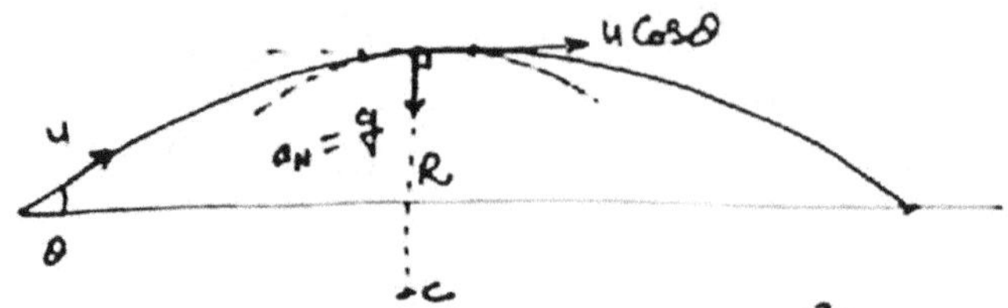


at topmost pt of trajectory $a_N = g = \frac{(u\cos\theta)^2}{R} \Rightarrow R = \frac{u^2\cos^2\theta}{g}$ Ans.

Alt. method:

$R = \frac{\left[1+\left(\frac{dy}{dx}\right)^2\right]^{3/2}}{|d^2y/dx^2|}$

eq$^n$ → $y = x\tan\theta - \frac{gx^2}{2u^2\cos^2\theta}$

slope → $\frac{dy}{dx} = \tan\theta - \frac{2gx}{2u^2\cos^2\theta}\bigg|_{Top} = 0$

$\frac{d^2y}{dx^2} = -\frac{g}{u^2\cos^2\theta}$

$R = \frac{1}{g/u^2\cos^2\theta} = \frac{u^2\cos^2\theta}{g}$ Ans.

Scan for Video Explanation

Web Reference at www.physicsgalaxy.com

**Module-12: Solved Example-7**

**Ex :** A particle moves in a circular path of radius 1 m with an angular speed

$$\omega = 2t^2 + 1 \text{ rad/sec}$$

Find the angle between total acceleration and normal acceleration at t = 1 sec.

Sol$^n$: Ang. acc$^n$ of particle $\alpha = \frac{d\omega}{dt} = 4t$ rad/s$^2$

tangential acc$^n$ of Part $a_t = R\alpha = 4t\big|_{t=1s} = 4$ m/s$^2$

Normal acc$^n$ of Part $a_N = \omega^2 R = (2t^2+1)^2 R$

$= 4t^4 + 4t^2 + 1\big|_{t=1sec}$

$= 9$ m/s$^2$

$\tan\theta = \frac{a_t}{a_N} = \frac{4}{9}$

$\theta = \tan^{-1}\left(\frac{4}{9}\right)$ Ans.

Scan for Video Explanation

Web Reference at www.physicsgalaxy.com

**Module-13: Solved Example-8**

**Ex :** A stone is thrown horizontally from top of a tower at 10 m/s. Find the radius of curvature of its trajectory at the end of 3 sec after the projection. (Take g = 10 m/s$^2$)

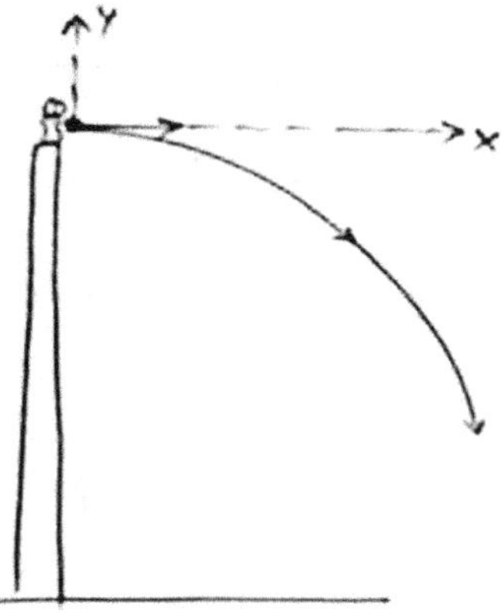


Sol$^n$:

Eq$^n$ of Traj $y = -\frac{gx^2}{2u^2}$

Rad. of Curvature $r = \frac{\left[1+\left(\frac{dy}{dx}\right)^2\right]^{3/2}}{\left|\frac{d^2y}{dx^2}\right|}$

$\frac{dy}{dx} = -\frac{gx}{u^2} = -\frac{10(30)}{(10)^2} = -3$

$\frac{d^2y}{dx^2} = -\frac{g}{u^2} = -\frac{10}{100} = -\frac{1}{10}$

at $t = 3$ sec
$x = 10t = 30$ m.

$r = \frac{[1+9]^{3/2}}{1/10} = 100\sqrt{10}$ m Ans.

Web Reference at www.physicsgalaxy.com | Scan for Video Explanation

**Module-14: Horizontal Circular Motion of a Conical Pendulum**

# Horizontal Circular Motion of a Pendulum Bob:

Conical Pendulum.

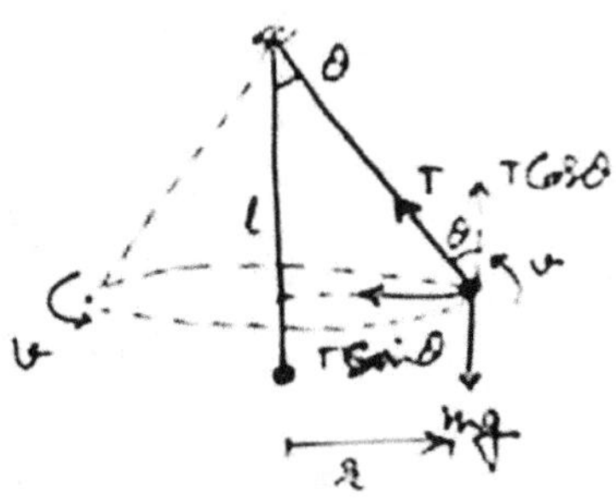


for uniform and Stable Horizontal Circular Motion

$T\sin\theta = \frac{mv^2}{r}$ —— (1)

$T\cos\theta = mg$. —— (2)

$\frac{(1)}{(2)} \Rightarrow \boxed{\tan\theta = \frac{v^2}{rg}} \Rightarrow \boxed{v = \sqrt{rg\tan\theta}}$

$\sqrt{(1)^2+(2)^2} \Rightarrow \boxed{T = m\sqrt{\frac{v^4}{r^2} + g^2}}$

Web Reference at www.physicsgalaxy.com | Scan for Video Explanation

**Module-15: Concept of Centrifugal Force**

# Concept of Centrifugal force :

→ Circular Motion is an accelerated Motion

→ A ref. frame moving in a circular path will be a non- inertial refrence frame.

Pseudo force
$F = \left(\frac{mv^2}{R}\right)$ [Centrifugal force]

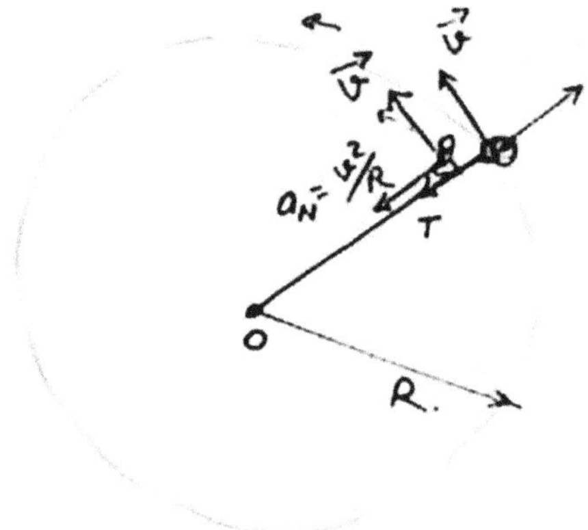


FBD of Body
w.r.to revolving observer

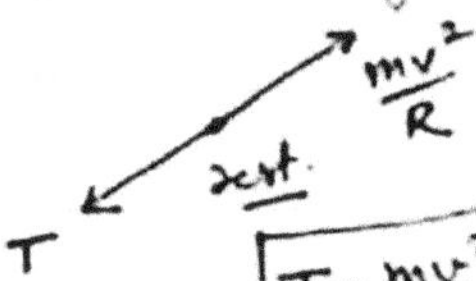


$$T = \frac{mv^2}{R}$$ (Ans)

in the ref frame of obs.

Scan for Video Explanation

Web Reference at www.physicsgalaxy.com

**Module-15: Concept of Centrifugal Force**

Centrifugal force :

→ It is a Pseudo force

→ It appears to that observer which moves in circular path along with the body in C.M.

→ It appears to act in radially outward direction

→ magnitude $F = \frac{mv^2}{R}$.

Scan for Video Explanation

Web Reference at www.physicsgalaxy.com

**Module-15: Concept of Centrifugal Force**

# Analysis of Conical Pendulum by Centrifugal Force:

FBD of bob as seen by an observer in rotating ref frame,

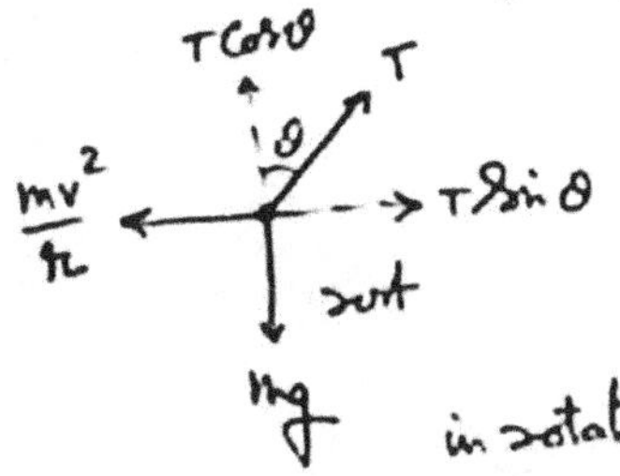


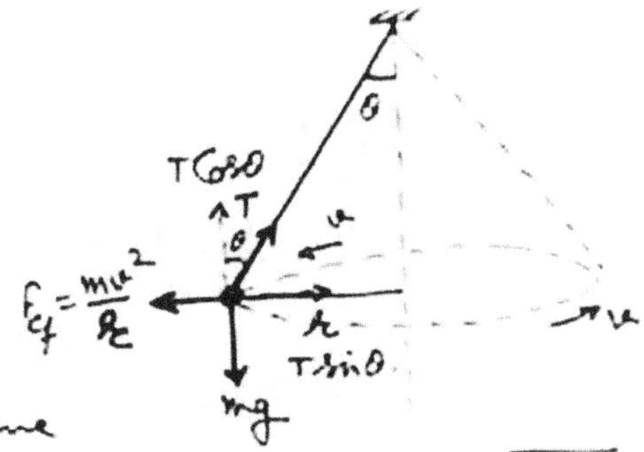


in rotating ref frame as body is at rest

$$T \sin\theta = \frac{mv^2}{R} \quad \text{—(1)}$$

$$T \cos\theta = mg \quad \text{—(2)}$$

$$\tan\theta = \frac{v^2}{Rg}$$

$$v = \sqrt{Rg\tan\theta}$$

$$T = m\sqrt{\frac{v^4}{R^2} + g^2}$$

Web Reference at www.physicsgalaxy.com | Scan for Video Explanation

**Module-16: Turning of an Automobile on a Road**

# Skidding of an automobile while taking a turn:

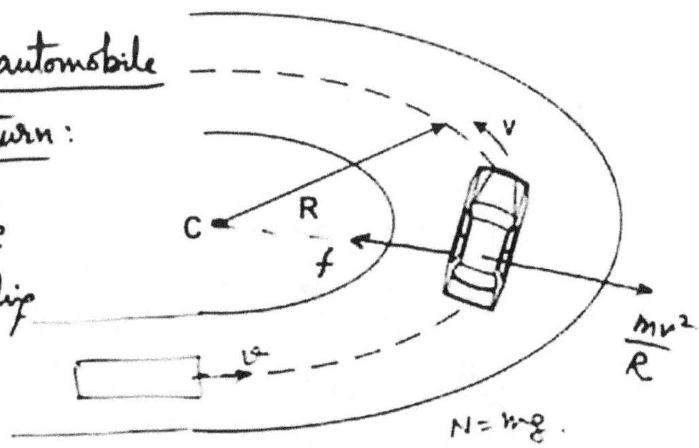


for safe turning without skidding

$$f = \frac{mv^2}{R}$$

↑ static friction

$N = mg$.

for safe turning

$$v < \sqrt{\mu Rg}.$$

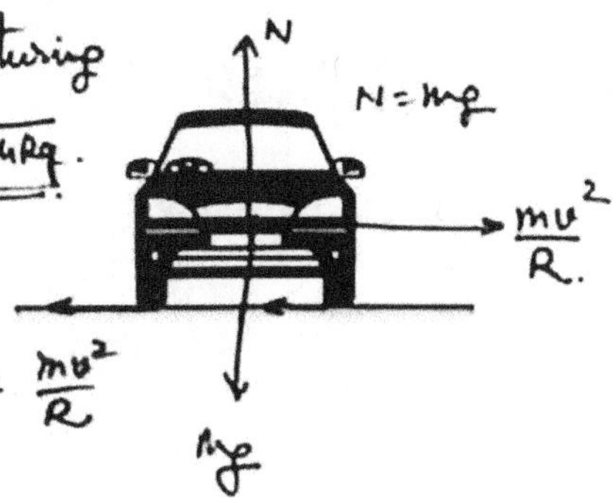


for skidding

$$\frac{mv^2}{R} \geq \mu N$$

$$\frac{mv^2}{R} \geq \mu mg$$

$$v \geq \sqrt{\mu Rg}$$

$$f = \frac{mv^2}{R}$$

Web Reference at www.physicsgalaxy.com | Scan for Video Explanation

**Module-16: Turning of an Automobile on a Road**

# Skidding of an automobile while taking a turn on Banked Road:

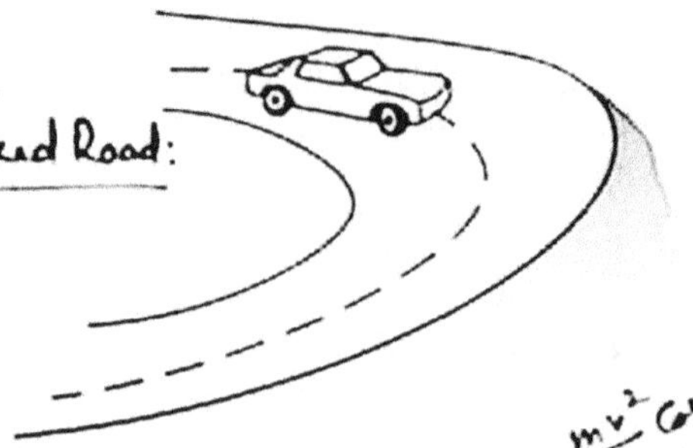

Case-I:

If $mg\sin\theta = \frac{mv^2}{R}\cos\theta$

$\Rightarrow$ No skidding tendency

$$\tan\theta = \frac{v^2}{Rg}$$

$$\theta = \tan^{-1}\left(\frac{v^2}{Rg}\right)$$

angle of banking suitable for a speed $v$.

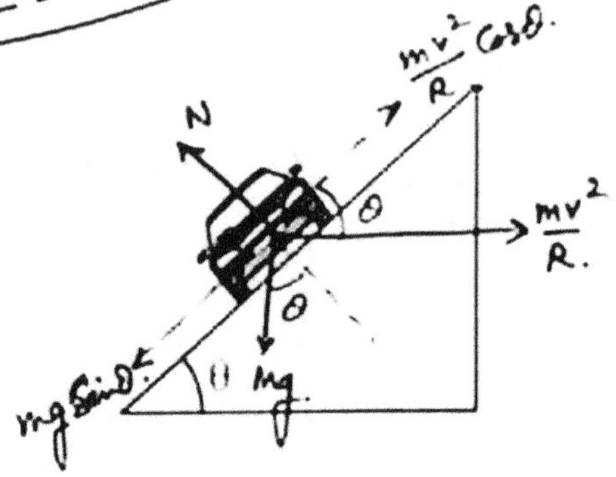


Scan for Video Explanation

Web Reference at www.physicsgalaxy.com

**Module-16: Turning of an Automobile on a Road**

Case-II:

If $\frac{mv^2}{R}\cos\theta > mg\sin\theta$.

f will act in downward dir such that for safe turning

$$f = \frac{mv^2}{R}\cos\theta - mg\sin\theta$$

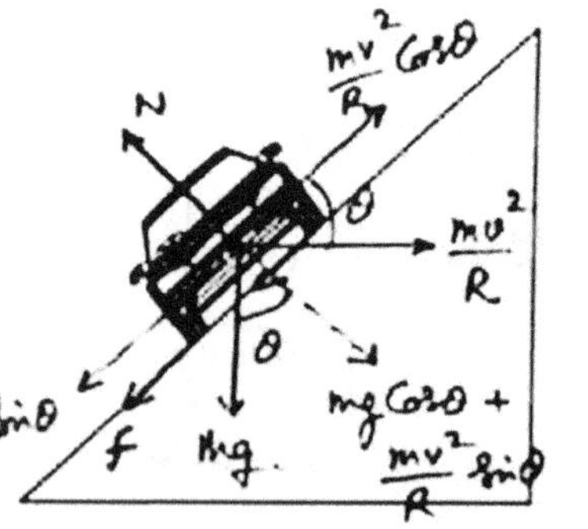


Car will skid upward if

$$\left(\frac{mv^2}{R}\cos\theta - mg\sin\theta\right) = \mu\left(mg\cos\theta + \frac{mv^2}{R}\sin\theta\right)$$

$$v = \sqrt{\frac{Rg(\sin\theta + \mu\cos\theta)}{(\cos\theta - \mu\sin\theta)}}$$

max speed limit for safe turning

Scan for Video Explanation

Web Reference at www.physicsgalaxy.com

**Module-16: Turning of an Automobile on a Road**

Case-III

If $mg\sin\theta > \frac{mv^2}{R}\cos\theta$

f will act in upward dir such that

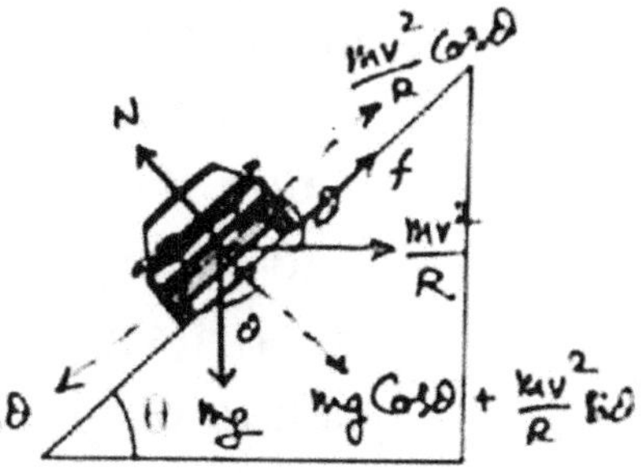


Static fr $f = mg\sin\theta - \frac{mv^2}{R}\cos\theta$

$N = mg\cos\theta + \frac{mv^2}{R}\sin\theta$

Car will skid downward if

$$\left(mg\sin\theta - \frac{mv^2}{R}\cos\theta\right) = \mu\left(mg\cos\theta + \frac{mv^2}{R}\sin\theta\right)$$

$$v = \sqrt{\frac{Rg(\sin\theta - \mu\cos\theta)}{\cos\theta + \mu\sin\theta}}$$

minimum speed limit for safe turning if θ > Angle of Repose.

Web Reference at www.physicsgalaxy.com | Scan for Video Explanation

**Module-17: Solved Example-9**

**Ex :** A horizontal disc which is free to rotate about its vertical central axis is initially at rest. At a distance r from axis, a coin is placed on disc and the friction coefficient between coin and disc is μ. Find the angular speed at which disc should be rotated so that coin will start slipping.

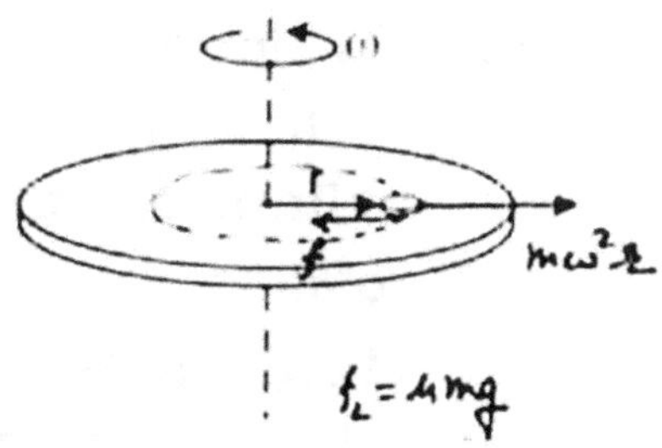


Sol$^n$:

When coin is at rest on disc

$$f = m\omega^2 r$$

The coin will slip when $f = \mu mg$.

$$\mu mg = m\omega^2 r \Rightarrow \boxed{\omega \geq \sqrt{\frac{\mu g}{r}}}\ \text{Ans}$$

Web Reference at www.physicsgalaxy.com | Scan for Video Explanation

**Module-18: Solved Example-10**

**Ex :** In a horizontal disc of radius R a smooth groove is cut along its radius as shown and a small bead is placed at a distance a from the centre of disc. If disc is set in rotation at angular speed ω, the bead will slide out. Find the speed with which bead will slide out from the edge of disc.

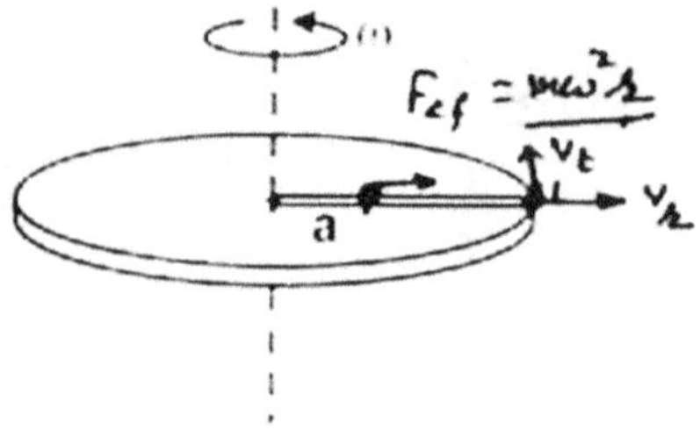


Sol$^n$: When bead will move off the edge of disc its tangential speed $v_t = R\omega$

w.r.to disc if a is the acc$^n$ of bead we use $m\omega^2 r = ma \Rightarrow v\frac{dv}{dr} = \omega^2 r$

$$\int_0^{v_r} v\,dv = \int_a^R \omega^2 r\,dr \rightarrow \frac{v_r^2}{2} = \frac{\omega^2}{2}(R^2 - a^2)$$

$$v_r = \omega\sqrt{R^2 - a^2}$$

$$v_{bead} = \sqrt{v_t^2 + v_r^2} = \sqrt{2\omega^2 R^2 - \omega^2 a^2}$$ Ans.

Scan for Video Explanation | Web Reference at www.physicsgalaxy.com

**Module-19: Solved Example-11**

**Ex :** A block B is supported at the vertical surface of a hollow cylindrical drum which is rotating about its vertical axis at an angular speed ω. Find the minimum friction coefficient between block and wall of cylinder so that block will not fall during circular motion.

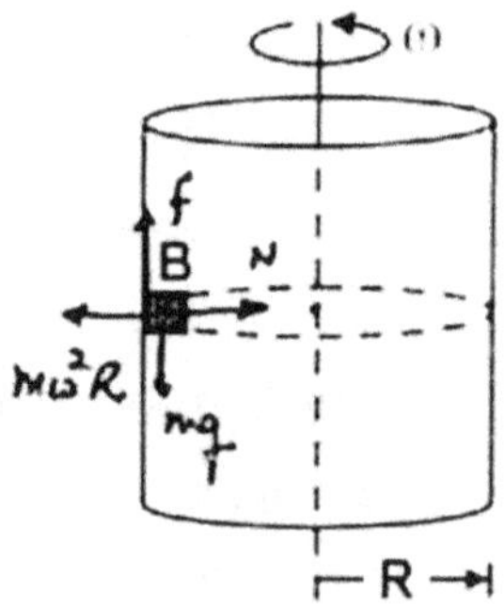


$f_L = \mu N$.

Sol$^n$: Normal Reaction on Block is $N = m\omega^2 R$

for block not to slide down $mg < \mu N$

$$mg < \mu m\omega^2 R$$

or $\mu > \frac{g}{\omega^2 R}$ Ans.

Scan for Video Explanation | Web Reference at www.physicsgalaxy.com

**Module-20: Solved Example-12**

**Ex :** On a smooth horizontal table a block is attached to a point C through a string as shown and it is set in circular motion with a uniform angular speed such that tension in string is $T_0$. Now length of string and angular speed of revolution are doubled find new tension in the string.

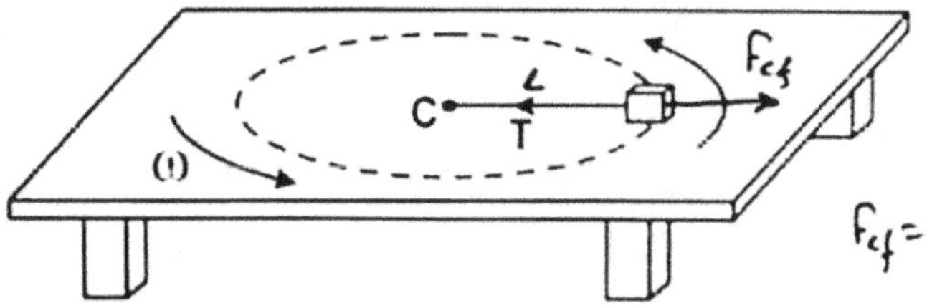


$f_{cf} = m\omega^2 l$

**Sol$^n$:**

Initially $T_0 = m\omega^2 l$

Final state $T = m(2\omega)^2(2l) = 8m\omega^2 l = 8T_0$ Ans.

Web Reference at www.physicsgalaxy.com | Scan for Video Explanation

**Module-21: Vertical Circular Motion of a Pendulum Bob**

# Vertical Circular Motion of a Pendulum Bob:

Using W-E eq$^n$ we get.

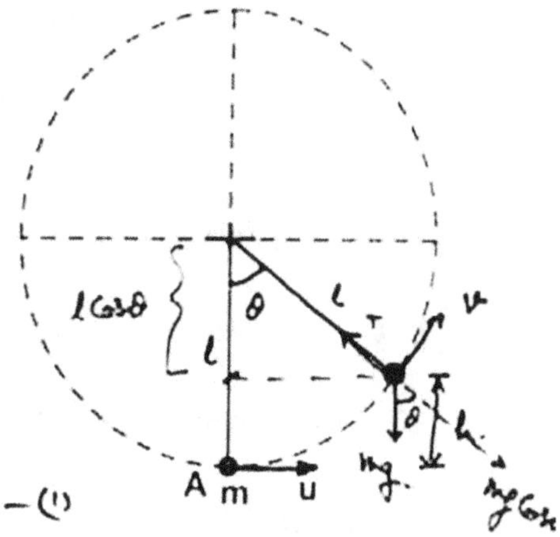


$k_i \qquad\qquad k_f$

$$\frac{1}{2}mu^2 - mgl(1-\cos\theta) = \frac{1}{2}mv^2$$

$$\boxed{v = \sqrt{u^2 - 2gl(1-\cos\theta)}} \quad -(1)$$

$h = l(1-\cos\theta)$

mg

$$T - mg\cos\theta = \frac{mv^2}{l}$$

$$T = mg\cos\theta + \frac{m}{l}\left[u^2 - 2gl(1-\cos\theta)\right]$$

$$\boxed{T = \frac{mu^2}{l} - 2mg + 3mg\cos\theta} \quad -(2)$$

Web Reference at www.physicsgalaxy.com | Scan for Video Explanation

**Module-21: Vertical Circular Motion of a Pendulum Bob**

# 'Angular Amplitude' of Pendulum:

from eq^n (1)

$$v = \sqrt{u^2 - 2gl(1-\cos\theta)}$$

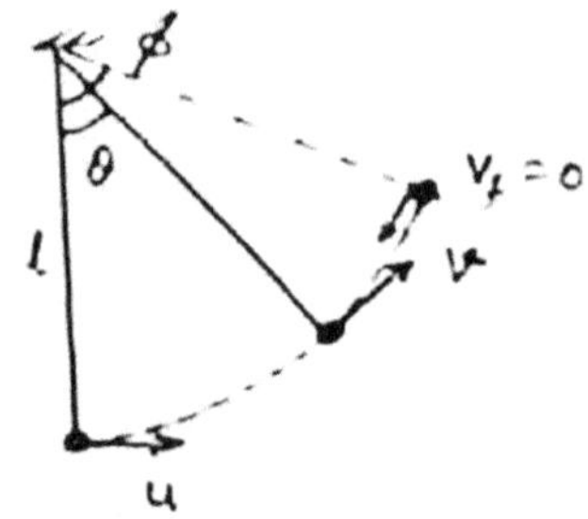


If $\phi$ is the angular amplitude

$\Rightarrow$ at $\theta = \phi$, $v = 0$

$$0 = \sqrt{u^2 - 2gl(1-\cos\phi)}$$

$$0 = u^2 - 2gl + 2gl\cos\phi$$

$$\cos\phi = \left(\frac{2gl - u^2}{2gl}\right) \Rightarrow \boxed{\phi = \cos^{-1}\left(\frac{2gl - u^2}{2gl}\right)}$$

Ang. amplitude.

Scan for Video Explanation

Web Reference at www.physicsgalaxy.com

**Module-21: Vertical Circular Motion of a Pendulum Bob**

# Angle where string of Pendulum will slack:

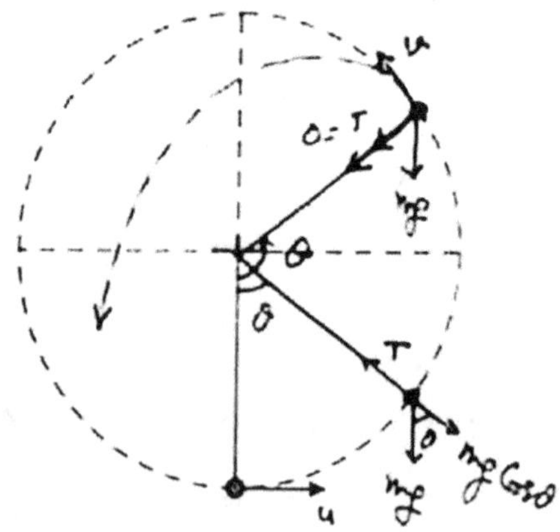


at $\theta < \pi/2$, $T \neq 0$

at $\theta > \pi/2$ it is possible that at some angle T becomes 0.

from eq^n (2)

$$T = \frac{mu^2}{l} - 2mg + 3mg\cos\theta$$

let at $\theta = \alpha$, $T = 0 \Rightarrow$

$$0 = \frac{mu^2}{l} - 2mg + 3mg\cos\alpha.$$

$$\cos\alpha = \left(\frac{2gl - u^2}{3gl}\right) \Rightarrow \boxed{\alpha = \cos^{-1}\left(\frac{2gl - u^2}{3gl}\right)}$$

Scan for Video Explanation

Web Reference at www.physicsgalaxy.com

**Module-22: Cases of Projection of a Pendulum Bob**

# Cases of Projection of a Pendulum Bob:

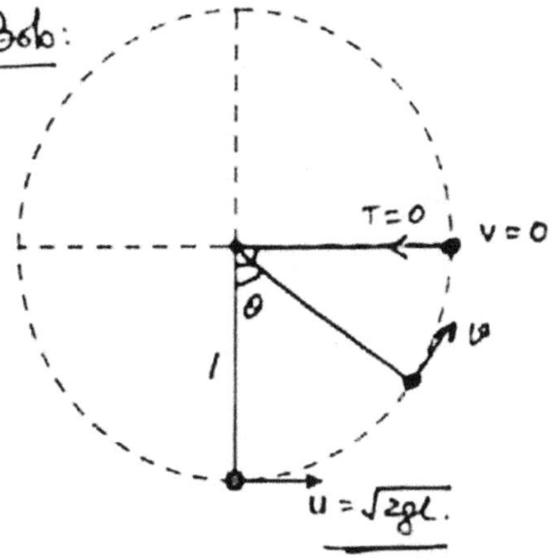


Case I: $u = \sqrt{2gl}$

Ang. Amp.

$$\phi = \cos^{-1}\left(\frac{2gl - u^2}{2gl}\right)$$
$$= \cos^{-1}(0) = \pi/2$$

let at $\theta = \alpha$, $T = 0$

$$\Rightarrow \quad \alpha = \cos^{-1}\left(\frac{2gl - u^2}{3gl}\right)$$
$$= \cos^{-1}(0) = \pi/2.$$

At $u = \sqrt{2gl}$, vel of bob & tension in string becomes zero simultaneously when $\theta = \pi/2$ / string becomes horizontal.



**Module-22: Cases of Projection of a Pendulum Bob**

Case-II: $u = \sqrt{4gl}$

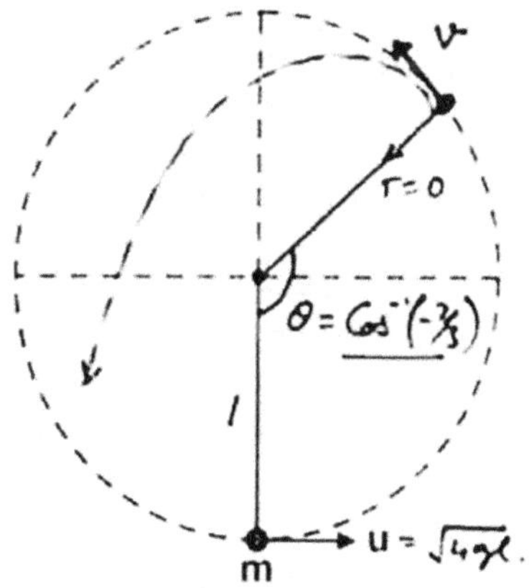


Ang. Amp. $\phi = \cos^{-1}\left(\frac{2gl - u^2}{2gl}\right)$
$$= \cos^{-1}(-1)$$
$$= \pi$$

At $\theta = \alpha$ if $T = 0$

$$\alpha = \cos^{-1}\left(\frac{2gl - u^2}{3gl}\right)$$
$$= \cos^{-1}(-2/3) = 131^\circ.$$

NOTE: If $u = \sqrt{4gl}$ then this speed is sufficient to carry the bob to the topmost pt of vertical circle. but tension in string is not supporting beyond $\theta = \cos^{-1}(-2/3)$.

**Module-22: Cases of Projection of a Pendulum Bob**

Case-III : $u = \sqrt{5gl}$

Ang. Amp.:

$$\phi = \cos^{-1}\left(\frac{2gl - u^2}{2gl}\right)$$

$$= \cos^{-1}\left(-3/2\right)$$

(not possible)

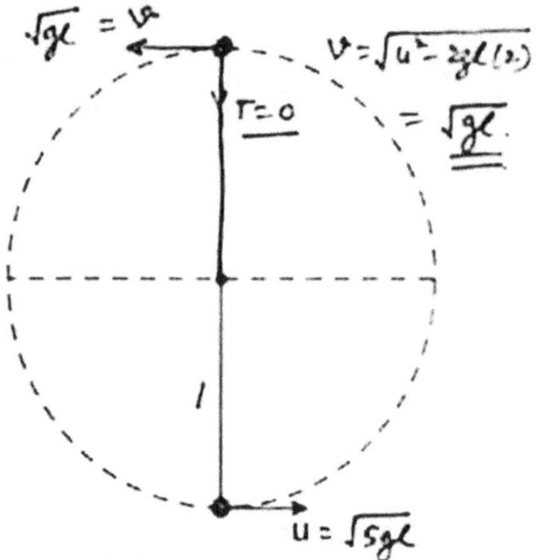


During CM $v \neq 0$.

let at $\theta = \alpha$, $T = 0 \Rightarrow \alpha = \cos^{-1}\left(\frac{2gl - u^2}{3gl}\right)$

$$= \cos^{-1}(-1) = \pi.$$

At bottommost pos. if $u = \sqrt{5gl}$, it is the min speed required to complete vertical circular motion.

Scan for Video Explanation

Web Reference at www.physicsgalaxy.com

**Module-22: Cases of Projection of a Pendulum Bob**

at $\theta = \pi$ then from eqn(1)

$$v = \sqrt{u^2 - 2gl(1-(-1))}$$

$$= \sqrt{gl}$$

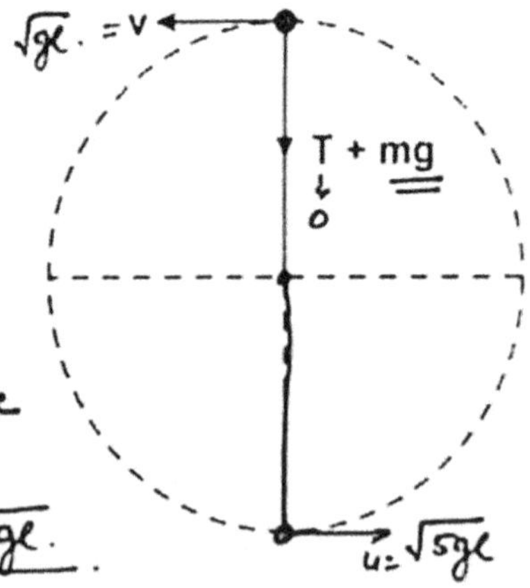


at $T = 0$ in top most pos. we use

$$mg = \frac{mv^2}{l} \longrightarrow v = \sqrt{gl}.$$

* Condition to complete vertical circular path by a bob:

$T = 0$ at the top most pt

$T \geq 0$.

Scan for Video Explanation

Web Reference at www.physicsgalaxy.com

**Module-23: Vertical Circular Motion of a Pendulum with a Rod**

# Vertical Circular Motion of a Pendulum with light rigid rod:

If $u = \sqrt{4gl}$, in case of a light rigid rod, bob will complete the circle as rod will never slack.

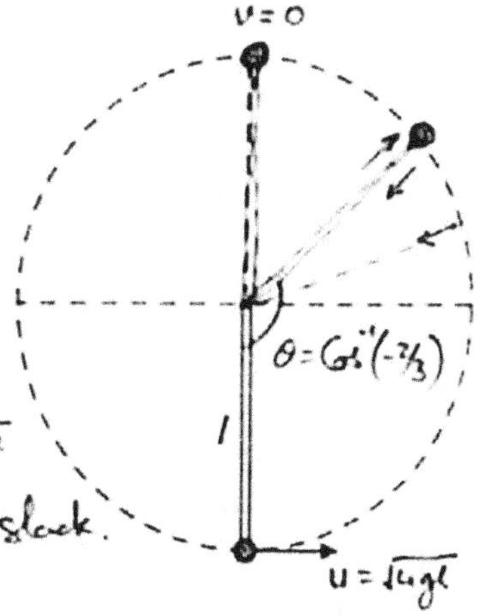


NOTE: After $\theta = \cos^{-1}(-2/3)$ the dir of force by rod on bob changes from inward to outward.

Web Reference at www.physicsgalaxy.com | Scan for Video Explanation

**Module-24: Circular Motion inside a Vertical Circular Track**

# Circular Motion of a body inside a vertical circular track:

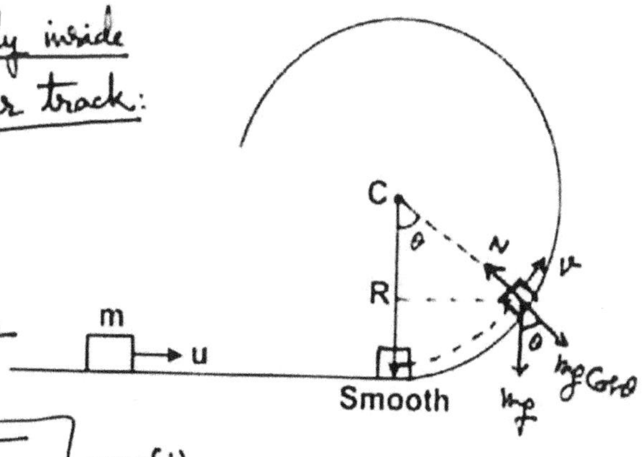


using WE eq$^n$.

$$\underset{k_i}{\frac{1}{2}mu^2} - mgR(1-\cos\theta) = \underset{k_f}{\frac{1}{2}mv^2}$$

$$\boxed{v = \sqrt{u^2 - 2gR(1-\cos\theta)}} \quad \text{—(1)}$$

$h = R(1-\cos\theta)$

at ang. pos $\theta$.

$$N - mg\cos\theta = \frac{mv^2}{R}$$

$$N = mg\cos\theta + \frac{m}{R}\left(u^2 - 2gR(1-\cos\theta)\right)$$

$$\boxed{N = \frac{mu^2}{R} - 2mg + 3mg\cos\theta} \quad \text{—(2)}$$

Web Reference at www.physicsgalaxy.com | Scan for Video Explanation

**Module-24: Circular Motion inside a Vertical Circular Track**

# Cases of Projection:

→ at $u = \sqrt{2gR}$, body will reach pt B & return

→ at $u = \sqrt{4gR}$, body will leave the circular track at $\theta = \cos^{-1}(-2/3)$ & follow proj motion.

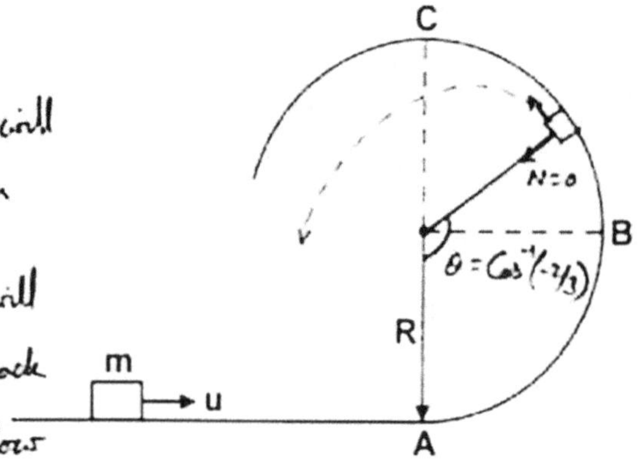


→ at $u = \sqrt{5gR}$, at top most pt $N = 0$ & $v = \sqrt{gR}$ & body will Complete the cir track.

→ at $\sqrt{5gR} > u > \sqrt{2gR}$, body will leave the track ($N=0$) somewhere betwⁿ pt B & C.

Scan for Video Explanation

Web Reference at www.physicsgalaxy.com

**Module-25: Vertical Circular Motion inside a Tube**

# VCM of a body inside a Tube:

at pos. B.

$$v = \sqrt{u^2 - 2gR(1-\cos\theta)}$$

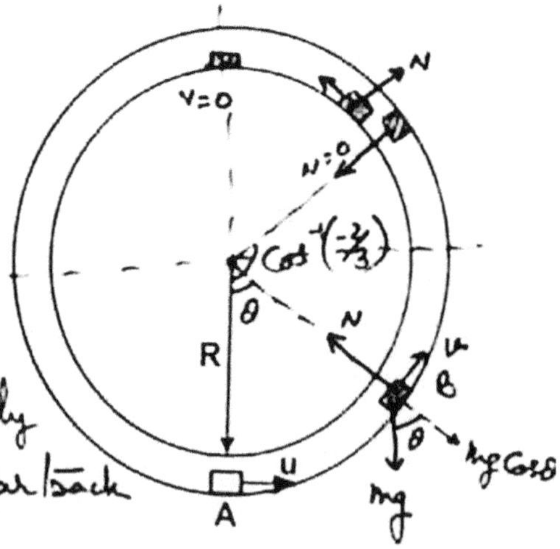


→ As path is inside a tube, body will never leave the circular track

→ If $u = \sqrt{4gR}$ body will loose Contact with the outer wall of tube at $\theta = \cos^{-1}(-2/3)$ and it takes support from the inner wall and Continue the circular motion.

Scan for Video Explanation

Web Reference at www.physicsgalaxy.com

**Module-26: Vertical Circular Motion outside a Track**

# VCM Outside a circular track:

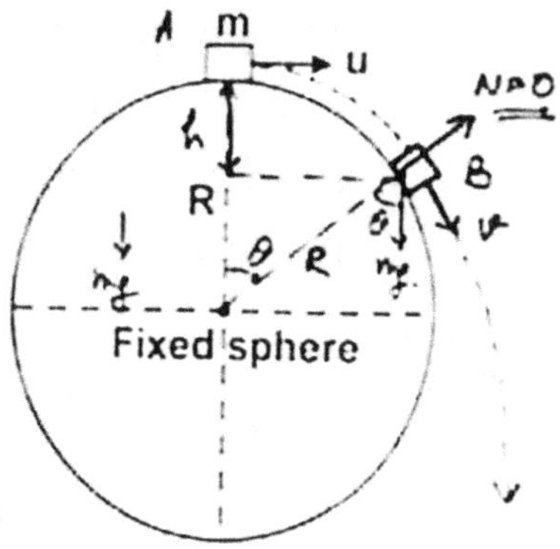


W-E eqⁿ from A to B.

$$\frac{1}{2}mu^2 + mgR(1-\cos\theta) = \frac{1}{2}mv^2$$

$$\boxed{v = \sqrt{u^2 + 2gR(1-\cos\theta)}} \quad -(1)$$

at pt B.

$$mg\cos\theta - N = \frac{mv^2}{R}$$

$$N = mg\cos\theta - \frac{mv^2}{R}$$

$$N = mg\cos\theta - \frac{m}{R}\left[u^2 + 2gR(1-\cos\theta)\right]$$

$$\boxed{N = 3mg\cos\theta - 2mg - \frac{mu^2}{R}} \quad -(2)$$



**Module-26: Vertical Circular Motion outside a Track**

→ at some pt $\theta = \beta$ on track $N=0$ and body will leave the circular track

from eqⁿ(2)

$$0 = 3mg\cos\beta - 2mg - \frac{mu^2}{R}$$

$$\cos\beta = \left(\frac{2gR + u^2}{3gR}\right)$$

$$\beta = \cos^{-1}\left(\frac{2gR + u^2}{3gR}\right)$$

**Module-27: Solved Example-13**

**Ex :** A body is released from rest from top of a fixed spherical surface such that it starts sliding down. Find the acceleration of body when it is at a position at an angle 30° from the initial position.

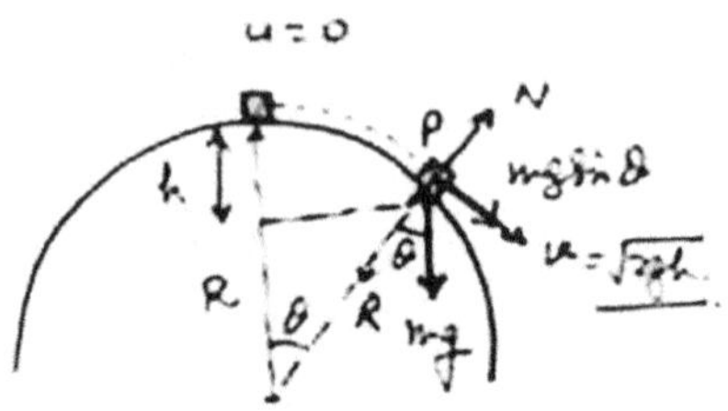


Solⁿ:

$$mgR(1-\cos\theta) = \frac{1}{2}mv^2$$

$$v = \sqrt{2gR(1-\cos\theta)}$$

normal accⁿ at pt P is

$$a_N = \frac{v^2}{R} = 2g(1-\cos 30^\circ) = 2g\left(1-\frac{\sqrt{3}}{2}\right) = g(2-\sqrt{3})$$

$$h = R(1-\cos\theta)$$

tangential accⁿ at pt P is $a_t = g\sin\theta = g/2$

Total accⁿ

$$a_T = \sqrt{a_t^2 + a_N^2} = \sqrt{g^2/4 + g^2(2-\sqrt{3})^2} = g\sqrt{1/4 + (4+3-2\sqrt{3})}$$

Ans.

Scan for Video Explanation

Web Reference at www.physicsgalaxy.com

**Module-28: Solved Example-14**

**Ex :** Figure shows a loop track of radius r. A block B starts from an inclined plane connected to the track at a distance h above the top of the track. Find the minimum value of h for successful looping.

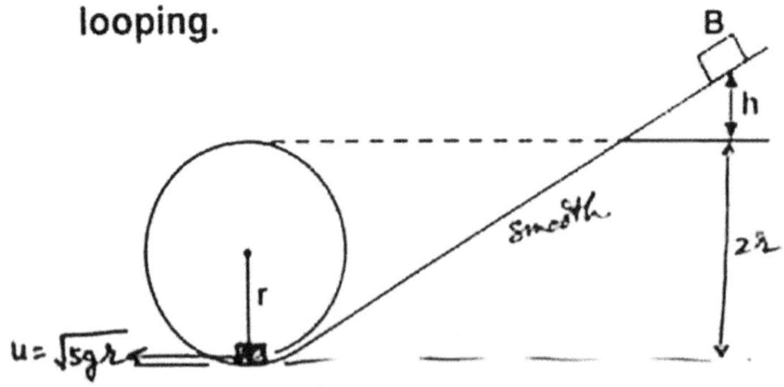


Solⁿ:

$$\frac{1}{2}mu^2 = mg(2r+h) \Rightarrow u = \sqrt{2g(2r+h)}$$

just to complete the loop $\sqrt{2g(2r+h)} = \sqrt{5gr}$

$$4gr + 2gh = 5gr \Rightarrow h = r/2$$

Ans.

Scan for Video Explanation

Web Reference at www.physicsgalaxy.com

**Module-29: Solved Example-15**

**Ex :** A simple pendulum bob (length = $l$) is projected horizontally with speed $u = \sqrt{gl}$ from the bottom position. Find the speed of particle and the inclination of string with the vertical at the instant when tension in string is equal to the weight of bob.

Sol$^n$: at an angle $\theta$ from initial position

$$mg = T = \frac{mu^2}{l} - 2mg + 3mg\cos\theta$$

$$mg = mg - 2mg + 3mg\cos\theta$$

$$\cos\theta = 2/3 \Rightarrow \theta = \cos^{-1}(2/3) \text{ Ans.}$$

$$v = \sqrt{u^2 - 2gl(1-\cos\theta)} = \sqrt{gl - 2gl(1-2/3)}$$

$$v = \sqrt{gl/3} \text{ Ans.}$$

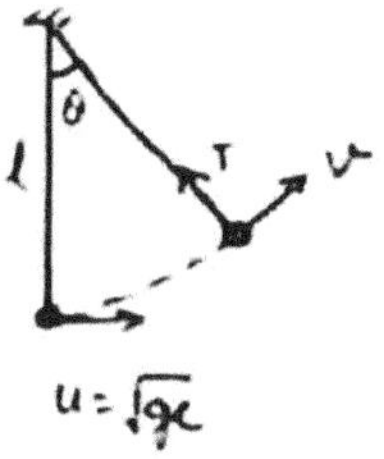


Web Reference at www.physicsgalaxy.com | Scan for Video Explanation

**Module-30: Solved Example-16**

**Ex :** A simple pendulum oscillates in a vertical plane. When it passes through the mean position, tension in string is 3 times the weight of bob. What is the maximum angular displacement of string from vertical during oscillations.

Sol$^n$: at mean position we use

$$h = l - l\cos\theta$$

$$T = mg + \frac{mv^2}{l}$$

$$T = mg + \frac{m}{l}(2gl(1-\cos\theta))$$

$$3mg = mg + 2mg - 2mg\cos\theta$$

$$\cos\theta = 0 \Rightarrow \theta = 90^\circ \text{ Ans.}$$

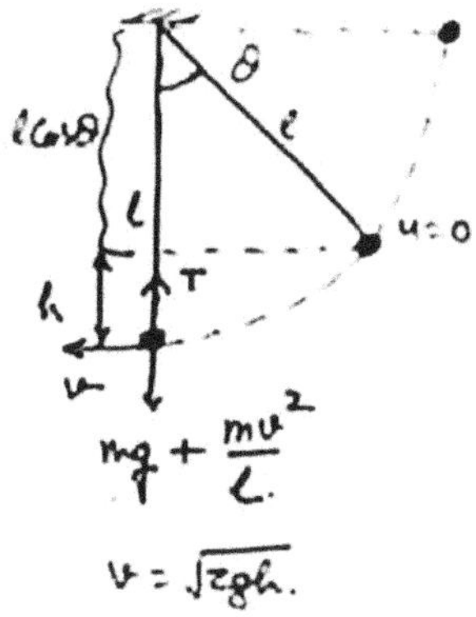


Web Reference at www.physicsgalaxy.com | Scan for Video Explanation

**Module-31: Solved Example-17**

**Ex :** A bucket tied at the end of a 1.6 m long rod is whirled in a vertical circle with constant speed. What should be the minimum speed so that water from the bucket does not spill, when the bucket is at the height position. (Take g = 10 m/s$^2$)

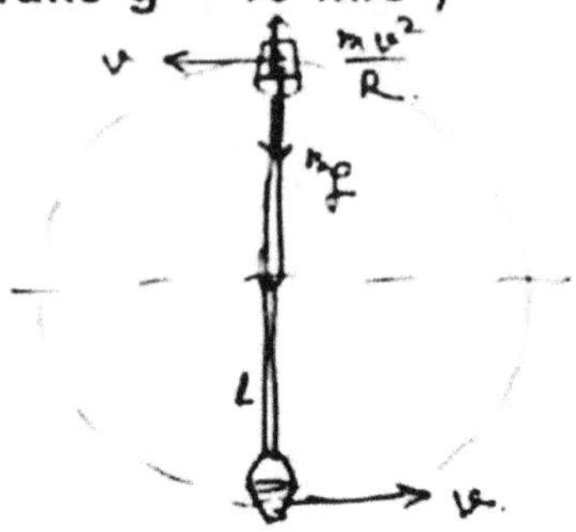

**Sol$^n$:** at the topmost pt water will not spill if

$$\frac{mv^2}{l} = mg$$

$$v = \sqrt{gl}$$

$$= \sqrt{10 \times 1.6} = 4 \text{ m/s. Ans.}$$

Scan for Video Explanation

Web Reference at www.physicsgalaxy.com

# Chapter 9
# System of Particle-I

## Lecture Notes Modules


**Module-1:** Centre of Mass
**Module-2:** How to Locate Centre of Mass
**Module-3:** Centre of Mass of Multiple Objects
**Module-4:** Solved Example-1
**Module-5:** Solved Example-2
**Module-6:** Solved Example-3
**Module-7:** Centre of Mass of Subtractive Objects
**Module-8:** Centre of Mass of Continuous Extended Bodies
**Module-9:** Centre of Mass of a Uniform Half Ring
**Module-10:** Centre of Mass of a Uniform Half Disc
**Module-11:** Centre of Mass of a Uniform Hollow Hemisphere
**Module-12:** Centre of Mass of a Uniform Solid Hemisphere
**Module-13:** Centre of Mass of a Uniform Hollow Cone
**Module-14:** Centre of Mass of a Uniform Solid Cone
**Module-15:** Solved Example-4
**Module-16:** Solved Example-5
**Module-17:** Solved Example-6
**Module-18:** Displacement of Centre of Mass
**Module-19:** Velocity and Acceleration of Centre of Mass
**Module-20:** Solved Example-7
**Module-21:** Solved Example-8

**Module-1: Centre of Mass**

# Centre of Mass:

It is a pt associated with a SoP where all the masses of system are supposed to be Concentrated.

→ For uniform bodies → COM is located at Geometric Centre of bodies.

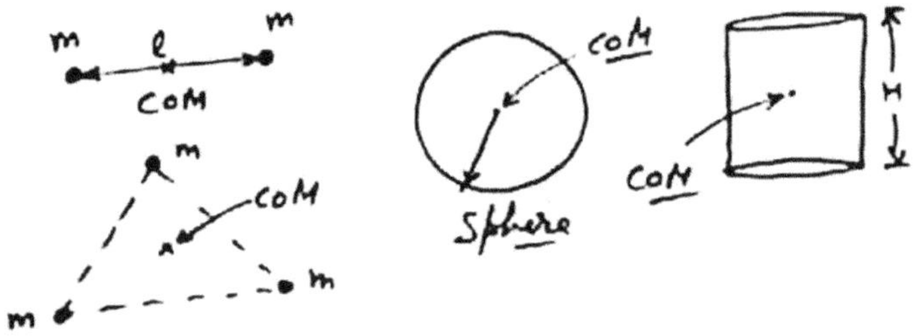


Scan for Video Explanation

Web Reference at www.physicsgalaxy.com

**Module-1: Centre of Mass**

# Types of System of Particles:

Every physical System in nature ] ← System of Particle

Based on relative location of particles in a system SoP can be classified in two Categories -

(1) Discrete object System

(2) Continuous object System.

Scan for Video Explanation

Web Reference at www.physicsgalaxy.com

**Module-1: Centre of Mass**

DOS : → SOP with several independent objects

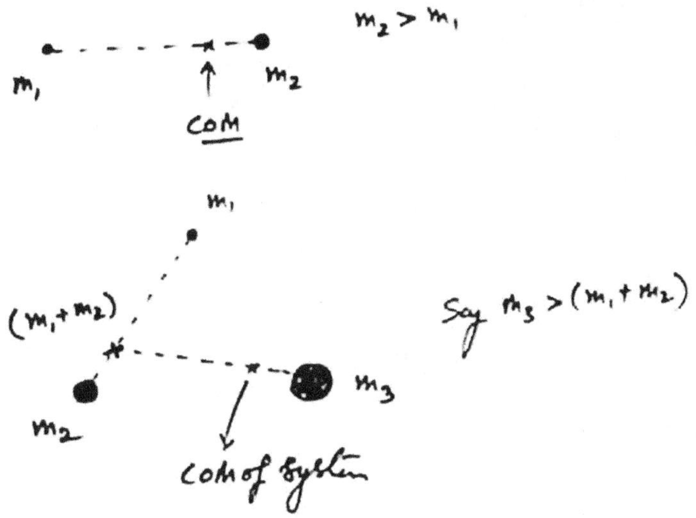


Web Reference at www.physicsgalaxy.com

Scan for Video Explanation

**Module-1: Centre of Mass**

COS. → SOP with all particles linked together to form an extended body.

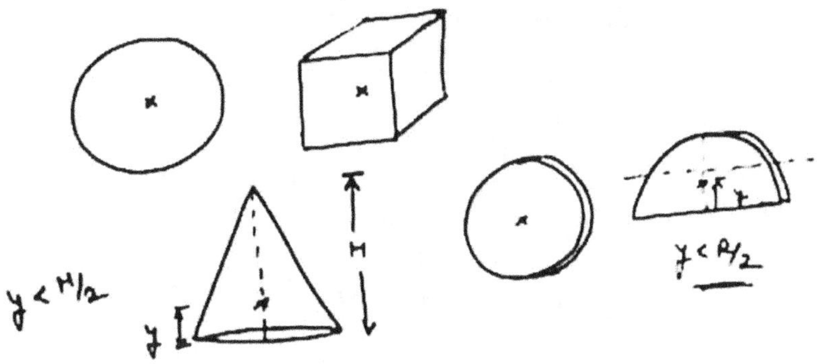


Web Reference at www.physicsgalaxy.com

Scan for Video Explanation

**Module-2: How to Locate Centre of Mass**

# How to locate Centre of Mass of a SOP:

## Mass Moment:

For a given pt mass its mass moment is always calculated w.r.to a Refrence pt.

Mass moment of "m" w.r.t pt O is given as-

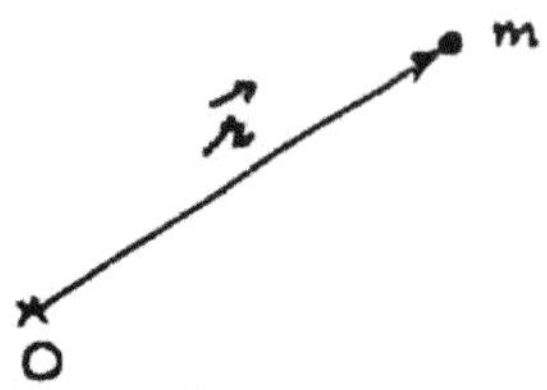


$$\vec{Z}_{mO} = m\vec{r}$$

Scan for Video Explanation

Web Reference at www.physicsgalaxy.com

**Module-2: How to Locate Centre of Mass**

# Centre of Mass in terms of Mass Moments:

In a SOP, Centre of Mass is defined as a point about which mass moments of all particles of SOP is 0.

$$\sum_{i=1}^{N} m_i \vec{r}_i = 0$$

$m_i$ ← mass of each particle

$\vec{r}_i$ ← position vec of $i^{th}$ particle w.r.to COM.

Scan for Video Explanation

Web Reference at www.physicsgalaxy.com

**Module-2: How to Locate Centre of Mass**

# COM of a Two Particle System:

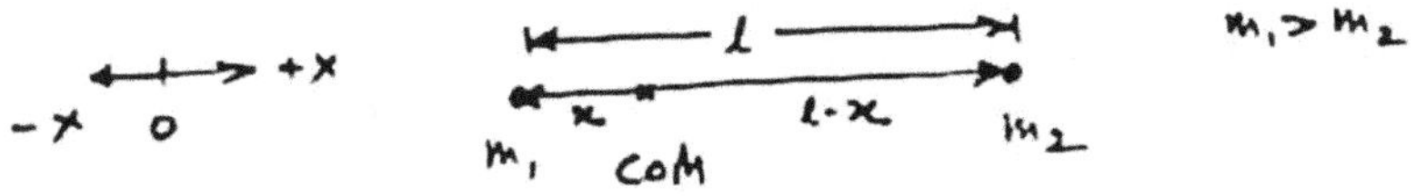


by mass-moment definition of COM we use

$$\sum_{i=1}^{2} m_i \vec{r}_i = 0$$

$$m_2(l-x)\hat{i} + m_1(-x)\hat{i} = 0$$

$$m_2 l = (m_1 + m_2)x$$

$$\boxed{x = \frac{m_2 l}{m_1 + m_2}}$$

Web Reference at www.physicsgalaxy.com | Scan for Video Explanation

**Module-2: How to Locate Centre of Mass**

**Ex :** Find location of center of mass of this system from 2 kg mass.

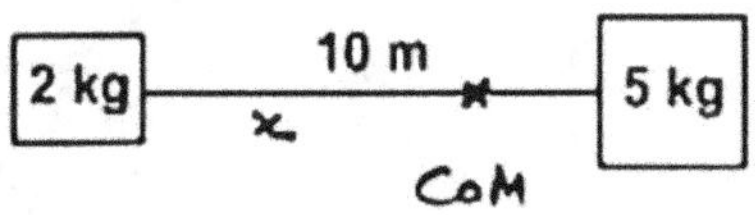


Sol$^n$:

$$x = \frac{5 \times 10}{2+5} = \frac{50}{7}\text{ m. Ans.}$$

Web Reference at www.physicsgalaxy.com | Scan for Video Explanation

**Module-2: How to Locate Centre of Mass**

**Ex :** Find co-ordinates of centre of mass in the system shown in figure.

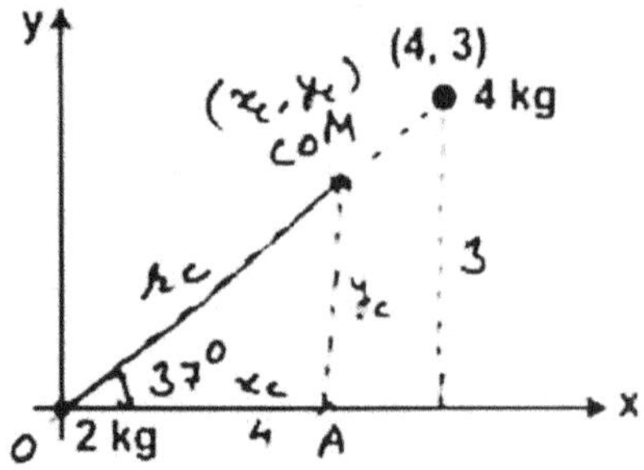


**Sol$^n$:**

$$r_c = \frac{m_2 r}{m_1 + m_2}$$

$$= \frac{4 \times 5}{6} = \frac{10}{3}\,m.$$

co-ordinates of COM are

$$x_c = r_c \cos 37^0 = \frac{10}{3} \times \frac{4}{5} = \frac{8}{3}\,m$$

$$y_c = r_c \sin 37^0 = \frac{10}{3} \times \frac{3}{5} = 2$$

$$COM \rightarrow (8/3, 2) \text{ Ans.}$$

Scan for Video Explanation

Web Reference at www.physicsgalaxy.com

**Module-3: Centre of Mass of Multiple Objects**

# Centre of Mass of a SoP with Multiple Objects:

By definition of Centre of Mass

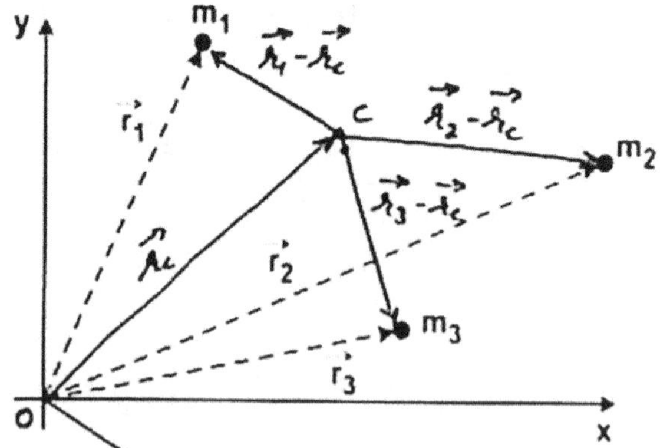


$$m_1(\vec{r}_1 - \vec{r}_c) + m_2(\vec{r}_2 - \vec{r}_c) + m_3(\vec{r}_3 - \vec{r}_c) = 0$$

$$(m_1 + m_2 + m_3)\vec{r}_c = m_1\vec{r}_1 + m_2\vec{r}_2 + m_3\vec{r}_3$$

$$\boxed{\vec{r}_c = \frac{m_1\vec{r}_1 + m_2\vec{r}_2 + m_3\vec{r}_3}{m_1 + m_2 + m_3}} = \frac{\sum_{i=1}^{N} m_i \vec{r}_i}{\sum_{i=1}^{N} m_i}$$

Scan for Video Explanation

Web Reference at www.physicsgalaxy.com

**Module-3: Centre of Mass of Multiple Objects**

$$\vec{r}_1 = x_1\hat{i} + y_1\hat{j}$$

$$\vec{r}_2 = x_2\hat{i} + y_2\hat{j}$$

$$\vdots \qquad \vdots$$

$$\vec{r}_c = \frac{m_1\vec{r}_1 + m_2\vec{r}_2 + \cdots}{m_1 + m_2 + \cdots} = \frac{m_1(x_1\hat{i} + y_1\hat{j}) + m_2(x_2\hat{i} + y_2\hat{j}) + \cdots}{m_1 + m_2 + \cdots}$$

$$x_c\hat{i} + y_c\hat{j} = \left(\frac{m_1x_1 + m_2x_2 + \cdots}{m_1 + m_2 + \cdots}\right)\hat{i} + \left(\frac{m_1y_1 + m_2y_2 + \cdots}{m_1 + m_2 + \cdots}\right)\hat{j}$$

$$x_c = \frac{\sum_{i=1}^{N} m_i x_i}{\sum_{i=1}^{N} m_i} \quad ; \quad y_c = \frac{\sum_{i=1}^{N} m_i y_i}{\sum_{i=1}^{N} m_i}$$

$$\left[z_c = \sum_{i=1}^{N} m_i z_i \Big/ \sum_{i=1}^{N} m_i\right]$$



**Module-4: Solved Example-1**

Ex : Figure shows a co-ordinate system in which 8 uniform cubes of side a are placed as shown. If masses of cubes 1, 2, 3 & 4 is m each & 2m for remaining cubes, find location of centre of mass of this system.

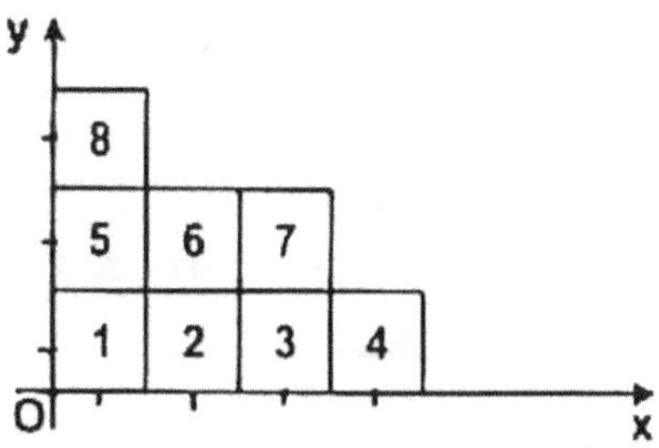


Sol$^n$:

$$x_c = \frac{\Sigma m_i x_i}{\Sigma m_i} = \frac{5m \times a/2 + 3m \times \frac{3a}{2} + 3m \times \frac{5a}{2} + m \times \frac{7a}{2}}{12m}$$

$$= \frac{18}{12}a = \frac{3a}{2}$$

$$y_c = \frac{\Sigma m_i y_i}{\Sigma m_i} = \frac{4m \times a/2 + 6m \times 3a/2 + 2m \times \frac{5a}{2}}{12m}$$

$$= \frac{16}{12}a = \frac{4}{3}a.$$

COM $\rightarrow \left(\frac{3a}{2}, \frac{4}{3}a\right)$ Ans.

**Module-5: Solved Example-2**

Ex : Figure shows a uniform plate of shape E as shown. Find the co-ordinates of centre of mass of this structure.

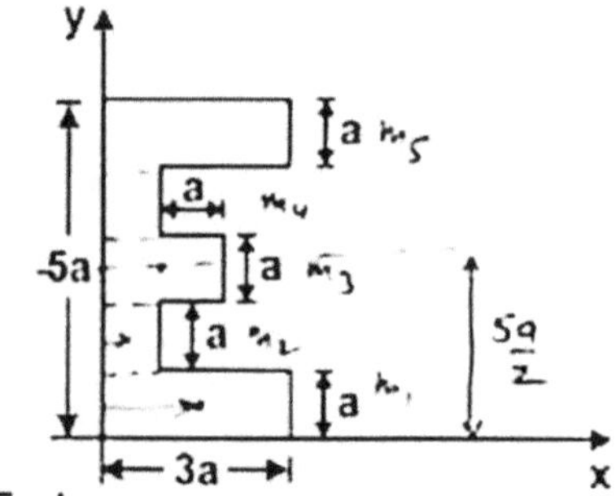


Sol$^n$: As plate is 2D object its mass is uniform so it is proportional to its area.

$y_c = 5a/2$

$m_1 = 3a^2$

$m_2 = a^2$

$m_3 = 2a^2$

$m_4 = a^2$

$m_5 = 3a^2$

$$x_c = \frac{m_1x_1 + m_2x_2 + \ldots}{m_1 + m_2 + \ldots}$$

$$x_c = \frac{2 \times 3a^2 \times \frac{3a}{2} + a^2 \times a/2 + 2a^2 \times a + a^2 \times a/2}{10a^2}$$

$$x_c = \frac{12a}{10} = 6a/5 \qquad \text{CoM} \rightarrow \left(\frac{6a}{5}, \frac{5a}{2}\right)$$

Scan for Video Explanation

Web Reference at www.physicsgalaxy.com

**Module-6: Solved Example-3**

Ex : Figure shows a uniform two dimensional T-shaped structure. Find co-ordinates of centre of mass of this object.

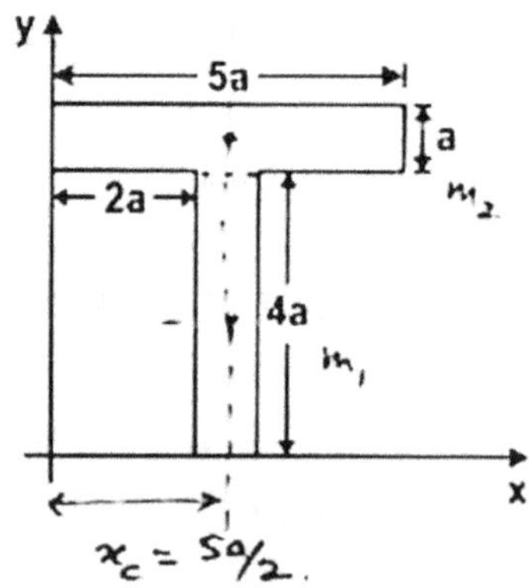


$x_c = 5a/2$

Sol$^n$:

$m_1 = 4a^2$

$m_2 = 5a^2$

$$y_c = \frac{m_1y_1 + m_2y_2}{m_1 + m_2}$$

$$y_c = \frac{4a^2 \times 2a + 5a^2 \times 9a/2}{9a^2} = \frac{61a}{18}$$

$$\text{CoM} \rightarrow \left(\frac{5a}{2}, \frac{61a}{18}\right)$$

Scan for Video Explanation

Web Reference at www.physicsgalaxy.com

**Module-7: Centre of Mass of Subtractive Objects**

# Centre of Mass of Subtractive Objects:

↓

A part is cut.

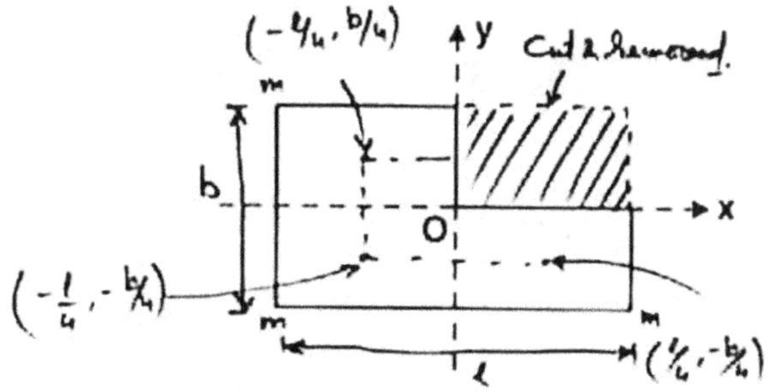


Ques: Find COM of the uniform plate shown in fig if 1/4th of its part is cut & removed as shown.

Soln: COM of remaining object is

$$x_c = \frac{m(-l/4) + m(-l/4) + m(l/4)}{3m} = -l/12$$

$$y_c = \frac{m(b/4) + m(-b/4) + m(-b/4)}{3m} = -b/12$$

$$COM = (-l/12, -b/12) \text{ Ans.}$$



**Module-7: Centre of Mass of Subtractive Objects**

Ex : Figure shows a uniform disc of radius R with centre $C_1$ and from this disc a small disc of radius R/2 is cut with centre at $C_2$ as shown. Find the distance of centre of mass from $C_1$ of the remaining part of disc.

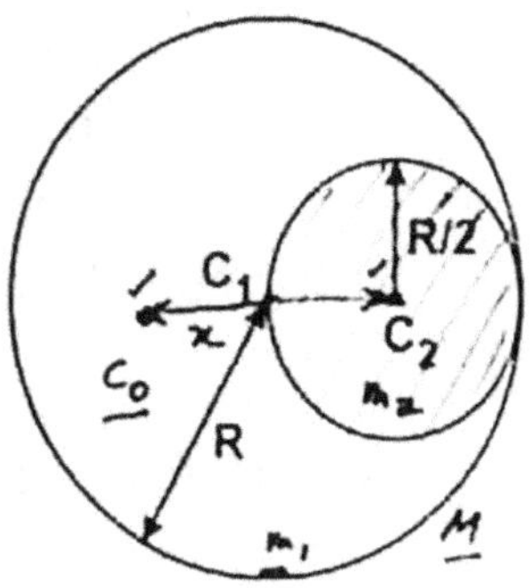


$$m_2 = \pi(R/2)^2 = \frac{\pi R^2}{4} = \frac{M}{4}$$

$$m_1 = M - m_2 = \frac{3M}{4}$$

Soln: By property of mass moments we use

$$m_2(R/2) - m_1 x = 0$$

$$\frac{M}{4} \cdot R/2 = \frac{3M}{4} \cdot x \Rightarrow x = R/6 \text{ Ans.}$$

**Module-8: Centre of Mass of Continuous Extended Bodies**

# Centre of Mass of Continuous Extended Objects:

for discrete object System

for a SOP

$$\vec{r}_c = \frac{\sum_{i=1}^{N} m_i \vec{r}_i}{\sum_{i=1}^{N} m_i}$$

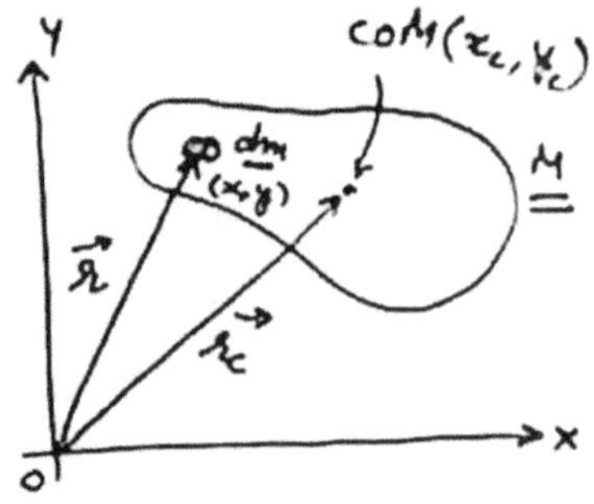


$$\vec{r}_c = \frac{\int dm\,\vec{r}}{\int dm} = \frac{1}{M}\int dm\,\vec{r}$$

Co-ordinates of COM of body can be given as

$$x_c = \frac{1}{M}\int dm\,x \quad \& \quad y_c = \frac{1}{M}\int dm\,y$$

Scan for Video Explanation

Web Reference at www.physicsgalaxy.com

**Module-8: Centre of Mass of Continuous Extended Bodies**

**Ex :** Figure shows a rod AB of length L. Linear mass density of this rod is non-uniform and it varies with the distance x from end A of rod as

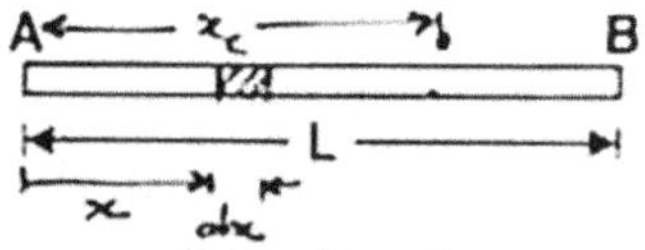


$$\lambda = Cx^2 \text{ kg/m}$$

Where C is a positive constant. Find distance of centre of mass of this rod from end A of rod.

Sol$^n$:

$$x_c = \frac{1}{M}\int dm\,x$$

$$= \frac{1}{M}\int_0^L \lambda dx . x$$

$dm = \lambda dx$

$M = \int dm = \int_0^L C x^2 dx$

$M = \frac{CL^3}{3}$

$$x_c = \frac{1}{M}\int_0^L C x^3 dx = \frac{CL^4}{4M} = \frac{CL^4}{4\left(\frac{CL^3}{3}\right)} = \frac{3L}{4} \text{ Ans.}$$

Scan for Video Explanation

Web Reference at www.physicsgalaxy.com

**Module-9: Centre of Mass of a Uniform Half Ring**

# Centre of Mass of a Uniform Half Ring:

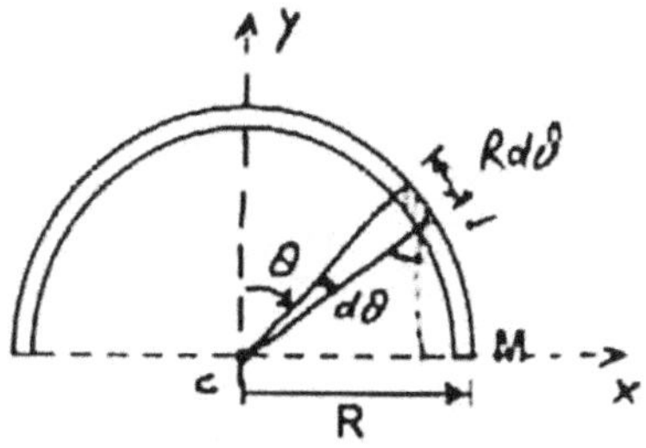


First we consider an elemental mass dm on Ring.

$$dm = \frac{M}{\pi}.d\theta$$

y co-ordinate of dm is $y = R\cos\theta$

y-co-ordinate of COM is $y_c = \frac{1}{M}\int dm\, y = \frac{1}{M}\int_{-\pi/2}^{+\pi/2}\frac{M}{\pi}d\theta\,(R\cos\theta)$

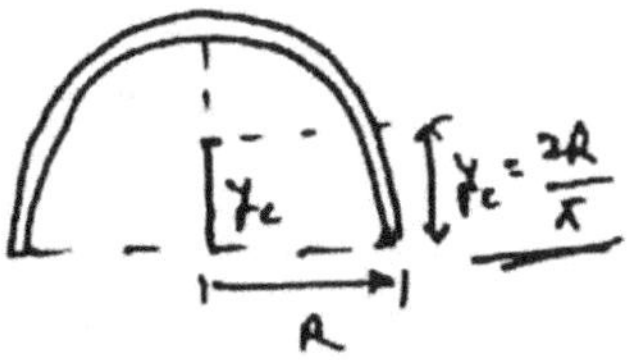


$$y_c = \frac{R}{\pi}\int_{-\pi/2}^{+\pi/2}\cos\theta\, d\theta = \frac{R}{\pi}\left[\sin\theta\right]_{-\pi/2}^{+\pi/2}$$

$$y_c = \frac{R}{\pi}\left[1+1\right] = \frac{2R}{\pi}$$

Web Reference at www.physicsgalaxy.com

Scan for Video Explanation

**Module-10: Centre of Mass of a Uniform Half Disc**

# Centre of Mass of a Uniform Half Disc:

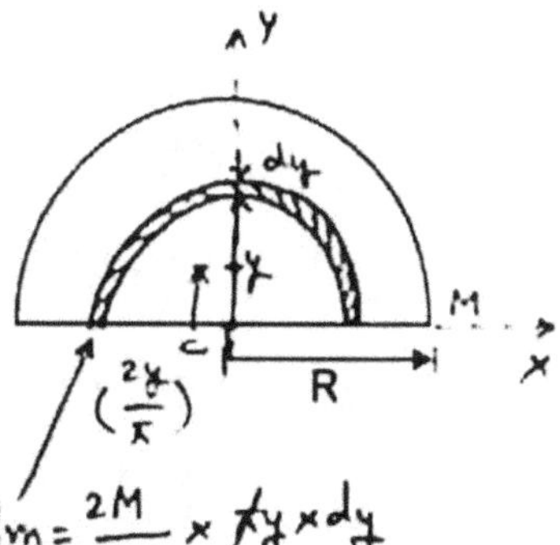


Here we take an elemental half ring of mass dm.

$$dm = \frac{2M}{\pi R^2}\times \pi y \times dy = \frac{2M}{R^2}.y\,dy$$

$$y_c = \frac{1}{M}\int dm\, y$$

$$= \frac{1}{M}\int_0^R \frac{2M}{R^2}.y\,dy\times\left(\frac{2y}{\pi}\right) = \frac{4}{\pi R^2}\int_0^R y^2 dy$$

$$= \frac{4}{\pi R^2}\left[\frac{y^3}{3}\right]_0^R = \frac{4}{\pi R^2}\times\frac{R^3}{3}$$

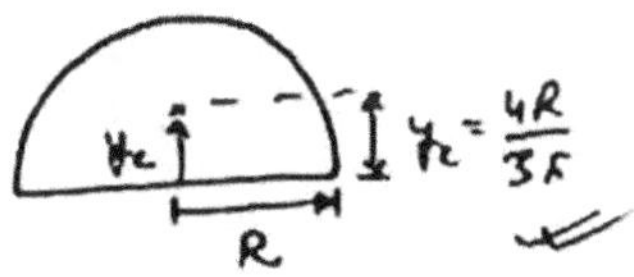


$$\left[y_c = \frac{4R}{3\pi}\right]$$

Web Reference at www.physicsgalaxy.com

Scan for Video Explanation

**Module-11: Centre of Mass of a Uniform Hollow Hemisphere**

# Centre of Mass of a Hollow Hemisphere:

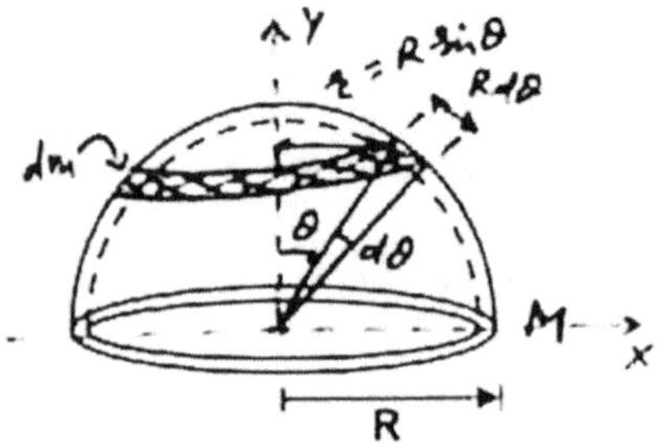


Here we Consider an elemental strip ring of mass dm.

$$dm = \frac{M}{2\pi R^2} \times 2\pi R \sin\theta . R d\theta$$

$$dm = M \sin\theta d\theta$$

$$y_c = \frac{1}{M}\int dm\, y$$

$$= \frac{1}{M}\int_0^{\pi/2} M \sin\theta d\theta . (R\cos\theta) \times \frac{2}{2}$$

$$= \frac{R}{2}\int_0^{\pi/2} \sin 2\theta\, d\theta = \frac{R}{2}\left[-\frac{\cos 2\theta}{2}\right]_0^{\pi/2}$$

$$= \frac{R}{2}\left[\frac{1}{2}+\frac{1}{2}\right] = R/2$$

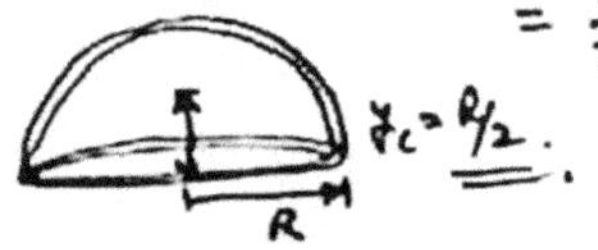


Scan for Video Explanation | Web Reference at www.physicsgalaxy.com

**Module-12: Centre of Mass of a Uniform Solid Hemisphere**

# Centre of Mass of a Solid Hemisphere:

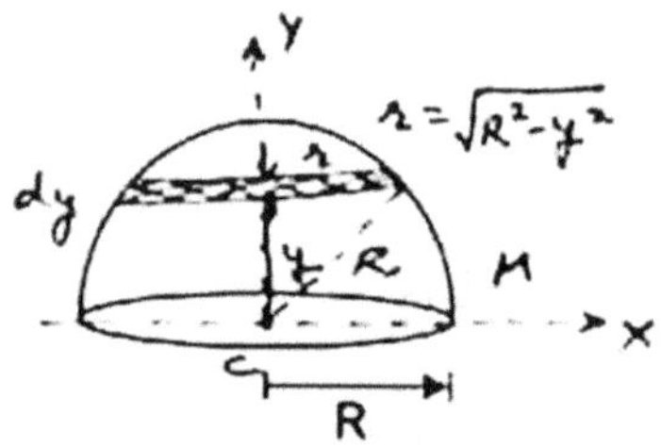


Here we Consider an elemental disc of mass dm

$$dm = \frac{3M}{2\pi R^3} \times \pi r^2 dy$$

$$= \frac{3M}{2\pi R^3} \times \pi (R^2 - y^2) dy$$

$$y_c = \frac{1}{M}\int dm\, y$$

$$= \frac{1}{M}\int_0^R \frac{3M}{2R^3}(R^2-y^2)\,dy \times y$$

$$= \frac{3}{2R^3}\int_0^R (R^2 y - y^3)\,dy$$

$$= \frac{3}{2R^3}\left[\frac{R^2 y^2}{2} - \frac{y^4}{4}\right]_0^R = \frac{3}{2R^3}\left[\frac{R^4}{2} - \frac{R^4}{4}\right] = \frac{3R}{8}$$

Scan for Video Explanation | Web Reference at www.physicsgalaxy.com

**Module-13: Centre of Mass of a Uniform Hollow Cone**

# Centre of Mass of a Uniform Hollow Cone:

we consider a strip ring of mass dm as an element

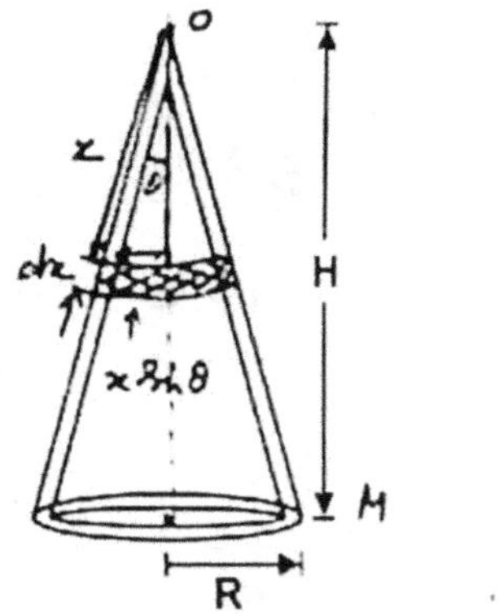


$$y_c = \frac{1}{M}\int dm\, y$$

$$= \frac{1}{M}\int_0^{\sqrt{R^2+H^2}} \frac{2M}{H^2+R^2}\cdot x\,dx\cdot\left[x\left(\frac{H}{\sqrt{H^2+R^2}}\right)\right] \quad (\cos\theta)$$

$$= \frac{2H}{(H^2+R^2)^{3/2}}\int_0^{\sqrt{R^2+H^2}} x^2\,dx$$

$$= \frac{2H}{(H^2+R^2)^{3/2}}\left[\frac{x^3}{3}\right]_0^{\sqrt{R^2+H^2}}$$

$$dm = \frac{M}{\pi R\sqrt{H^2+R^2}}\times 2\pi x\sin\theta\, dx$$

$$dm = \frac{M}{R\sqrt{H^2+R^2}}\cdot 2x\frac{R}{\sqrt{H^2+R^2}}\, dx$$

$$dm = \frac{2M}{H^2+R^2}\cdot x\,dx.$$

Web Reference at www.physicsgalaxy.com

Scan for Video Explanation

**Module-13: Centre of Mass of a Uniform Hollow Cone**

$$y_c = \frac{2H}{(H^2+R^2)^{3/2}}\times\frac{(R^2+H^2)^{3/2}}{3}$$

$$y_c = \frac{2H}{3}$$

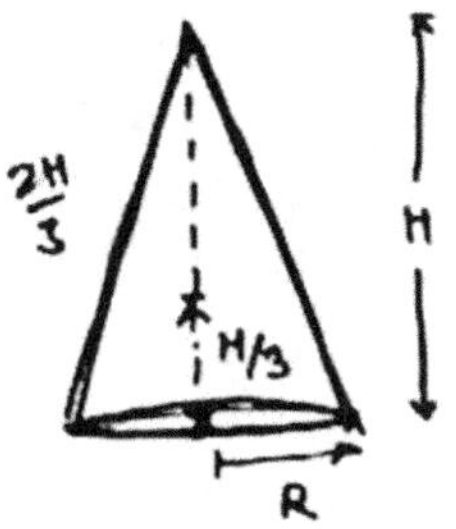


$$y_{c_{base}} = H/3.$$

Web Reference at www.physicsgalaxy.com

Scan for Video Explanation

## Module-14: Centre of Mass of a Uniform Solid Cone

# Centre of Mass of a Uniform Solid Cone:

We consider an elemental disc of mass dm.

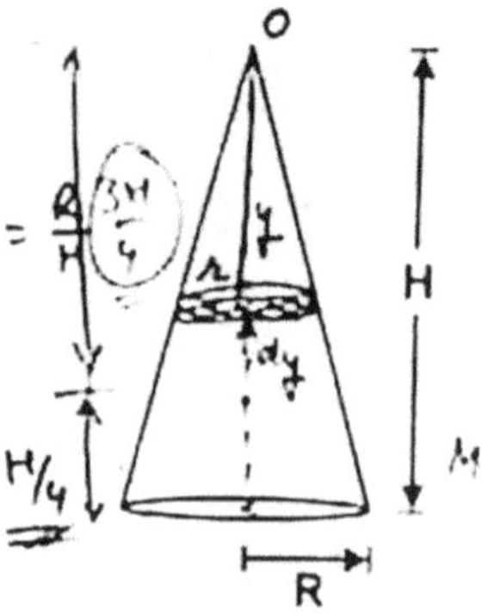


$$\frac{r}{y} = \frac{R}{H}$$

$$dm = \frac{3M}{\pi R^2 H} \times \pi r^2 dy$$

$$dm = \frac{3M}{\pi R^2 H} \times \pi \left(\frac{Ry}{H}\right)^2 dy$$

$$dm = \frac{3My^2}{H^3} dy$$

$$y_c = \frac{1}{M}\int dm\, y$$

$$= \frac{1}{M}\int_0^H \frac{3My^2}{H^3} . dy \times y$$

$$= \frac{3}{H^3}\int_0^H y^3 dy = \frac{3}{H^3} . \left[\frac{y^4}{4}\right]_0^H$$

$$= \frac{3}{H^3} . \frac{H^4}{4} = \frac{3H}{4}$$

Scan for Video Explanation

Web Reference at www.physicsgalaxy.com

## Module-15: Solved Example-4

**Ex :** Find centre of mass of a uniformly dense object which consist of a hemisphere joined at the base of a cone as shown.

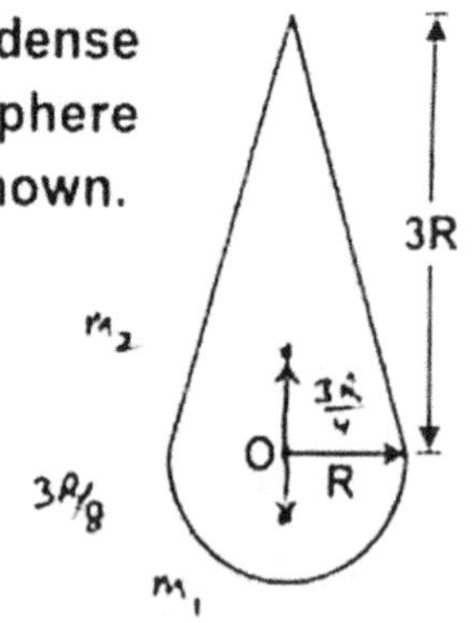


**Sol:**

$$m_1 = \frac{2}{3}\pi R^3 \rightarrow \frac{2}{3}m$$

$$m_2 = \frac{1}{3}\pi R^2 (3R) = \pi R^3 \rightarrow m$$

$$y_c = \frac{m_2\left(\frac{3R}{4}\right) - m_1\left(\frac{3R}{8}\right)}{m_1 + m_2}$$

$$y_c = \frac{m\left(\frac{3R}{4}\right) - \frac{2}{3}m . \frac{3R}{8}}{\frac{5m}{3}}$$

$$y_c = \frac{3R}{10} \text{ Ans.}$$

Scan for Video Explanation

Web Reference at www.physicsgalaxy.com

**Module-16: Solved Example-5**

**Ex :** Figure shows a square plate of side a from which a semicircular disc is removed as shown. Find the centre of mass of the remaining object.

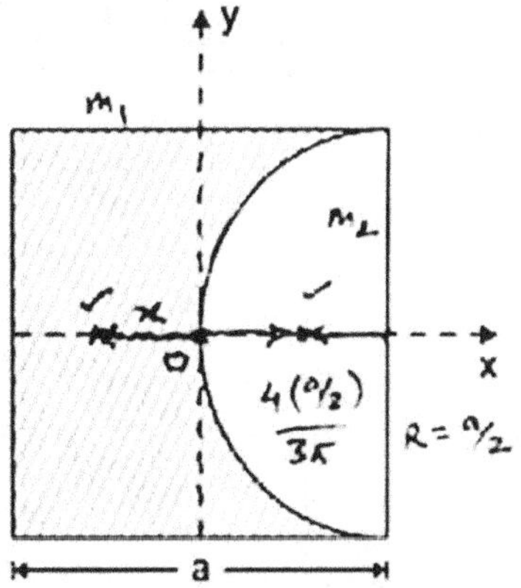


Sol$^n$:

COM of $m_1 \rightarrow -x$

COM of $m_2 \rightarrow \frac{a}{2} - \frac{2a}{3\pi}$

$m_1 + m_2 = a^2$

$m_2 = \frac{\pi (a/2)^2}{2} = \frac{\pi a^2}{8}$

$m_1 = a^2 - \frac{\pi a^2}{8}$

$$m_2\left(\frac{a}{2} - \frac{2a}{3\pi}\right) = m_1 x$$

$$x = \frac{m_2 a}{m_1}\left(\frac{1}{2} - \frac{2}{3\pi}\right)$$

$$x = \frac{a(3\pi - 4)}{6(8-\pi)} \quad \text{Ans}$$

Web Reference at www.physicsgalaxy.com | Scan for Video Explanation

**Module-17: Solved Example-6**

**Ex :** From a uniform cylinder of height 2R and radius R, a hemisphere is cut as shown. Find centre of mass of the remaining object.

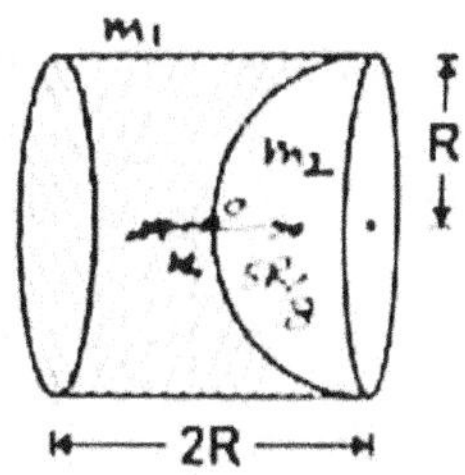


Sol$^n$:

$$m_2 = \frac{2}{3}\pi R^3 \rightarrow m$$

$$m_1 = \pi R^2(2R) - \frac{2}{3}\pi R^3 = \frac{4}{3}\pi R^3 \rightarrow 2m$$

by def. of COM using mass moments

$$m_2(5R/8) - m_1 x = 0$$

$$x = \frac{5R}{8} \times \frac{m_2}{m_1} = \frac{5R}{16} \quad \text{Ans}$$

Web Reference at www.physicsgalaxy.com | Scan for Video Explanation

**Module-18: Displacement of Centre of Mass**

# Displacement of Centre of Mass of a SOP:

for a given SOP

$$\vec{r}_c = \frac{m_1\vec{r}_1 + m_2\vec{r}_2 + ---}{m_1 + m_2 + ---}$$

If $\vec{\Delta r}_1, \vec{\Delta r}_2, ---$ are the displacements in position of masses $m_1, m_2, m_3 ---$ then disp in pos. of COM can be given as.

$$\vec{\Delta r}_c = \frac{m_1\vec{\Delta r}_1 + m_2\vec{\Delta r}_2 + ---}{m_1 + m_2 + ----}$$

Scan for Video Explanation

Web Reference at www.physicsgalaxy.com

**Module-18: Displacement of Centre of Mass**

**Ex :** A body of A mass 5 kg is placed at (5, 0) and another body B of mass 10 kg is placed at (8, 0). If A is displaced to (7, 0) and B is displaced to (12, 0), find displacement of centre of mass of this system.

**Sol$^n$:**

5kg: $\Delta x_1 = 2$ units

10kg: $\Delta x_2 = 4$ units

$$\Delta x_c = \frac{m_1\Delta x_1 + m_2\Delta x_2}{m_1 + m_2}$$

$$= \frac{5\times 2 + 10\times 4}{15} = \frac{50}{15} = \frac{10}{3} \text{ units Ans.}$$

Scan for Video Explanation

Web Reference at www.physicsgalaxy.com

**Module-19: Velocity and Acceleration of Centre of Mass**

# Velocity & Acceleration of Centre of Mass of a SoP:

As we know for a SoP, p.v of COM is given as

$$\vec{r}_c = \left(\frac{m_1\vec{r}_1 + m_2\vec{r}_2 + \cdots}{m_1 + m_2 + \cdots}\right)$$

velocity of COM can be given as-

$$\left[\vec{v}_c = \frac{d\vec{r}_c}{dt} = \left(\frac{m_1\vec{v}_1 + m_2\vec{v}_2 + \cdots}{m_1 + m_2 + \cdots}\right) = \frac{\vec{P}_{system}}{\sum_{i=1}^{n} m_i}\right]$$

$$\left[\text{vel of COM} = \frac{\text{Total momentum of SoP}}{\text{Sum of masses.}}\right]$$

**Web Reference at www.physicsgalaxy.com** | **Scan for Video Explanation**

**Module-19: Velocity and Acceleration of Centre of Mass**

Acceleration of COM of a SoP

$$\vec{a}_c = \frac{d\vec{v}_c}{dt} = \left(\frac{m_1\vec{a}_1 + m_2\vec{a}_2 + \cdots}{m_1 + m_2 + \cdots}\right)$$

$$\left[\vec{a}_c = \frac{\vec{F}_{ext}}{\sum_{i=1}^{n} m_i}\right]$$

$$\left[\text{Acc}^n \text{ of COM} = \frac{\text{Net External force on a SoP}}{\text{Sum of masses.}}\right]$$

**Web Reference at www.physicsgalaxy.com** | **Scan for Video Explanation**

**Module-20: Solved Example-7**

**Ex :** A body of mass 4 kg moves along +X direction at speed 20 m/s. Find the speed with which a 10 kg mass should move along +Y direction such that their centre of mass travels along line x = y. Assume both particle starts at the same time from origin.

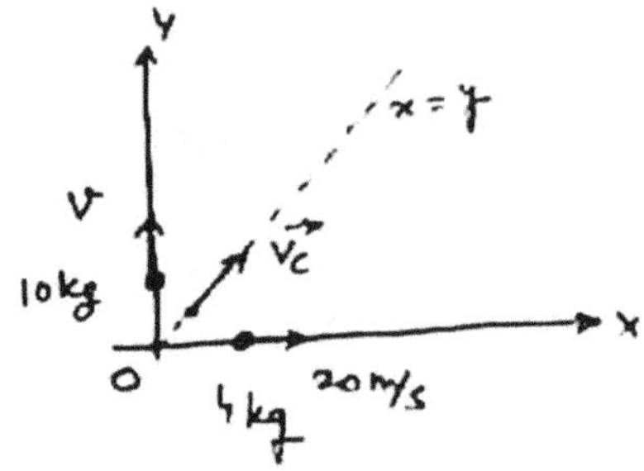


Solⁿ:

As we know

$$\vec{v}_c = \frac{m_1\vec{v}_1 + m_2\vec{v}_2}{m_1 + m_2}$$

$$\vec{v}_c = \frac{4(20\hat{i}) + 10(v\hat{j})}{14}$$

for $\vec{v}_c$ to be along line x = y ⇒ $80 = 10v$

$v = 8$ m/s Ans.

Scan for Video Explanation

Web Reference at www.physicsgalaxy.com

**Module-21: Solved Example-8**

**Ex :** Two blocks of masses 3 kg and 5 kg are connected by an ideal spring and placed on a smooth surface as shown. A force of 40 N is applied on 3 kg block and it is observed that at an instant acceleration of 3 kg mass is 2 m/s$^2$. Find acceleration of 5 kg mass at this instant.

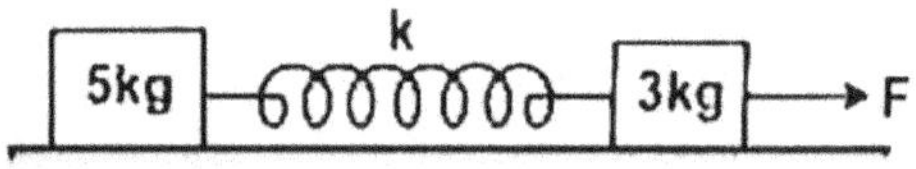


Solⁿ: As F is the only ext force present

accⁿ of COM $a_c = \frac{F}{8} = \frac{40}{8} = 5$ m/s$^2$

As we know $a_c = \frac{m_1a_1 + m_2a_2}{m_1 + m_2}$ ⇒ $5 = \frac{3\times 2 + 5a}{8}$

$5a = 40 - 6 = 34$

$a = \frac{34}{5}$ m/s$^2$ Ans.

Scan for Video Explanation

Web Reference at www.physicsgalaxy.com

# Chapter 10
# System of Particle-II

## Lecture Notes Modules


**Module-1:** Conservation of Momentum
**Module-2:** Solved Example-1
**Module-3:** Solved Example-2
**Module-4:** Solved Example-3
**Module-5:** Effect of Internal Forces
**Module-6:** Solved Example-4
**Module-7:** Solved Example-5
**Module-8:** Solved Example-6
**Module-9:** Solved Example-7
**Module-10:** Solved Example-8
**Module-11:** Impulse & Conservation of Momentum
**Module-12:** Solved Example-9
**Module-13:** Solved Example-10
**Module-14:** Solved Example-11
**Module-15:** Collisions
**Module-16:** Types of Collisions Based on Material Properties
**Module-17:** Collision of Bodies of Different Materials
**Module-18:** Coefficient of Restitution
**Module-19:** Head-on Collisions
**Module-20:** Solved Example-12
**Module-21:** Partial Elastic Head-on Collision
**Module-22:** Solved Example-13
**Module-23:** Solved Example-14
**Module-24:** Solved Example-15
**Module-25:** Solved Example-16
**Module-26:** Solved Example-17
**Module-27:** Oblique Collision of a Ball on a Flat Surface
**Module-28:** Oblique Collision in a 2D Plane
**Module-29:** Solved Example-18
**Module-30:** Solved Example-19
**Module-31:** Solved Example-20
**Module-32:** Mass Variation Cases
**Module-33:** Super Elastic Collision

**Module-1: Conservation of Momentum**

# Law of Conservation of Momentum:

for a SoP

$$\vec{V}_{CM} = \frac{\vec{P}_{system}}{\Sigma m_i} \quad \& \quad \vec{a}_{CM} = \frac{\vec{F}_{ext}}{\Sigma m_i}$$

if on a SoP $\vec{F}_{ext} = 0 \Rightarrow \vec{a}_{CM} = 0 \Rightarrow \vec{V}_{CM} = \text{Constant}$

$$\Downarrow$$

$$\vec{P}_{system} = \text{Constant}$$

[ "If on a SoP, no external force is acting, the total momentum of SoP remain Conserved" ]

Scan for Video Explanation

Web Reference at www.physicsgalaxy.com

**Module-1: Conservation of Momentum**

if $\vec{F}_{ext} = 0 \Rightarrow \vec{P}_{system} = \text{Constant}$

= Recoiling of a Gun: before firing $\vec{P}_s = 0$

by Cons. of Momentum we write

before shot     after shot

$$0 = mv_B - Mv_G$$

$$mv_B = Mv_G$$

Scan for Video Explanation

Web Reference at www.physicsgalaxy.com

**Module-1: Conservation of Momentum**

# Adding a mass on a moving mass:

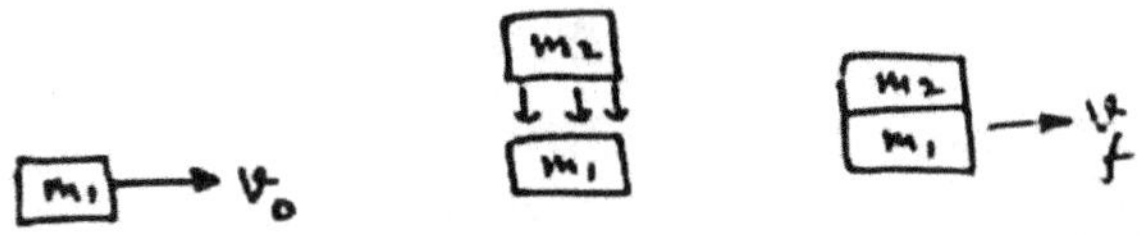


No ext. force.

by Cons. of momentum

$$m_1 v_0 = (m_1 + m_2) v_f$$

find speed of blocks is $v_f = \frac{m_1 v_0}{m_1 + m_2}$.

Web Reference at www.physicsgalaxy.com | Scan for Video Explanation

**Module-2: Solved Example-1**

Ex : A bullet of mass 5 kg is fired toward a 20 kg block at speed 30 m/s placed on smooth ground. It is observed that the bullet penetrates the block and comes out at a speed 10 m/s. Find the speed acquired by the block in this process.

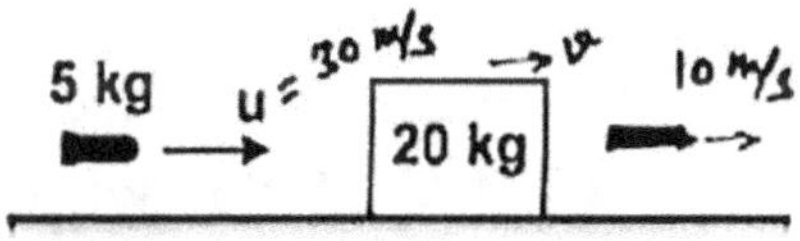


**Sol$^n$:** no $F_{ext}$ is present.

⇒ by Cons. of momentum we use

$$5 \times 30 = 20v + 5 \times 10$$

$$v = \frac{100}{20} = 5 \text{ m/s} \quad \text{Ans.}$$

Web Reference at www.physicsgalaxy.com | Scan for Video Explanation

**Module-3: Solved Example-2**

**Ex :** A pendulum bob of mass 2 kg is hanging from a string of length 1 m as shown. A bullet of mass 0.5 kg moving at 10 m/s strikes the bob and gets embedded into it. Find the maximum deflection angle from vertical to which the string will be deflected.

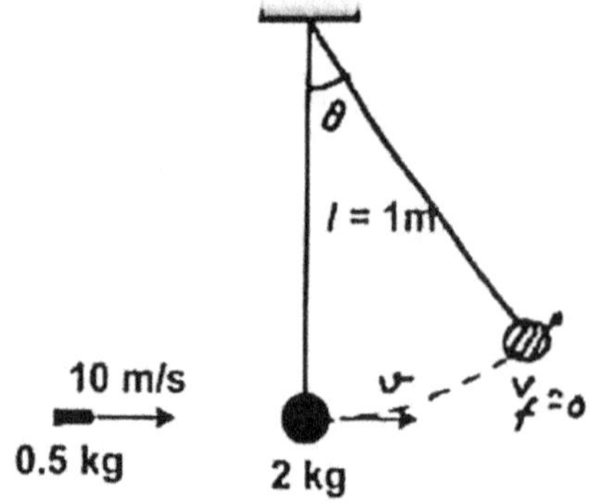


$g = 10\,m/s^2$

Sol$^n$: no $F_{ext}$ is present.

$$0.5 \times 10 = 2.5 \times v \Rightarrow v = 2\,m/s$$

As we know for any amplitude $\cos\theta = \frac{2gl - u^2}{2gl}$.

$$\cos\theta = \frac{2\times 10 - 4}{2\times 10} = \frac{16}{20} = \frac{4}{5}$$

$$\theta = \cos^{-1}(4/5) = 37^\circ \text{ Ans.}$$

Scan for Video Explanation



**Module-4: Solved Example-3**

**Ex :** A shell is fired from a canon with a speed of 100 m/s at 60° with horizontal. At the highest point of its trajectory, the shell explodes into two equal fragments. One of these moves along horizontal backward direction at 50 m/s, find the speed of other fragment at the time of explosion.

Sol$^n$:

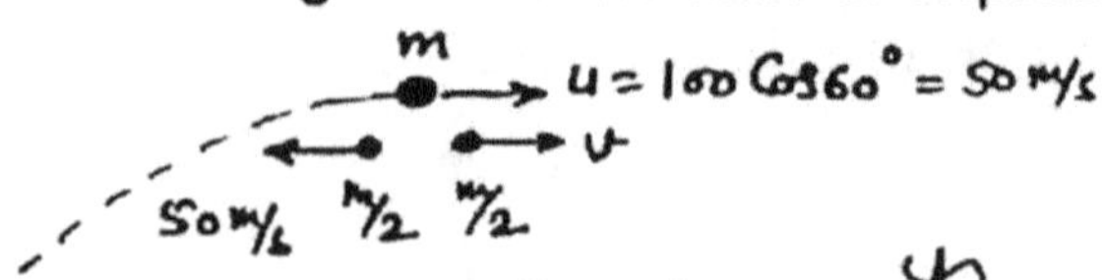


As $F_{ext} = 0$ in horiz dir. ⇒ by Cons of momentum.

$$mu = \frac{m}{2}v - \frac{m}{2}(50)$$

$$v = 2u + 50 = 150\,m/s \text{ Ans.}$$

Scan for Video Explanation

**Module-5: Effect of Internal Forces**

# Effect of Internal Forces:

If in a SOP, Centre of mass of System is at rest.

$\Rightarrow \vec{V}_{CM} = 0 = \text{Const.}$

& if no external forces are acting on the SOP then COM of System will remain at rest.

OR..

v. imp.

Only Internal forces of a System Can never Cause any motion in COM of a SOP if it is at rest.

Web Reference at www.physicsgalaxy.com

Scan for Video Explanation

**Module-5: Effect of Internal Forces**

# Practical Illustration:

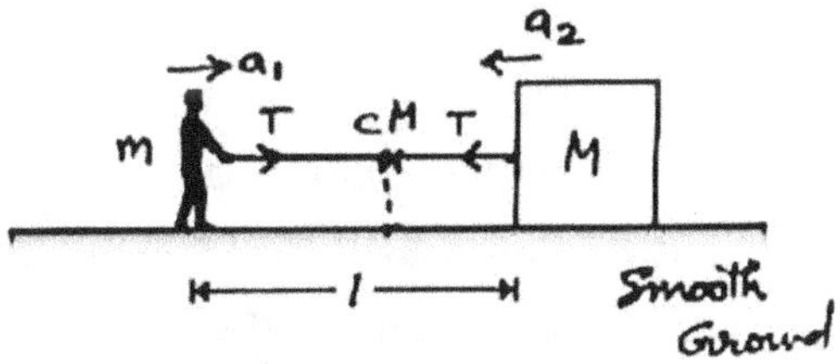


Here net ext force on System = 0.

COM → rest.

$ma_1 = Ma_2$

$a_{COM} = 0.$

in this situation boy & block both are displaced in such a way that they meet at the location of their COM.

Web Reference at www.physicsgalaxy.com

Scan for Video Explanation

**Module-5: Effect of Internal Forces**

Ex : A monkey holds an umbrella and this system is floating in air at rest due to buoyancy on balloon. If monkey starts climbing the string at speed u relative to string, find the speed with which balloon will move.

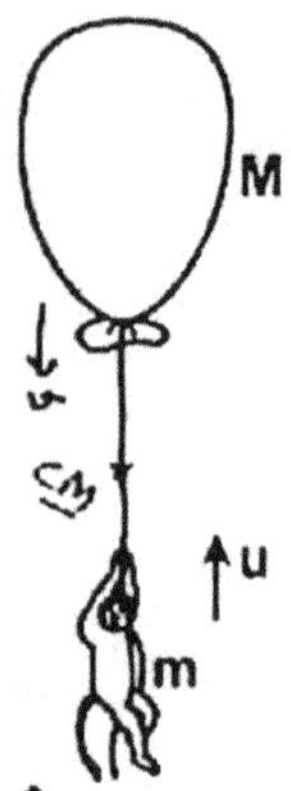


Sol$^n$:

Initial momentum = 0

Final momentum = $m(u-v) - Mv = 0$

$\left(v = \frac{mu}{M+m}\right)$ Ans.

Scan for Video Explanation

Web Reference at www.physicsgalaxy.com

**Module-6: Solved Example-4**

Ex : Figure shows a uniform boat of length 4m and mass 50 kg. At the edges of boat two friends Joc and Mac are standing. If they run to meet each other at centre of boat, find the displacement of boat when they meet. Ignore friction between boat and water.

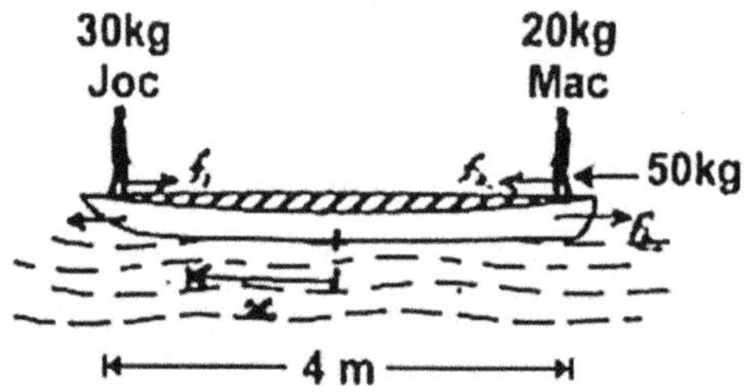


Sol$^n$:

$$\Delta x_{com} = 0 = \frac{m_1\Delta x_1 + m_2\Delta x_2 + m_3\Delta x_3}{m_1 + m_2 + m_3}$$

$$0 = \frac{30\times(2-x) - 20\times(2+x) - 50\times x}{100}$$

$$6 - 3x - 4 - 2x - 5x = 0 \Rightarrow 10x = 2$$

$x = 0.2\,m$ Ans.

Scan for Video Explanation

Web Reference at www.physicsgalaxy.com

**Module-7: Solved Example-5**

**Ex :** A flat car of mass M is at rest on a frictionless floor with a child of mass m standing at the edge. If child jumps off from the car toward right with initial velocity u, with respect to car, find the velocity of car after the jump.

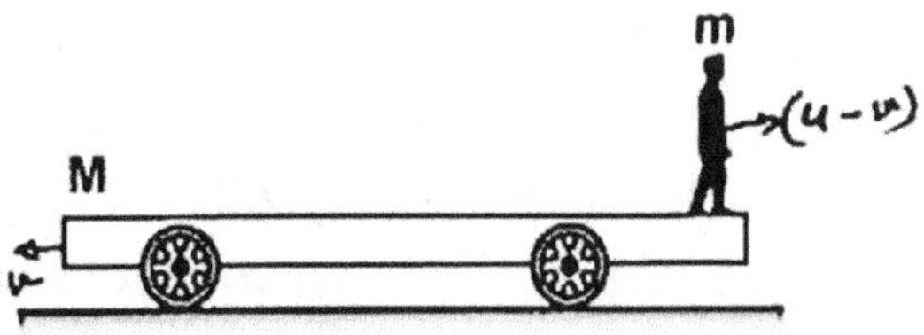


Sol$^n$. by Cons. of momentum, we use.

$$0 = m(u - v) - Mv$$

$$v = \frac{mu}{m+M} \quad \text{Ans.}$$

Web Reference at www.physicsgalaxy.com | Scan for Video Explanation

**Module-8: Solved Example-6**

**Ex :** A flat car of mass M with a child of mass m is moving at speed $v_1$. The child jumps in the direction of motion of car with a velocity u with respect to car. Find the final velocities of the child and that of car after jump.

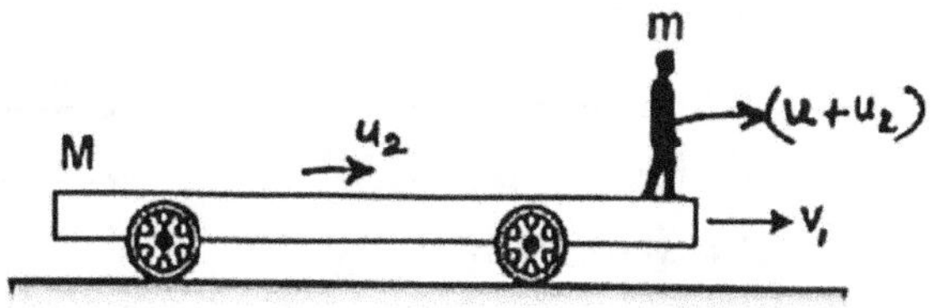


Sol$^n$. by Cons. of momentum

$$(M+m)v_1 = m(u+u_2) + Mu_2.$$

vel of Car $\quad u_2 = \dfrac{(M+m)v_1 - mu}{m+M.}$ Ans.

vel of child $\quad u + u_2 = \dfrac{(M+m)v_1 + Mu}{m+M.}$ Ans.

Web Reference at www.physicsgalaxy.com | Scan for Video Explanation

**Module-9: Solved Example-7**

**Ex :** A shell at rest at origin explodes into three fragments of masses 1kg, 2kg and m kg. The fragments of masses 1kg and 2 kg fly off with speeds 12 m/s along X-axis and 8 m/s along Y-axis respectively. If m kg flies off with speed 40 m/s then find the total mass of the shell.

Sol$^n$:

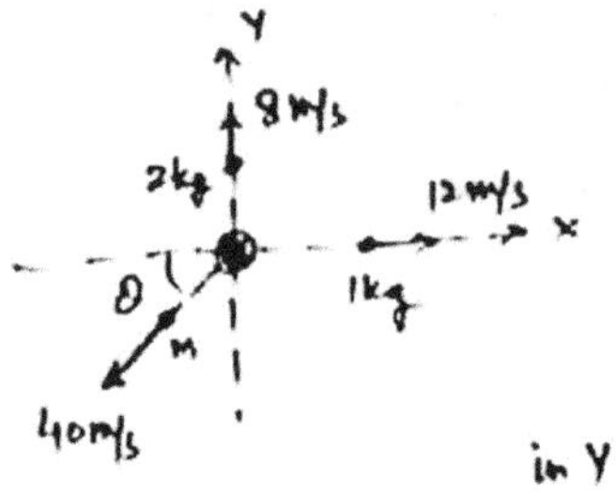


Initial momentum = 0

In x dir.

$m(40\cos\theta) = 1\times 12$ —(1)

in Y dir $m(40\sin\theta) = 2\times 8 = 16$ —(2)

$(1)^2 + (2)^2 \Rightarrow m^2(40)^2 = (12)^2 + (16)^2$

$m^2 = \frac{144+256}{1600} = \frac{1}{4}$

$m = \frac{1}{2} = 0.5$ kg.

$M_{shell} = 1 + 2 + 0.5$ kg

$= 3.5$ kg Ans.

Scan for Video Explanation

Web Reference at www.physicsgalaxy.com

**Module-10: Solved Example-8**

**Ex :** Two identical bars are connected by an ideal spring of natural length $l_0$, as shown. A constant horizontal force F starts acting on one of the bars as shown. Find the maximum and minimum separation between the bars during their motion.

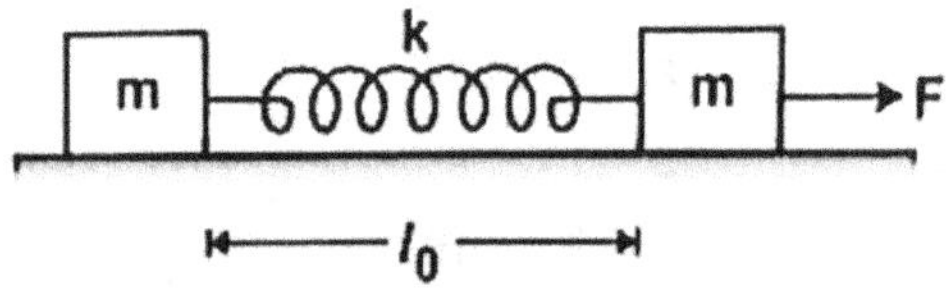


Sol$^n$: acc$^n$ of CoM $a_{CM} = \frac{F}{2m}$

min sep $= l_0$.

FBD of bars w.r.t. CoM.

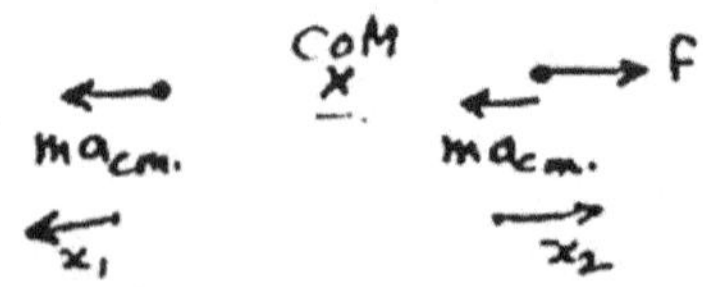


by energy Cons.

$ma_{cm}\cdot x_1 + (F - ma_{cm})x_2 = \frac{1}{2}k(x_1+x_2)^2$

$m\left(\frac{F}{2m}\right)x_1 + \left\{F - m\left(\frac{F}{2m}\right)\right\}x_2 = \frac{1}{2}k(x_1+x_2)^2$

$F\frac{(x_1+x_2)}{2} = \frac{1}{2}k(x_1+x_2)^2 \Rightarrow x_1+x_2 = \frac{F}{k}$ Ans.

Scan for Video Explanation

Web Reference at www.physicsgalaxy.com

**Module-11: Impulse & Conservation of Momentum**

# Impulse & Conservation of Momentum:

As we know $F = \frac{dP}{dt}$

$$\left.\begin{aligned} dP &= F\,dt \\ \Delta P &= \int F\,dt \end{aligned}\right] \text{Impulse}$$

Impulse ↓ momentum imparted by a force in a given time.

(Impulse = Force × time duration)

**Work**
- for Const force $W = F \cdot \Delta s$
- for var. force $W = \int F\,dx$

**Impulse**
- $J = F\Delta t$ (for a Constant F)
- $J = \int F\,dt$ (for a time varying F)

Web Reference at www.physicsgalaxy.com | Scan for Video Explanation

**Module-11: Impulse & Conservation of Momentum**

# Impulse - Momentum Equation:

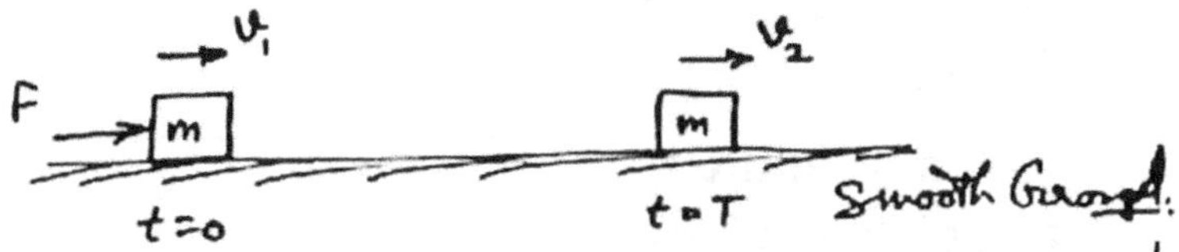


Initial Momentum + Impulse Imparted = Final momentum

$$mv_1 + Ft = mv_2$$

External force ⇒ Continuous change in momentum.

Web Reference at www.physicsgalaxy.com | Scan for Video Explanation

**Module-11: Impulse & Conservation of Momentum**

**Ex :** A body of mass 10 kg is pulled from rest with a time varying force $F = 2t^2$ N on a rough surface having friction coefficient $\mu = 0.2$. Find the speed of block after 6 seconds.

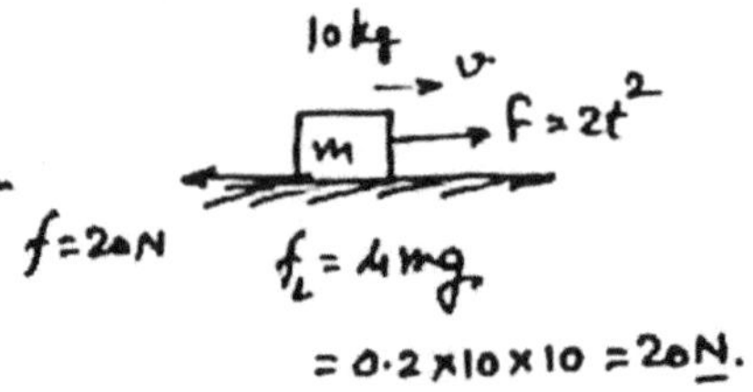


Sol$^n$:

block will start sliding when

$$2t^2 = 20$$

$$t = \sqrt{10} \text{ sec}$$

Impulse by ext force $J_1 = \int_{\sqrt{10}}^{6} 2t^2\,dt = \left(\frac{2t^3}{3}\right)_{\sqrt{10}}^{6}$

$$= \frac{2}{3}[216 - 10\sqrt{10}] \text{ N-s}$$

Impulse by fr. $J_2 = -20(6-\sqrt{10})$

using Impulse-momentum eq$^n$ → $\frac{2}{3}[216 - 10\sqrt{10}] - 20[6-\sqrt{10}] = 10 v_f$

$$v_f = \left(24 + \frac{40}{3}\sqrt{10}\right)/10 = \frac{12}{5} + \frac{4}{3}\sqrt{10} \text{ m/s Ans.}$$

Scan for Video Explanation

Web Reference at www.physicsgalaxy.com

**Module-12: Solved Example-9**

**Ex :** A block A is released from rest from top of a wedge block of height h as shown in figure.

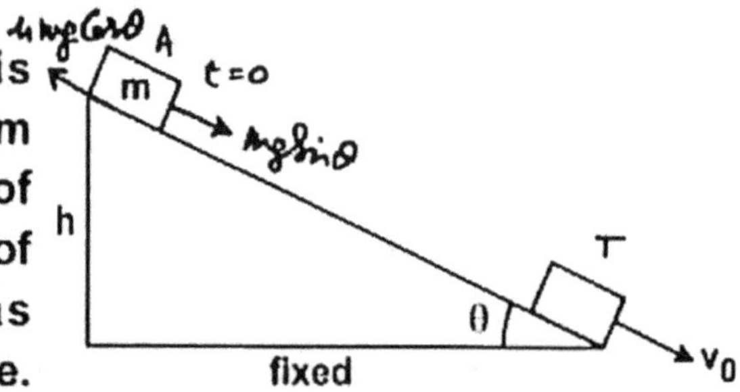


If velocity of block when it reaches the bottom of incline is $v_0$, find the time of sliding. Friction coefficient on inclined plane is $\mu$.

Sol$^n$: net downward force on block along incline is

$$F_{down} = mg\sin\theta - \mu mg\cos\theta$$

W-E th.: $(mg\sin\theta - \mu mg\cos\theta) \times h\,\text{Cosec}\,\theta = \frac{1}{2} m v_0^2$ —(1)

by Imp eq$^n$: $(mg\sin\theta - \mu mg\cos\theta) \times T = m v_0$ —(2)

$\frac{(2)}{(1)}$ $\frac{T}{h\,\text{Cosec}\,\theta} = \frac{2}{v_0} \Rightarrow T = \frac{2h\,\text{Cosec}\,\theta}{v_0}$ Ans.

Scan for Video Explanation

Web Reference at www.physicsgalaxy.com

**Module-13: Solved Example-10**

**Ex :** Figure shows a spring block system on which time varying force 5t is applied on 2kg mass. After 10 second, velocity of 3 kg mass is 30 m/s. Find velocity of 2 kg mass at this instant.

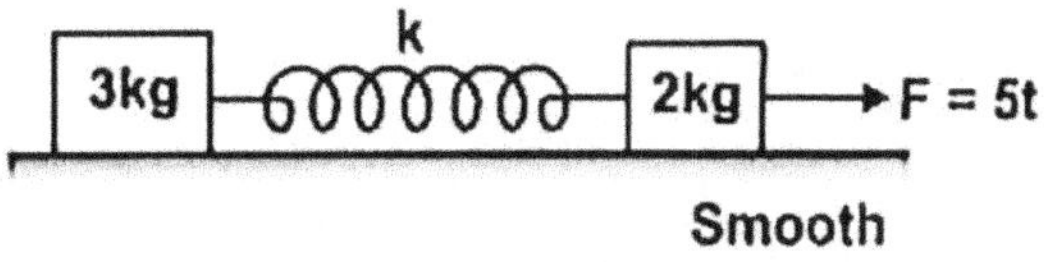


**Sol$^n$:** Total impulse by external force in 10 s is $J = \int_0^{10} F\,dt$

$= \int_0^{10} 5t\,dt = 250$ N·s.

$(\Sigma m)\, v_{CM} = 250$.

$v_{CM} = 250/5 = 50$ m/s.

$v_{CM} = \dfrac{m_1 v_1 + m_2 v_2}{m_1 + m_2} = \dfrac{3\times 30 + 2\times v_2}{5} = 50$

$\Rightarrow v_2 = \dfrac{160}{2} = 80$ m/s Ans.



**Module-14: Solved Example-11**

**Ex :** Figure shows a smooth L-shaped block of mass M placed on a smooth plane. If a small mass m is thrown on it with speed u as shown, find the minimum speed u such that it is able to reach the topmost point B on block.

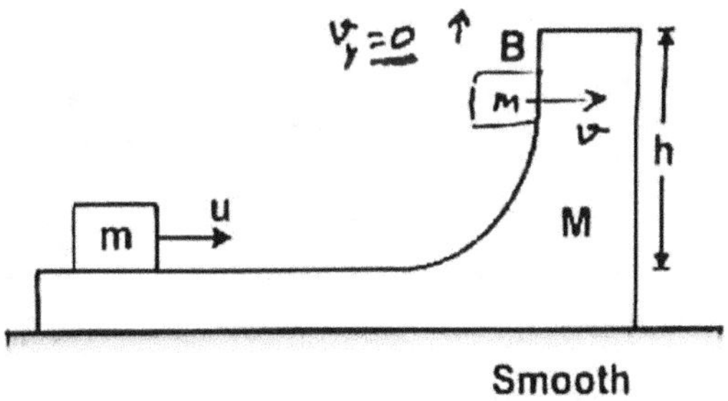


**Sol$^n$:** As no $F_{ext}$ is present $\Rightarrow$ $mu = (m+M)v$

$v = \dfrac{mu}{m+M}$

by W/E theorem -

$\frac{1}{2}mu^2 - mgh = \frac{1}{2}(m+M)\left(\dfrac{mu}{m+M}\right)^2 \Rightarrow \frac{1}{2}mu^2\left[1 - \dfrac{m}{m+M}\right] = mgh$

$u = \sqrt{\dfrac{2gh(M+m)}{M}}$ Ans.

**Module-15: Collisions**

# Collision:

It is a process of momentum transfer by internal forces of a System.

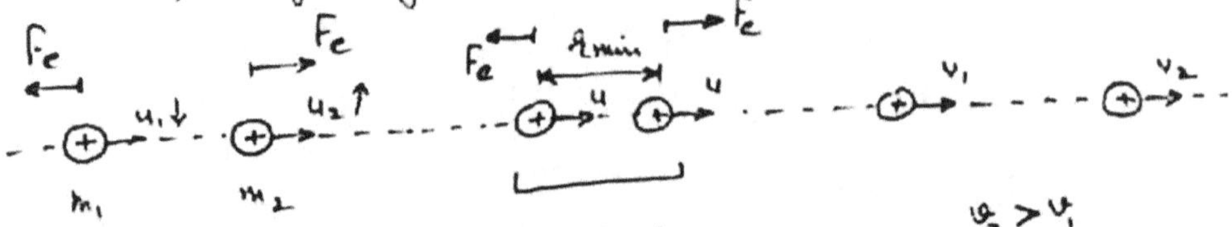


Say $u_1 > u_2$

$t = 0$

State of min KE during Collision

$v_2 > v_1$

$t = T$

Impact.

↓ physical Contact occurs

Scan for Video Explanation

Web Reference at www.physicsgalaxy.com

**Module-15: Collisions**

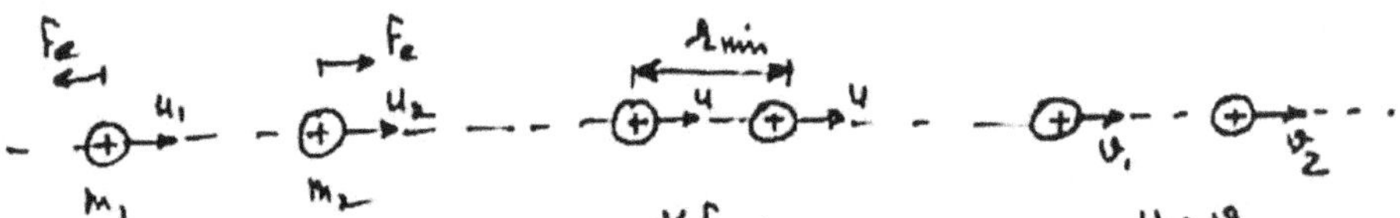


As no external force is present in the above case

⇒ Total momentum of System will remain Conserved.

$$m_1u_1 + m_2u_2 = m_1v_1 + m_2v_2$$

Scan for Video Explanation

Web Reference at www.physicsgalaxy.com

## Module-15: Collisions

# Impact between two bodies:

↓

Physical Contact

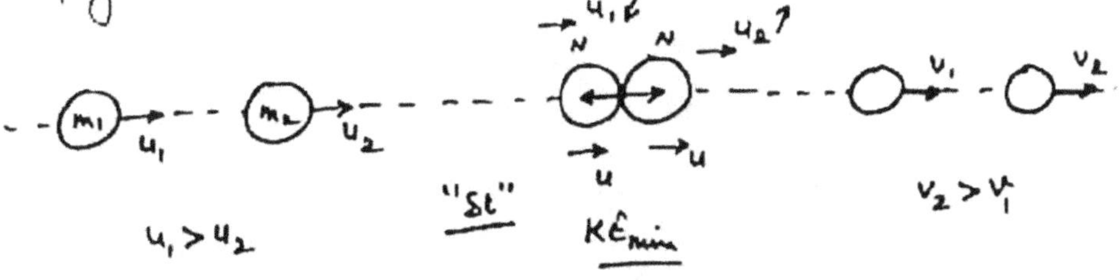


As no external force is present

$$m_1u_1 + m_2u_2 = m_1v_1 + m_2v_2$$

Web Reference at www.physicsgalaxy.com | Scan for Video Explanation

## Module-16: Types of Collisions Based on Material Properties

# Types of Collisions based on material properties:

Based on types of material, collisions can be classified in 3 categories - These are -

during Collision

$KE_f = KE_i$ (1) Elastic Collision → KE loss = 0

$KE_f < KE_i$ (2) Partial Elastic Collision → Some KE loss

$KE_f < KE_i$ (min) (3) Inelastic Collisions → KE loss → max.

Web Reference at www.physicsgalaxy.com | Scan for Video Explanation

**Module-16: Types of Collisions Based on Material Properties**

# Classification of Solids:

(1) Elastic material
(2) Partially elastic material
(2) Inelastic materials (plastic)

(1) Elastic Material: which deform on application of external force & regains original shape on removal of force.

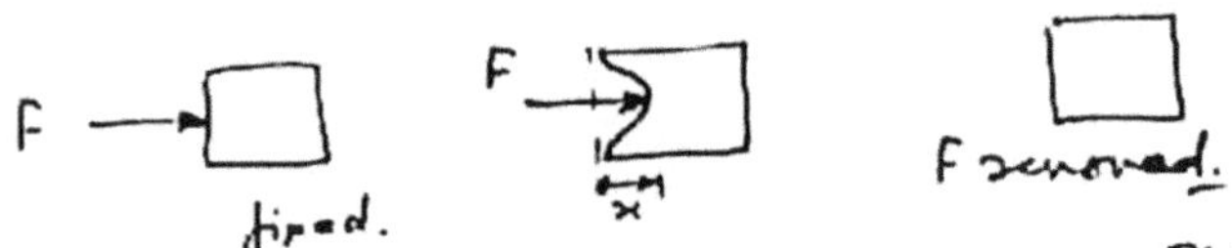


Deformation ⇒ PE is stored. → on removal of force this PE is released

Scan for Video Explanation

Web Reference at www.physicsgalaxy.com

**Module-16: Types of Collisions Based on Material Properties**

# Partially Elastic Material:

W/D in Deformation → partly stored in form of PE & remaining energy is dissipated.

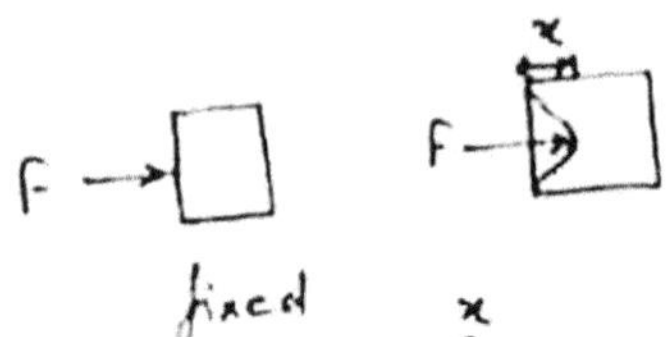


$PE < \int_0^x F\,dx$ → on removal of ext force only the stored PE will be released.

Scan for Video Explanation

Web Reference at www.physicsgalaxy.com

## Module-16: Types of Collisions Based on Material Properties

# Inelastic Materials:

W/D in deformation → Completely dissipated
&
No PE is stored
⇓
on removal of force
object remain in
deformed state.

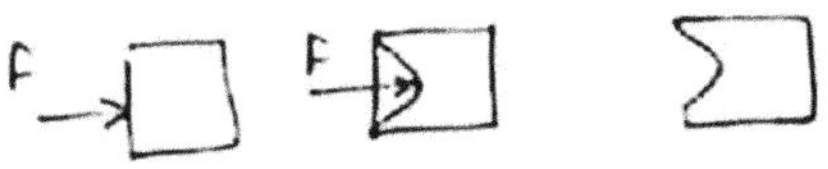


No restoring
force exist in
Inelastic materials

Web Reference at www.physicsgalaxy.com

Scan for Video Explanation

## Module-17: Collision of Bodies of Different Materials

# Collision of bodies of different materials:

$(u_1 > u_2)$

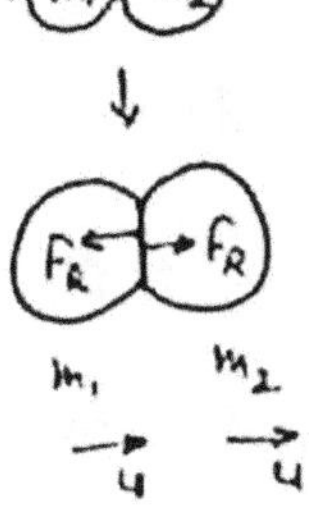


PE → max
KE → min

for elastic or
partially elastic
materials bodies will
separate.

$$m_1u_1 + m_2u_2 = (m_1 + m_2)u$$

$$\left[u = \frac{m_1u_1 + m_2u_2}{m_1 + m_2}\right]$$

inelastic material
along line of motion bodies do not
separate.

Web Reference at www.physicsgalaxy.com

Scan for Video Explanation

**Module-18: Coefficient of Restitution**

# Coefficient of Restitution:

When two bodies collide then this is a constant parameter which accounts for the elasticity of material of bodies.

def:

$$e = \frac{\text{velocity of separation along line of Impact}}{\text{velocity of approach along line of Impact}}$$

$u_1 > u_2$

$v_{ap} = u_1 - u_2$

$$e = \left(\frac{v_2 - v_1}{u_1 - u_2}\right)$$

$v_2 > v_1$

sep vel $v_{sep} = v_2 - v_1$

Scan for Video Explanation

Web Reference at www.physicsgalaxy.com

**Module-18: Coefficient of Restitution**

For different type of material of bodies value of 'e' is different -

for elastic collision $e = 1$

for partial elastic collision $0 < e < 1$

Inelastic collisions $e = 0 = \frac{v_{sep}}{v_{app}}$

↓

bodies do not separate along line of impact.

Scan for Video Explanation

Web Reference at www.physicsgalaxy.com

**Module-18: Coefficient of Restitution**

# Types of Collisions :

Based on material of bodies involved in collision. These are categories in 3 ways.

(1) Elastic Collisions : In such cases before & after collision no energy loss takes place in deformation of bodies.

$KE_{iB} = KE_{fA}$. $(e = 1)$

(2) Partially Elastic Collisions : In such cases some energy loss takes place during collision under deformation of bodies $KE_{iB} > KE_{fA}$ $[0 < e < 1]$

Web Reference at www.physicsgalaxy.com | Scan for Video Explanation

**Module-18: Coefficient of Restitution**

(3) Inelastic Collision :

In such cases loss in energy during collision will be maximum and after collision no restoring force exist for separation of bodies. $(e = 0)$

# Classification of Collisions based on motion direction :

Two Categories :

(1) Head-on Collisions 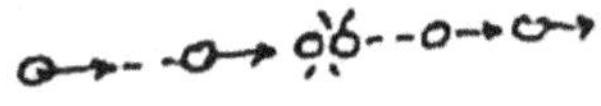

(2) Oblique Collisions

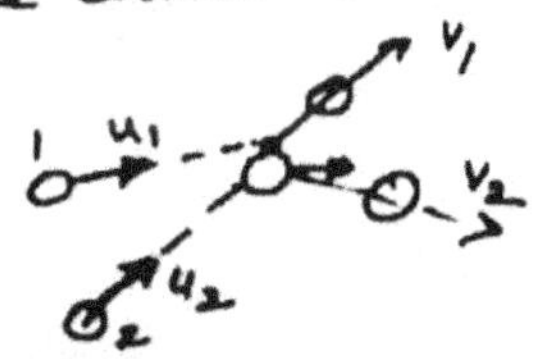


Web Reference at www.physicsgalaxy.com | Scan for Video Explanation

**Module-19: Head-on Collisions**

# Head-on Collisions :

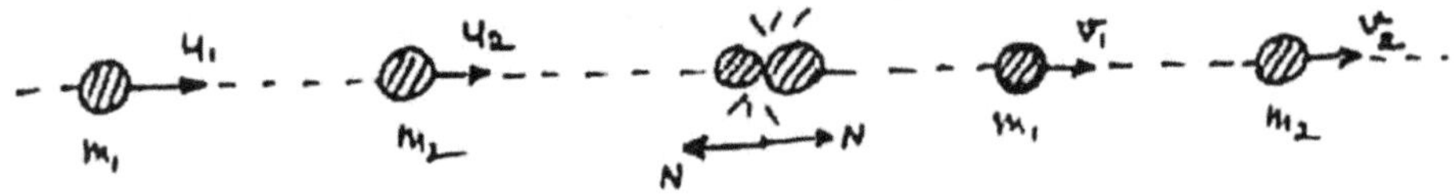


When line of motion of two bodies before and after collision are same it is termed as Head-on Collision.

As no ext force is present ⇒

$$m_1u_1 + m_2u_2 = m_1v_1 + m_2v_2 \quad \text{—(1)}$$

COR $e = 1.$

Scan for Video Explanation

Web Reference at www.physicsgalaxy.com

**Module-19: Head-on Collisions**

$$m_1u_1 + m_2u_2 = m_1v_1 + m_2v_2 \quad \text{—(1)}$$

$e = 1. = \frac{v_{sep}}{v_{app}}$

$$v_{sep} = v_{app}$$

$$v_2 - v_1 = u_1 - u_2 \quad \text{—(2)}$$

As collision is elastic ⇒ $KE_i = KE_f$.

$$\frac{1}{2}m_1u_1^2 + \frac{1}{2}m_2u_2^2 = \frac{1}{2}m_1v_1^2 + \frac{1}{2}m_2v_2^2 \quad \text{—(3)}$$

Solving any two eq$^n$ of (1), (2) & (3) we'll get values of $v_1$ and $v_2$ ....

Scan for Video Explanation

Web Reference at www.physicsgalaxy.com

**Module-19: Head-on Collisions**

from eq$^n$ (2)

$$v_2 = u_1 - u_2 + v_1$$

from eq$^n$ (1)

$$m_1u_1 + m_2u_2 = m_1v_1 + m_2(u_1 - u_2 + v_1)$$

On solving, we get

v. imp:

$$\vec{v}_1 = \left(\frac{m_1 - m_2}{m_1 + m_2}\right)\vec{u}_1 + \left(\frac{2m_2}{m_1 + m_2}\right)\vec{u}_2 \quad \text{—(A)}$$

$$\vec{v}_2 = \left(\frac{m_2 - m_1}{m_1 + m_2}\right)\vec{u}_2 + \left(\frac{2m_1}{m_1 + m_2}\right)\vec{u}_1 \quad \text{—(B)}$$

Web Reference at www.physicsgalaxy.com | Scan for Video Explanation

**Module-19: Head-on Collisions**

# Special Cases of Head-on elastic Collisions :

$$\vec{v}_1 = \left(\frac{m_1 - m_2}{m_1 + m_2}\right)\vec{u}_1 + \left(\frac{2m_2}{m_1 + m_2}\right)\vec{u}_2$$

$$\vec{v}_2 = \left(\frac{m_2 - m_1}{m_1 + m_2}\right)\vec{u}_2 + \left(\frac{2m_1}{m_1 + m_2}\right)\vec{u}_1$$

Case-I: If $m_1 = m_2$.

$\Rightarrow$ $\vec{v}_1 = \vec{u}_2$ & $\vec{v}_2 = \vec{u}_1$ } after collision vel are swapped.

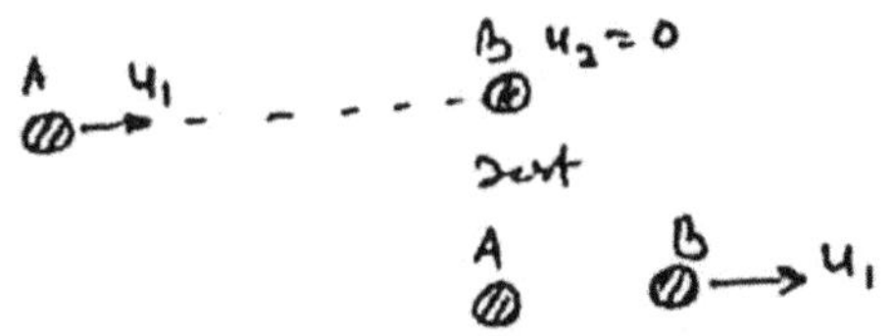


Web Reference at www.physicsgalaxy.com | Scan for Video Explanation

**Module-19: Head-on Collisions**

Case-II : If $m_1 >> m_2$

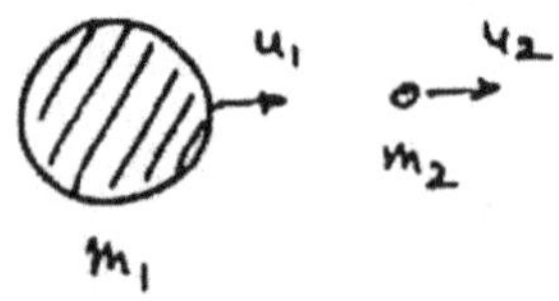


$$\vec{v}_1 = \left(\frac{m_1 - m_2}{m_1 + m_2}\right)\vec{u}_1 + \left(\frac{2m_2}{m_1 + m_2}\right)\vec{u}_2$$

$$\boxed{\vec{v}_1 = \vec{u}_1}$$

$$\vec{v}_2 = \left(\frac{m_2 - m_1}{m_1 + m_2}\right)\vec{u}_2 + \left(\frac{2m_1}{m_1 + m_2}\right)\vec{u}_1$$

$$\boxed{\vec{v}_2 = 2\vec{u}_1 - \vec{u}_2}$$

Scan for Video Explanation

Web Reference at www.physicsgalaxy.com

**Module-19: Head-on Collisions**

Case-III : If $m_2 >> m_1$

$$\vec{v}_1 = \left(\frac{m_1 - m_2}{m_1 + m_2}\right)\vec{u}_1 + \left(\frac{2m_2}{m_1 + m_2}\right)\vec{u}_2$$

$$\boxed{\vec{v}_1 = 2\vec{u}_2 - \vec{u}_1}$$

$$\vec{v}_2 = \left(\frac{m_2 - m_1}{m_1 + m_2}\right)\vec{u}_2 + \left(\frac{2m_1}{m_1 + m_2}\right)\vec{u}_1$$

$$\boxed{\vec{v}_2 = \vec{u}_2}$$

Scan for Video Explanation

Web Reference at www.physicsgalaxy.com

**Module-19: Head-on Collisions**

**Ex :** A horizontal platform is going up at a speed of 10 m/s and a small elastic ball falls on it at a speed of 5 m/s. Find the height to which ball will bounce after collision above the point of collision.

Sol^n:

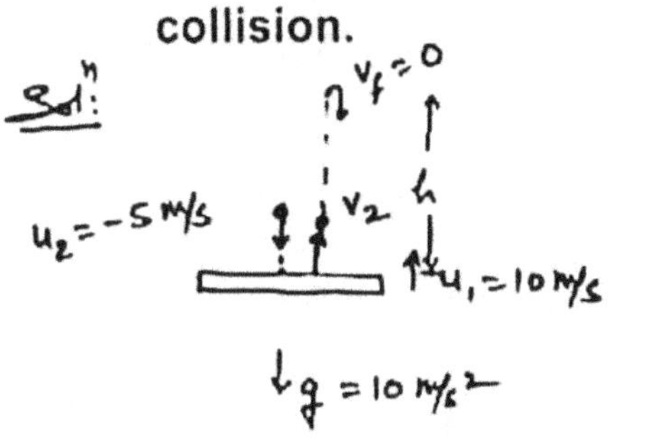


$$\vec{v}_2 = 2\vec{u}_1 - \vec{u}_2$$
$$= 2\times 10 - (-5)$$
$$= 25 \text{ m/s}$$

$$h = \frac{v_2^2}{2g} = \frac{(25)^2}{2\times 10} = \frac{625}{20} = 31.25 \text{ m. Ans.}$$

Web Reference at www.physicsgalaxy.com

Scan for Video Explanation

**Module-20: Solved Example-12**

**Ex :** A neutron moving at a speed v undergoes a head-on elastic collision with a neucleus of mass number A at rest. Find the ratio of kinetic energies of neutron after and before collision.

Sol^n:

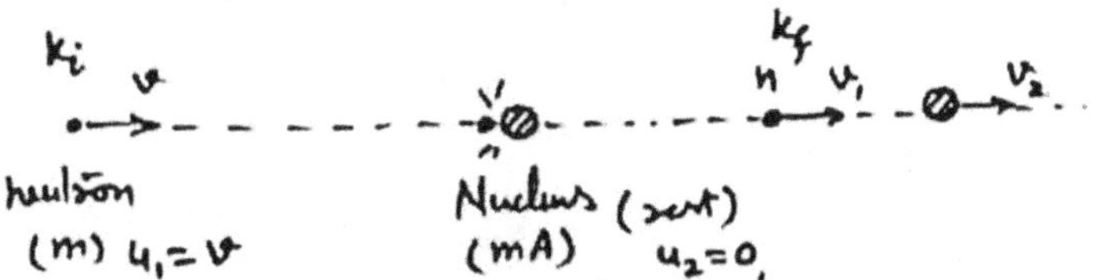


After head-on elastic collision speed of neutron is -

$$v_1 = \left(\frac{m_1 - m_2}{m_1 + m_2}\right) u_1 + \left(\frac{2m_2}{m_1 + m_2}\right) u_2$$

$$v_1 = \left(\frac{m - mA}{m + mA}\right) v = \left(\frac{1-A}{1+A}\right) v$$

$$r = \frac{K_f}{K_i} = \frac{\frac{1}{2} m v_1^2}{\frac{1}{2} m v^2} = \left(\frac{1-A}{1+A}\right)^2 \text{ Ans.}$$

Web Reference at www.physicsgalaxy.com

Scan for Video Explanation

**Module-21: Partial Elastic Head-on Collision**

# Partially elastic Head-on Collision: $(0<e<1)$

$v_{sep} = v_2 - v_1$

$v_{app} = u_1 - u_2$

by Cons. of momentum $\quad m_1u_1 + m_2u_2 = m_1v_1 + m_2v_2 \quad \text{—— (1)}$

using COR — $\quad v_2 - v_1 = e(u_1 - u_2) \quad \text{—— (2)}$

from eqⁿ (1) $\quad m_1u_1 + m_2u_2 = m_1v_1 + m_2(v_1 + eu_1 - eu_2)$

on Simplifying

$$\vec{v}_1 = \left(\frac{m_1 - em_2}{m_1 + m_2}\right)\vec{u}_1 + \left(\frac{m_2(1+e)}{m_1+m_2}\right)\vec{u}_2 \quad \text{—(A)}$$

$e=1$ ✓ &

$$\vec{v}_2 = \left(\frac{m_2 - em_1}{m_1 + m_2}\right)\vec{u}_2 + \left[\frac{m_1(1+e)}{m_1+m_2}\right]\vec{u}_1 \quad \text{—(B)}$$

Scan for Video Explanation

Web Reference at www.physicsgalaxy.com

**Module-21: Partial Elastic Head-on Collision**

# Inelastic Head-on Collision: $(e=0)$

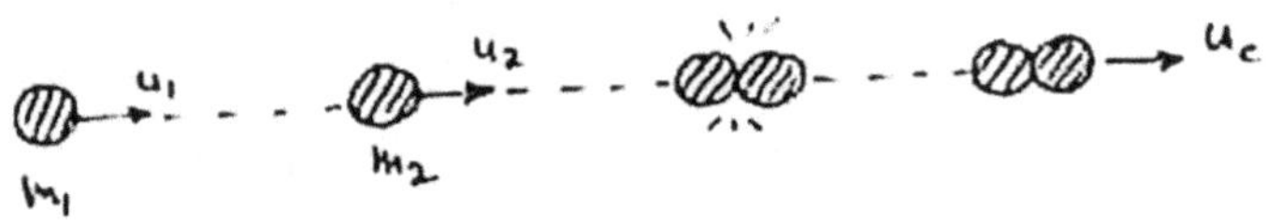


as $e=0 \Rightarrow v_{sep} = 0$

by Cons. of momentum

$$m_1u_1 + m_2u_2 = (m_1+m_2)u_c$$

$$\boxed{u_c = \frac{m_1u_1 + m_2u_2}{m_1+m_2}} \leftarrow v_{COM}.$$

Scan for Video Explanation

Web Reference at www.physicsgalaxy.com

**Module-21: Partial Elastic Head-on Collision**

Ex : A 3 kg ball moving at 10 m/s collides head-on inelastically with another ball of mass 2 kg moving toward the first ball at 20 m/s. After collision balls will stick to each other, find the amount of total energy loss in this collision.

Sol$^n$:

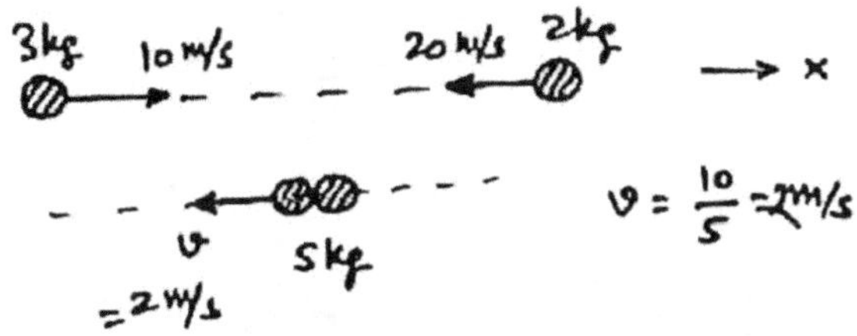


$\vec{P}_1 = 30\hat{i}$

$\vec{P}_2 = -40\hat{i}$

$-10\hat{i}$

$v = \frac{10}{5} = 2\,m/s$

$KE_i = \frac{1}{2}(3)(10)^2 + \frac{1}{2}(2)(20)^2 = 150 + 400 = 550\,J$

$KE_f = \frac{1}{2}(5)(2)^2 = 10\,J$

loss in energy $\Delta E_L = KE_i - KE_f$

$= 550 - 10 = 540\,J$ Ans.

Web Reference at www.physicsgalaxy.com | Scan for Video Explanation

**Module-22: Solved Example-13**

Ex : A ball is dropped on a floor from a height h. If coefficient of restitution is e, find the height to which ball will rise after rebound.

Sol$^n$:

using the speed eq$^n$

$v^2 = u^2 + 2as$

$0 = (e\sqrt{2gh})^2 - 2gh_1$

$h_1 = e^2 h$. Ans.

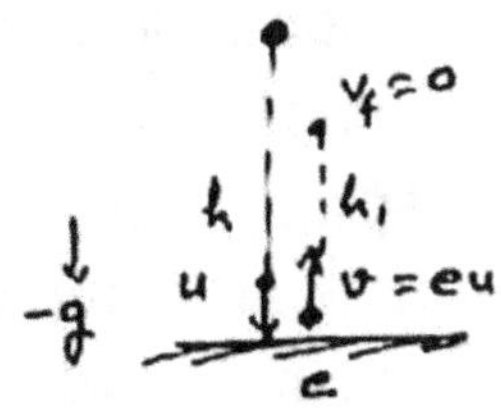


$u = \sqrt{2gh}$ $e = \frac{v}{u}$

$v = eu$

Web Reference at www.physicsgalaxy.com | Scan for Video Explanation

**Module-23: Solved Example-14**

**Ex :** A ball is released from a height h above a floor with coefficient of restitution $\underline{e}$ between floor & ball. Find the number of bounces ball will take before coming to rest. Also find the total time after which ball will come to rest.

Sol$^n$:

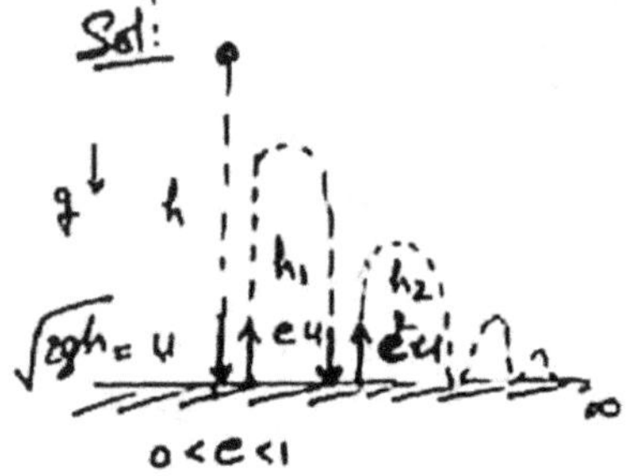


$$h_1 = \frac{(eu)^2}{2g} = e^2 h$$

$$h_2 = e^2 h_1 = e^4 h \qquad CR = e^2 < 1$$

$$h_3 = e^2 h_2 = e^6 h \qquad \text{GP with } CR < 1$$

$\Rightarrow$ Converging G.P.

$\Rightarrow$ $\infty$ no. of terms.

No. of bounces before Coming to rest $\rightarrow \infty$

Time of Motion $$T = \frac{u}{g} + \frac{2eu}{g} + \frac{2e^2u}{g} + \dots = \frac{u}{g}\left[1 + 2\left(\frac{e}{1-e}\right)\right]$$

$$= \sqrt{\frac{2h}{g}}\left(\frac{1+e}{1-e}\right) \text{ Ans.}$$

Scan for Video Explanation

Web Reference at www.physicsgalaxy.com

**Module-24: Solved Example-15**

**Ex :** Figure shows a U-shaped block of mass M placed on smooth ground in which an ideal spring of force constant k is attached as shown. If a ball of mass m moving at speed u strikes the spring, find the maximum fraction of kinetic energy which will be stored in spring in the process of collision.

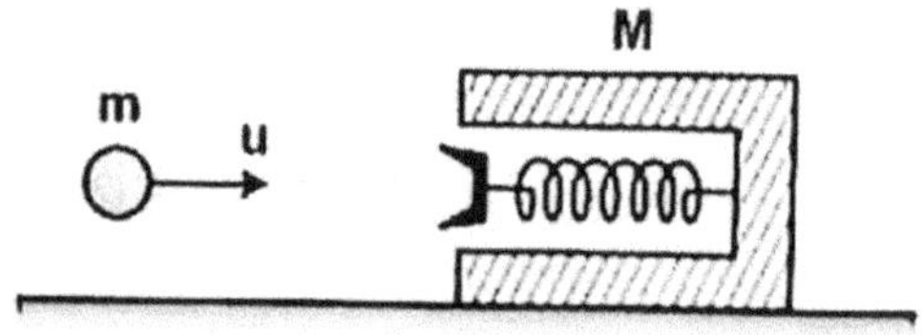


$(KE_{min}) \rightarrow v_f$

Sol$^n$: by Cons. of momentum $mu = (m+M) v_f \Rightarrow v_f = \frac{mu}{m+M}$

$$f = \frac{\Delta K_L}{K_i} = \frac{\frac{1}{2}mu^2 - \frac{1}{2}(m+M)v_f^2}{\frac{1}{2}mu^2} = 1 - \frac{m+M}{m}\cdot\frac{v_f^2}{u^2} = 1 - \frac{m}{m+M}$$

$$f = \frac{M}{m+M} \text{ Ans.}$$

Scan for Video Explanation

Web Reference at www.physicsgalaxy.com

**Module-25: Solved Example-16**

**Ex :** A ball of mass m moving at speed v makes a head-on collision with an identical ball at rest. The kinetic energy of the balls after the collision is three fourth of original. Find the coefficient of restitution.

Sol$^n$:

$m_1 = m_2 = m.$

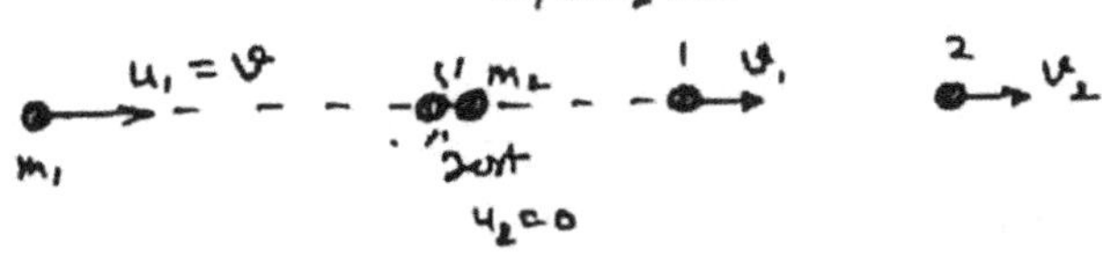


$$v_1 = \left(\frac{m_1 - em_2}{m_1 + m_2}\right) u_1 + \frac{m_2(1+e)}{m_1+m_2} \cdot u_2 = \left(\frac{1-e}{2}\right) v$$

$$v_2 = \left(\frac{m_2 - em_1}{m_1 + m_2}\right) u_2 + \frac{m_1(1+e)}{m_1+m_2} u_1 = \left(\frac{1+e}{2}\right) v$$

$$K_f = \frac{3}{4}\left(\frac{1}{2} m v^2\right)$$

$$\frac{1}{2} m \left(\frac{1-e}{2}\right)^2 v^2 + \frac{1}{2} m \left(\frac{1+e}{2}\right)^2 v^2 = \frac{3}{4}\left(\frac{1}{2} m v^2\right) \Rightarrow 2 + 2e^2 = 3$$

$$e = \frac{1}{\sqrt{2}} \text{ Ans.}$$

Web Reference at www.physicsgalaxy.com | Scan for Video Explanation

**Module-26: Solved Example-17**

**Ex :** A steel ball of mass m falls from a height on a horizontal surface. Find the cumulative momentum that the ball inparts to surface after numerous bounces, if coefficient of restitution is e.

Sol$^n$: Total momentum imparted to Surface can be given as-

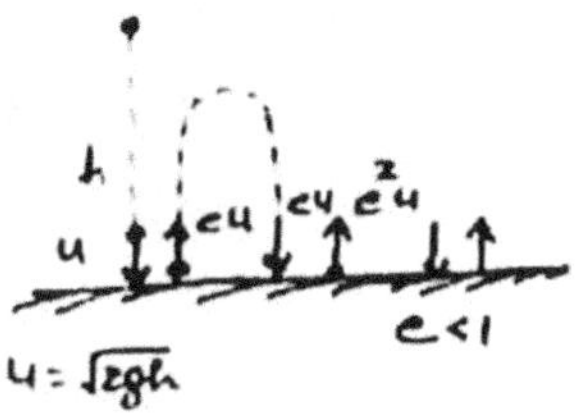


first hit $\Delta P_1 = mu + m(eu)$

Second hit $\Delta P_2 = m(eu) + m(e^2 u)$

$$\Delta P_T = mu + 2emu + 2e^2 mu + \cdots$$

$$= mu[1 + 2e + 2e^2 + \cdots] \quad \{\text{Conv. G.P.}\}$$

$$= mu\left[1 + \frac{2e}{1-e}\right] = mu\left(\frac{1+e}{1-e}\right) = m\sqrt{2gh}\left(\frac{1+e}{1-e}\right) \text{ Ans.}$$

Web Reference at www.physicsgalaxy.com | Scan for Video Explanation

**Module-27: Oblique Collision of a Ball on a Flat Surface**

# Oblique Collision of a ball on a flat surface:

Case-I : Elastic Collision

$e = 1$

$v_{app} = v_{sep}$

$u\cos\theta = v\cos\phi \quad -(1)$

$u\sin\theta = v\sin\phi \quad -(2)$

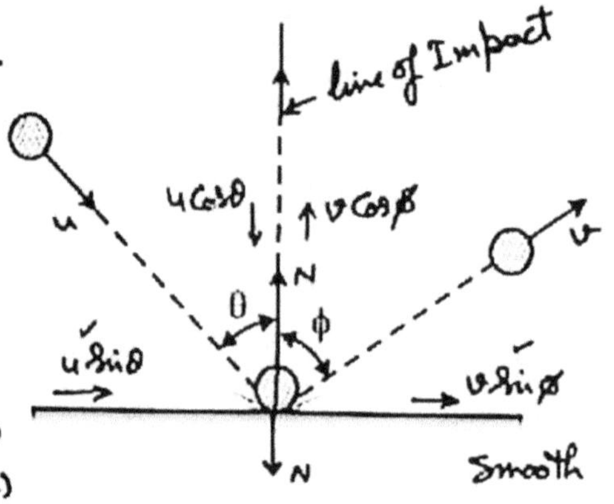


from (1) & (2) $\quad v = u \;\&\; \phi = \theta$

Angle of Reflection = Angle of Incidence.

Scan for Video Explanation

Web Reference at www.physicsgalaxy.com

**Module-27: Oblique Collision of a Ball on a Flat Surface**

Case-II: Partial Elastic Collision

$(0 < e < 1)$

Along line of impact

$v_{sep} = e\, v_{app}$

$v\cos\phi = e(u\cos\theta) \quad -(1)$

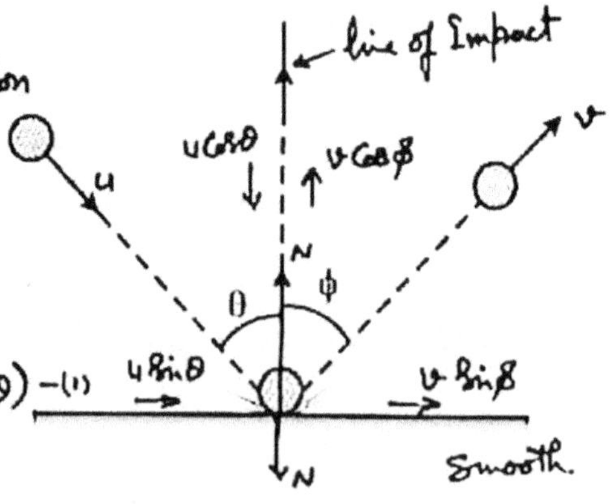


Along the surface

$v\sin\phi = u\sin\theta \quad -(2)$

$\sqrt{(1)^2 + (2)^2}$ $\quad \boxed{v = u\sqrt{\sin^2\theta + e^2\cos^2\theta}}$ $\quad$ for $e = 1 \Rightarrow \begin{bmatrix} v = u \\ \phi = \theta \end{bmatrix}$

$\frac{(2)}{(1)} \Rightarrow \boxed{\tan\phi = \frac{1}{e}\tan\theta}$

Scan for Video Explanation

Web Reference at www.physicsgalaxy.com

### Module-27: Oblique Collision of a Ball on a Flat Surface

**Case-III: Inelastic Collision ($e=0$)**

$\Rightarrow$ No separation

$v_{sep} = 0.$

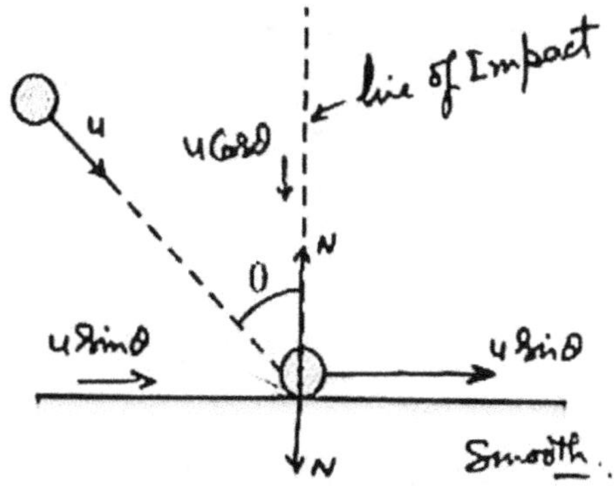


Web Reference at www.physicsgalaxy.com

Scan for Video Explanation

### Module-28: Oblique Collision in a 2D Plane

**# Oblique Collision in 2D Plane:**

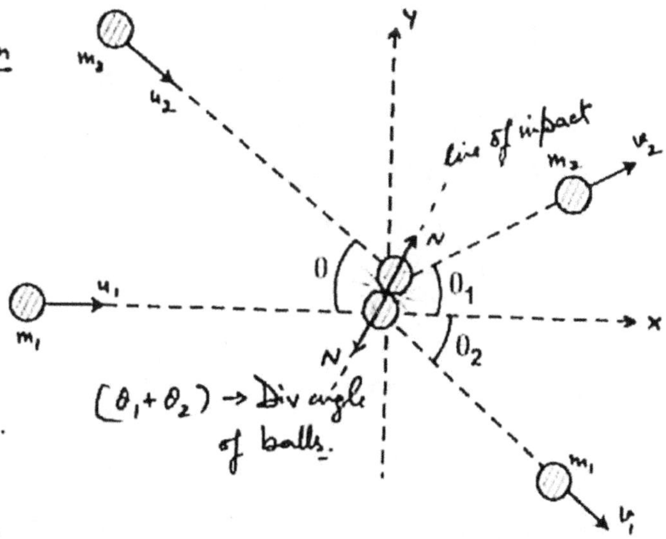


As no $F_{ext}$ is present, we use.

along x-dir $\quad m_1u_1 + m_2u_2\cos\theta = m_1v_1\cos\theta_1 + m_2v_2\cos\theta_2$

along y-dir $\quad m_2u_2\sin\theta = m_1v_1\sin\theta_1 + m_2v_2\sin\theta_2$

For elastic collisions – $\frac{1}{2}m_1u_1^2 + \frac{1}{2}m_2u_2^2 = \frac{1}{2}m_1v_1^2 + \frac{1}{2}m_2v_2^2$

Web Reference at www.physicsgalaxy.com

Scan for Video Explanation

**Module-28: Oblique Collision in a 2D Plane**

# Impact Parameter:

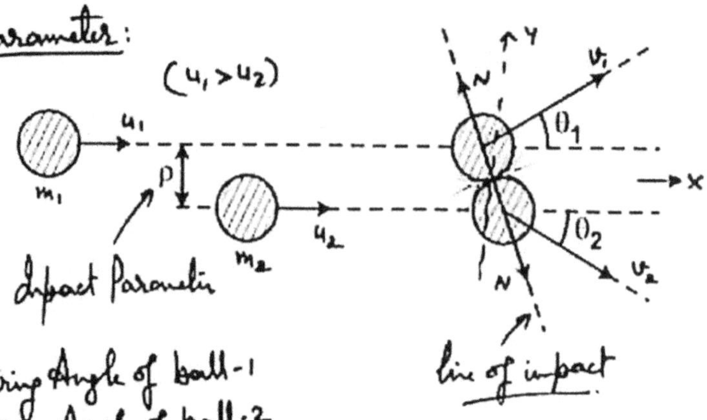


$\theta_1 \rightarrow$ Scattering Angle of ball-1

$\theta_2 \rightarrow$ Scattering Angle of ball-2

$\theta_1 + \theta_2 \rightarrow$ Divergence Angle of two balls after Collision

along x-dir $\quad m_1u_1 + m_2u_2 = m_1v_1\cos\theta_1 + m_2v_2\cos\theta_2$

along y-dir $\quad 0 = m_1v_1\sin\theta_1 - m_2v_2\sin\theta_2$

for elastic Collision $\quad \frac{1}{2}m_1u_1^2 + \frac{1}{2}m_2u_2^2 = \frac{1}{2}m_1v_1^2 + \frac{1}{2}m_2v_2^2$

Scan for Video Explanation

Web Reference at www.physicsgalaxy.com

**Module-28: Oblique Collision in a 2D Plane**

**Ex :** A ball of radius $R_1$ strikes another ball of same mass and radius $R_2$ at rest elastically. Find the divergence angle of the two balls after collision. Given that impact parameter $\rho < (R_1 + R_2)$.

Sol$^n$:

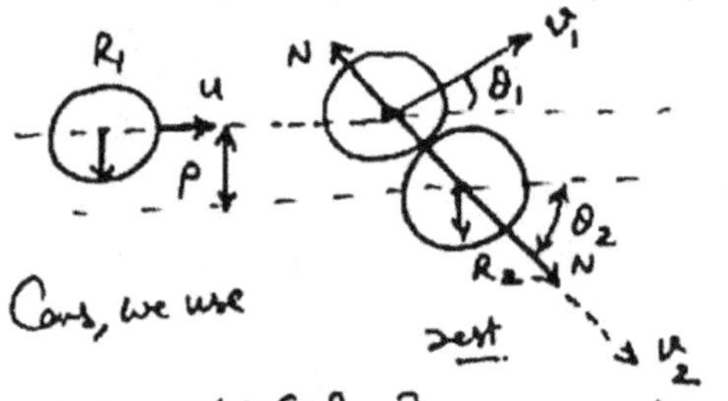


Div $= \theta_1 + \theta_2$

by momentum Cons, we use

$$\begin{bmatrix} mu = m v_1\cos\theta_1 + m v_2\cos\theta_2 \\ 0 = m v_1\sin\theta_1 - m v_2\sin\theta_2 \end{bmatrix} \rightarrow \text{Sq. \& adding}$$

$$u^2 = v_1^2 + v_2^2 + v_1v_2(\cos\theta_1\cos\theta_2 - \sin\theta_1\sin\theta_2)$$

$KE_i = KE_f$

$$\frac{1}{2}mu^2 = \frac{1}{2}mv_1^2 + \frac{1}{2}mv_2^2$$

$$\cos\theta_1\cos\theta_2 - \sin\theta_1\sin\theta_2 = 0$$

$$\cos(\theta_1 + \theta_2) = 0$$

$$\boxed{\theta_1 + \theta_2 = \pi/2} \text{ Ans.}$$

Scan for Video Explanation

Web Reference at www.physicsgalaxy.com

**Module-29: Solved Example-18**

**Ex :** A particle of mass m collides with a stationary particle of mass M is scattered at angle $\pi/2$ and M is recoiled at 30° to the initial direction of motion. Find the fraction of initial energy lost in collision if M = 5 m.

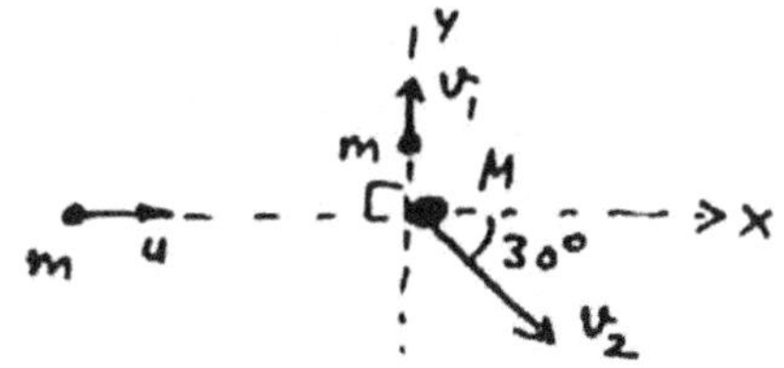


Sol$^n$: along x dir:

$mu = Mv_2 \cos 30°$ ($\cos 30° = \sqrt{3}/2$)

$v_2 = \dfrac{2u}{5\sqrt{3}}$ —(1)

along y-dir $\quad mv_1 = Mv_2 \sin 30° \Rightarrow v_1 = \dfrac{5}{2} v_2$ —(2)

$v_1 = \dfrac{5}{2}\left(\dfrac{2u}{5\sqrt{3}}\right) = \dfrac{u}{\sqrt{3}}$

fraction of energy lost in collision

$$f = \frac{\frac{1}{2}mu^2 - \left(\frac{1}{2}mv_1^2 + \frac{1}{2}Mv_2^2\right)}{\frac{1}{2}mu^2} = \frac{u^2 - v_1^2 - 5v_2^2}{u^2} = 1 - \frac{1}{3} - \frac{4}{15}$$

$= \dfrac{2}{5}$ Ans.

Web Reference at www.physicsgalaxy.com | Scan for Video Explanation

**Module-30: Solved Example-19**

**Ex :** A ball with a speed 9 m/s strikes another identical ball at rest such that after collision the direction of each ball makes an angle 30° with the original line of motion. Find the speeds of the two balls after collision. Is kinetic energy conserved in this collision.

Sol$^n$:

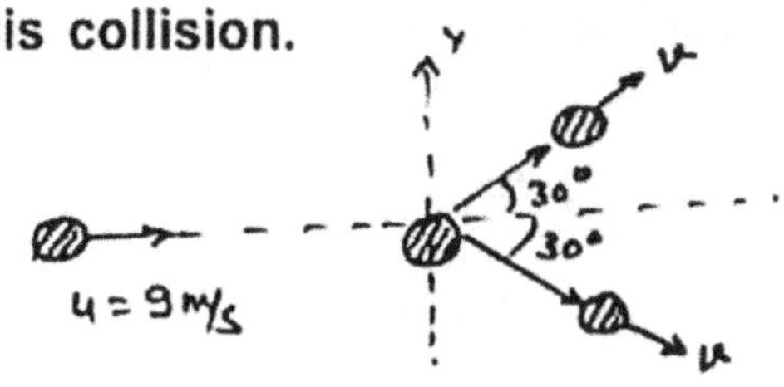


along x-dir $\quad mu = 2mv \cos 30° \Rightarrow v = \dfrac{u}{\sqrt{3}} = 3\sqrt{3}$ m/s Ans.

Divergence angle $\neq \pi/2$ (it is 60°)

This collision is NOT elastic $\Rightarrow$ KE will NOT be conserved. Ans.

Web Reference at www.physicsgalaxy.com | Scan for Video Explanation

**Module-31: Solved Example-20**

**Ex :** **A ball of mass m, moving at speed u along x-axis, strikes another ball of mass 2m kept at rest. First ball comes to rest after collision and the other breaks into two equal parts. One of these starts moving along y-axis with a speed v, what will be the velocity of the other part.**

Sol^n:

Along x-dir:

$mu = mv'\cos\theta$ —(1)

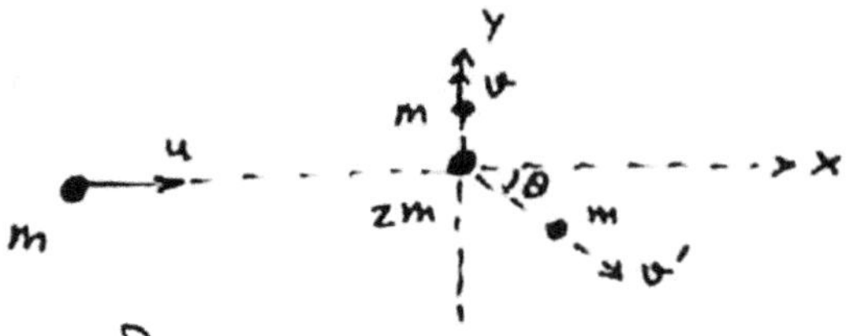


Along Y dir

$mv = mv'\sin\theta$ —(2) $\Big\} \Rightarrow \sqrt{(1)^2+(2)^2}$

$v'^2 = u^2 + v^2$

$v' = \sqrt{u^2+v^2}$ Ans.

Scan for Video Explanation | Web Reference at www.physicsgalaxy.com

**Module-32: Mass Variation Cases**

# Mass Variation Cases:

Whenever mass of a System (or bodies) changes with time during motion Such cases can be handled by Concept of Impulse and Conservation of momentum.

Two cases can be there -

Case-I: When external force is absent.

Case-II: When external force is present

Scan for Video Explanation | Web Reference at www.physicsgalaxy.com

**Module-32: Mass Variation Cases**

# Case-I: When $F_{ext}$ is not present

Hint → In such cases we use Cons. of momentum at an intermediate time of motion from $t$ to $t+dt$

Ex of Rocket Propulsion:

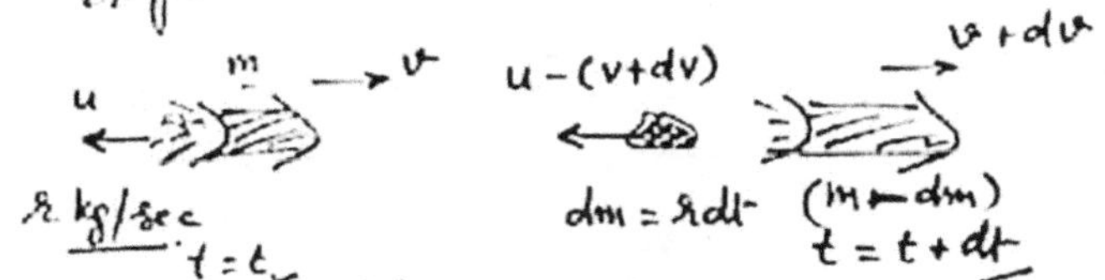


$u \leftarrow$ relative to rocket.

(t) (t+dt)

by Conservation of momentum → $mv = (m-dm)(v+dv) - dm(u-v-dv)$

**Web Reference at www.physicsgalaxy.com** | **Scan for Video Explanation**

**Module-32: Mass Variation Cases**

(t) (t+dt)

$$mv = (m-dm)(v+dv) - dm(u-v-dv)$$

$$mv = mv - dm\,v + m\,dv - dm\,u + dm\,v$$

$$m\frac{dv}{dt} = \frac{dm}{dt}u$$

$$ma = r\,u$$

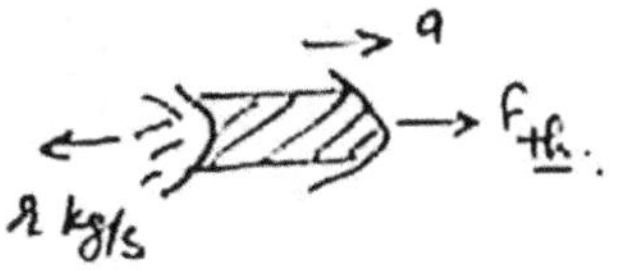


$$\boxed{F_{th} = r\,u}$$

**Web Reference at www.physicsgalaxy.com** | **Scan for Video Explanation**

**Module-32: Mass Variation Cases**

**Case-II:** When $F_{ext}$ is present

→ In such cases we use Impulse-momentum eq$^n$ for the motion of body at $t$ & $t+dt$.

**Ex:** Find vel of tank as time function in the given situation.

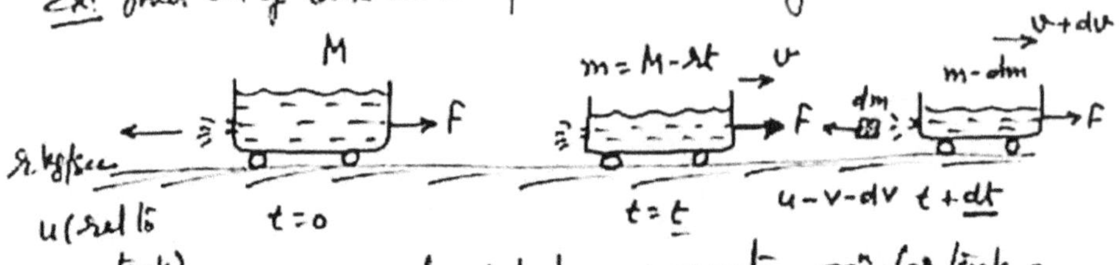


Here we write impulse-momentum eq$^n$ for tank -

$$mv + Fdt = (m - dm)(v + dv) - dm(u - v - dv)$$

$$mv + Fdt = mv - dm\,v + m\,dv - dm\,u + dm\,v$$

$$Fdt = m\,dv - dm\,u$$

Scan for Video Explanation

Web Reference at www.physicsgalaxy.com

**Module-32: Mass Variation Cases**

$$Fdt = m\,dv - dm\,u$$

$$m = M - rt$$
$$dm = r\,dt$$

$$Fdt = (M - rt)dv - ru\,dt$$

$$(F + ru)dt = (M - rt)dv$$

$$\int_0^v dv = \int_0^t (F + ru)\frac{dt}{(M - rt)}$$

$$v = (F + ru)\left[-\frac{1}{r}\ln(M - rt)\right]_0^t$$

$$v = \frac{(F + ru)}{r}\ln\left(\frac{M}{M - rt}\right) \quad \text{Ans.}$$

Scan for Video Explanation

Web Reference at www.physicsgalaxy.com

**Module-32: Mass Variation Cases**

Alternative method

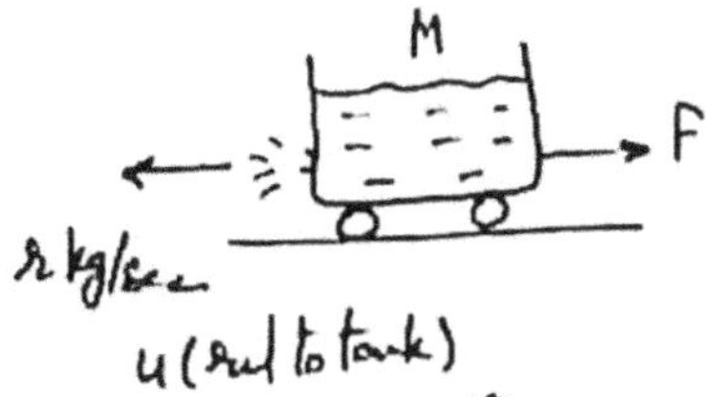


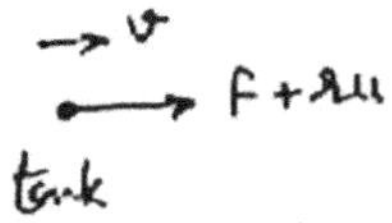


$$F + ru = m\frac{dv}{dt} \Rightarrow \int_0^v dv = \int_0^t \frac{dt}{(M - rt)}(F + ru)$$

$m = M - rt$

$v = - - - -$ ✓

Web Reference at www.physicsgalaxy.com | Scan for Video Explanation

**Module-33: Super Elastic Collision**

# Super elastic Collisions :

Collisions in which $e > 1$ are termed as Super elastic.

When $e = 1 \rightarrow$ no dissipation of energy

$e < 1 \rightarrow$ energy loss

$e = 0 \rightarrow$ max. energy loss.

$e > 1 \Rightarrow$ gain of energy.

$KE_f$ after Collision > $KE_i$ before Collision

$\Rightarrow e > 1$ happens only when in a Collision some energy is released. like the cases of explosion.

$v_{sep} > v_{app}$.

Web Reference at www.physicsgalaxy.com | Scan for Video Explanation

# Chapter 11
# Rigid Body Dynamics

## Lecture Notes Modules


**Module-1:** Rotational Motion
**Module-2:** Moment of Inertia
**Module-3:** Moment of Inertia of a Ring
**Module-4:** Moment of Inertia of a Disc
**Module-5:** Moment of Inertia of a Hollow Sphere
**Module-6:** Moment of Inertia of a Solid Sphere
**Module-7:** Moment of Inertia of a Hollow Cone
**Module-8:** Moment of Inertia of a Solid Cone
**Module-9:** Moment of Inertia of a Thin Rod
**Module-10:** Axes Theorem-Perpendicular Axes Theorem
**Module-11:** Axes Theorem-Parallel Axes Theorem
**Module-12:** Radius of Gyration
**Module-13:** Solved Example-1
**Module-14:** Solved Example-2
**Module-15:** Solved Example-3
**Module-16:** Solved Example-4
**Module-17:** Newton's Second Law in Rotational Motion
**Module-18:** Solved Example-5
**Module-19:** Solved Example-6
**Module-20:** Solved Example-7
**Module-21:** Solved Example-8
**Module-22:** Solved Example-9
**Module-23:** Angular Momentum
**Module-24:** Conservation of Angular Momentum
**Module-25:** Solved Example-10
**Module-26:** Solved Example-11
**Module-27:** Solved Example-12
**Module-28:** Solved Example-13
**Module-29:** Concept of Angular Impulse
**Module-30:** Kinetic Energy of a Rigid body in Rotational Motion
**Module-31:** Rolling Motion and Pure Rolling of a body on Inclined Plane
**Module-32:** Solved Example-14
**Module-33:** Solved Example-15
**Module-34:** Solved Example-16
**Module-35:** Solved Example-17
**Module-36:** Solved Example-18
**Module-37:** Analysis of Rolling with Sliding
**Module-38:** Solved Example-19
**Module-39:** Solved Example-20
**Module-40:** Work & Power in Rotational Motion
**Module-41:** Solved Example-21
**Module-42:** Solved Example-22

**Module-1: Rotational Motion**

# Rotational Motion:

It is the motion of a rigid body about an AOR such that different particles of body move in different circular paths with all the centres lie on the AOR.

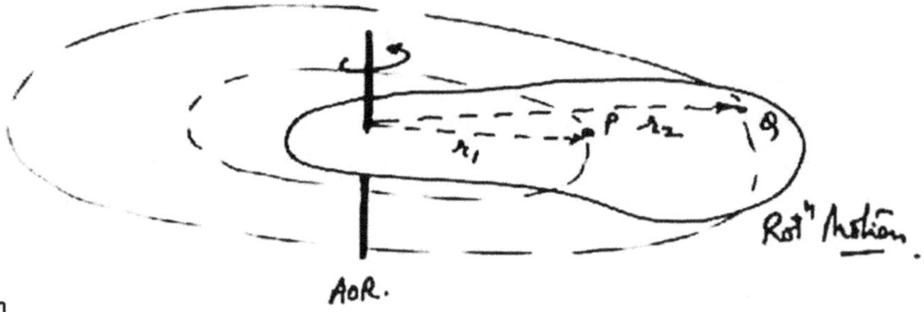


Scan for Video Explanation

Web Reference at www.physicsgalaxy.com

**Module-1: Rotational Motion**

# Various Sections of Rigid Body Dynamics:

→ Moment of Inertia
→ Newton's II law for Rot$^n$ Motion
→ Angular momentum & its Conservation
→ Kinetic Energy in Rot$^n$ Motion
→ Work & Power in Rot$^n$ Motion
→ Rolling Motion & its applications

——x——.

Scan for Video Explanation

Web Reference at www.physicsgalaxy.com

**Module-2: Moment of Inertia**

# Moment of Inertia:

It is the same property analogous to inertia in rot$^n$ motion

Inertia → mass. $m$.

$F = ma$

$a = \frac{F}{m}$ } for a given force $a \propto \frac{1}{m}$.

Similarly.

Moment of Inertia → mass & mass distribution w.r.to AOR.

$\tau = I\alpha$.

$\alpha = \frac{\tau}{I}$

$\alpha \propto \frac{1}{I}$ → change in state of rot$^n$ motion.

**Web Reference at www.physicsgalaxy.com**

**Scan for Video Explanation**

**Module-2: Moment of Inertia**

# Moment of Inertia of a point mass in Circular Motion:

MI of this particle is given as -

$$I = mr^2$$

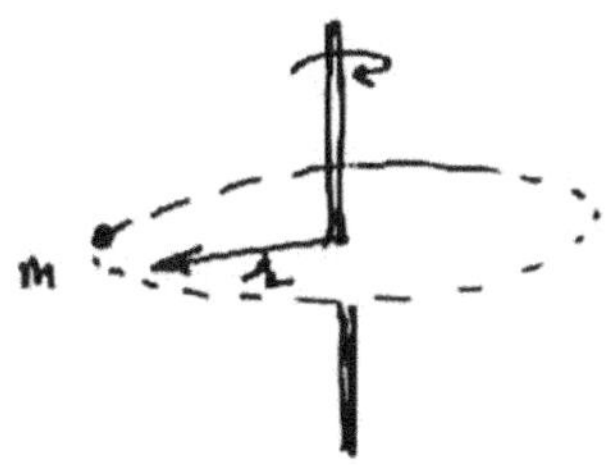


**Web Reference at www.physicsgalaxy.com**

**Scan for Video Explanation**

**Module-2: Moment of Inertia**

# MI of an extended body in Rotational Motion:

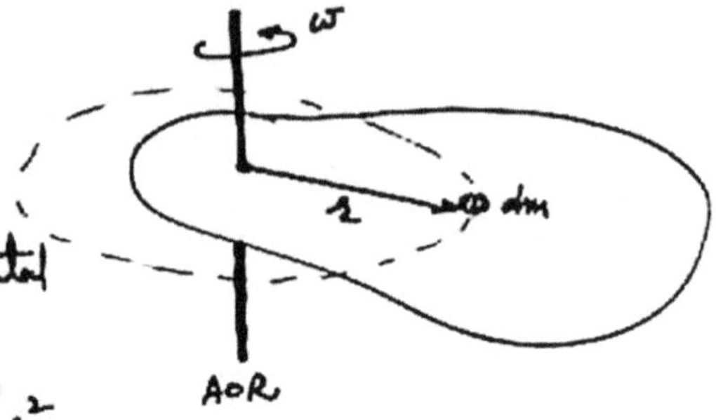


MI of the elemental mass dm is

$$dI = dm\,r^2$$

Total MI of whole body is $\boxed{I = \int dI = \int dm\,r^2}$

Valid for all rigid bodies

NOTE: MI depends on location of AOR for a rotating body.

Scan for Video Explanation

Web Reference at www.physicsgalaxy.com

**Module-3: Moment of Inertia of a Ring**

# Moment of Inertia of a Ring:

MI of elemental mass is

$$dI = dm\,R^2$$

$$I = \int dI = \int dm\,R^2$$

$$I = MR^2$$

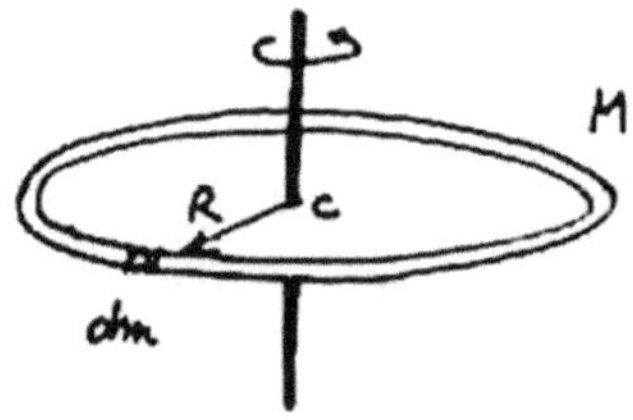


Scan for Video Explanation

Web Reference at www.physicsgalaxy.com

**Module-4: Moment of Inertia of a Disc**

# Moment of Inertia of a Disc:

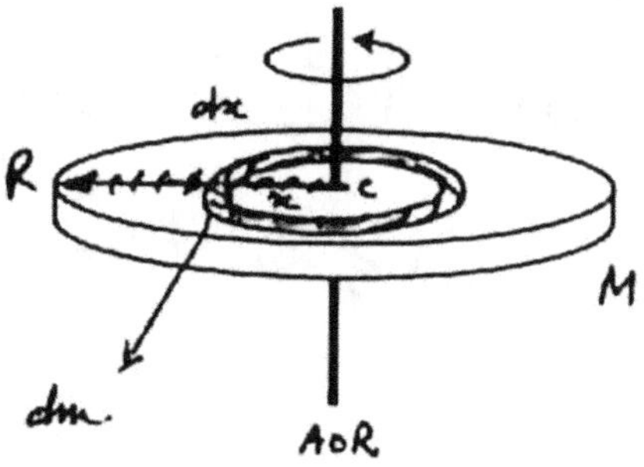


mass of elemental ring is

$$dm = \frac{M}{\pi R^2} \times 2\pi x\,dx$$

$$dm = \frac{M}{R^2} \cdot 2x\,dx$$

MI of the elemental ring $dI = dm\,x^2$

$$dI = \frac{2M}{R^2} \cdot x^3 dx$$

MI of whole disc $I = \int dI = \int_0^R \frac{2M}{R^2} \cdot x^3 dx$

$$= \frac{2M}{R^2}\left[\frac{x^4}{4}\right]_0^R = \frac{2M}{R^2}\left(\frac{R^4}{4} - 0\right)$$

$$\boxed{I = \frac{1}{2}MR^2}$$

Web Reference at www.physicsgalaxy.com | Scan for Video Explanation

**Module-5: Moment of Inertia of a Hollow Sphere**

# Moment of Inertia of a Hollow Sphere:

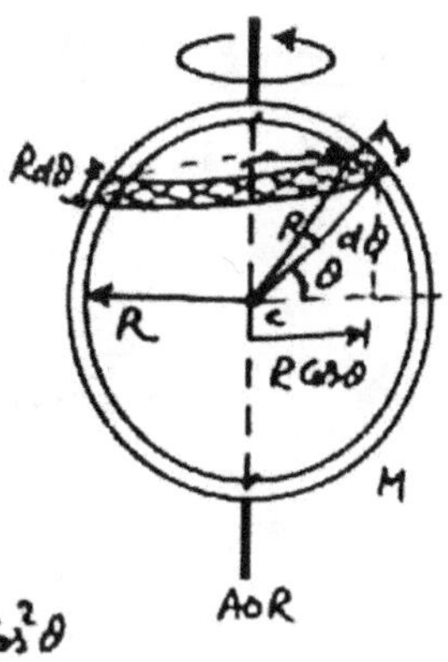


mass of element $dm = \frac{M}{4\pi R^2} \times 2\pi R \cos\theta \cdot R\,d\theta$

$$dm = \frac{1}{2} M \cos\theta\, d\theta$$

MI of the element is $dI = dm (R\cos\theta)^2$

$$dI = \left(\frac{1}{2} M \cos\theta\, d\theta\right) \cdot R^2 \cos^2\theta$$

$$dI = \frac{1}{2} MR^2 \cos^3\theta\, d\theta$$

MI of Hollow sphere is $I = \int dI = \int_{-\pi/2}^{+\pi/2} \frac{1}{2} MR^2 \cos^3\theta\, d\theta$

$$I = \frac{1}{2} mR^2 \int_{-\pi/2}^{+\pi/2} \cos^3\theta\, d\theta = \frac{2}{3} MR^2.$$

$t = \sin\theta$.

Web Reference at www.physicsgalaxy.com | Scan for Video Explanation

**Module-6: Moment of Inertia of a Solid Sphere**

# Moment of Inertia of a Solid Sphere:

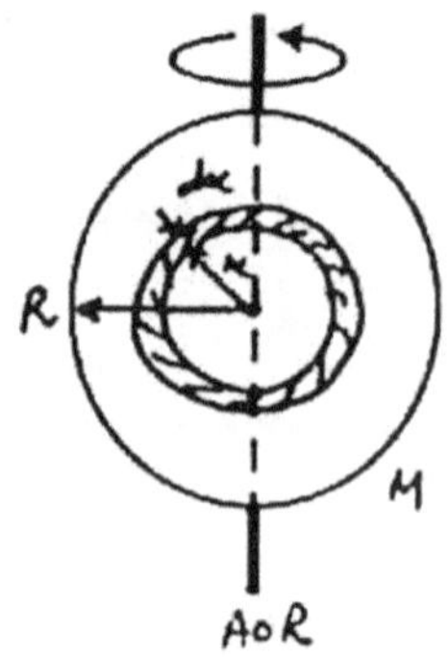


mass of element $dm = \frac{M}{\frac{4}{3}\pi R^3} \times 4\pi x^2 dx$

$$dm = \frac{3M}{R^3} \cdot x^2 dx$$

MI of the element is $dI = \frac{2}{3} dm\, x^2$

$$dI = \frac{2}{3}\left(\frac{3M}{R^3} \cdot x^2 dx\right) x^2$$

$$dI = \frac{2M}{R^3} \cdot x^4 dx$$

MI of Solid Sphere $I = \int dI = \frac{2M}{R^3} \int_0^R x^4 dx$

$$= \frac{2M}{R^3} \cdot \frac{R^5}{5} = \frac{2}{5} MR^2$$

Scan for Video Explanation

Web Reference at www.physicsgalaxy.com

**Module-7: Moment of Inertia of a Hollow Cone**

# Moment of Inertia of a Hollow Cone:

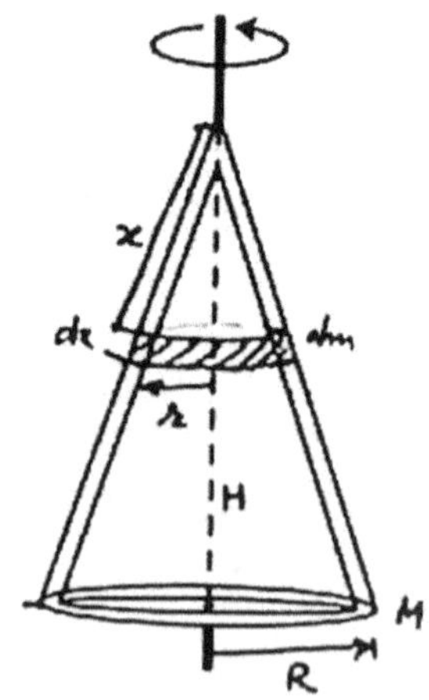


elemental mass $dm = \frac{M}{\pi R\sqrt{R^2+H^2}} \times 2\pi r\, dx$

$$dm = \frac{M}{\pi R\sqrt{R^2+H^2}} \times 2\pi \frac{R}{\sqrt{R^2+H^2}} \cdot x \times dx$$

✓ $dm = \frac{2Mx\, dx}{(R^2+H^2)}$

MI of element is $dI = dm\, r^2$

$$dI = \frac{2Mx}{(R^2+H^2)} dx \cdot \frac{R^2}{(R^2+H^2)} \cdot x^2$$

$$\frac{r}{x} = \frac{R}{\sqrt{R^2+H^2}} \cdot x$$

MI of H. Cone $I = \int dI = \frac{2MR^2}{(R^2+H^2)^2} \int_0^{\sqrt{R^2+H^2}} x^3 dx$

$$= \frac{2MR^2}{(R^2+H^2)^2} \times \frac{(R^2+H^2)^2}{4} = \frac{1}{2} MR^2$$

Scan for Video Explanation

Web Reference at www.physicsgalaxy.com

**Module-8: Moment of Inertia of a Solid Cone**

# Moment of Inertia of a Solid Cone:

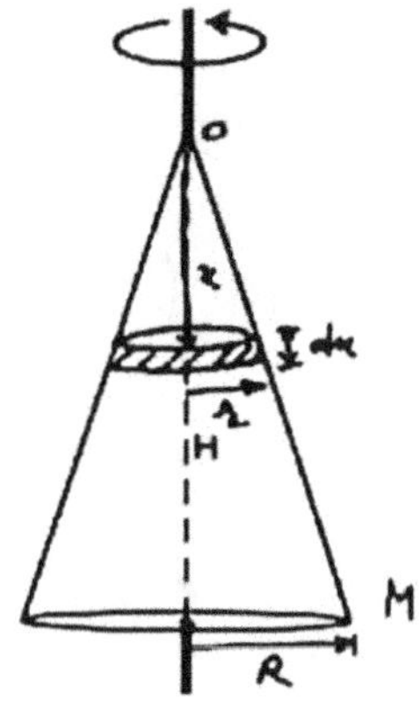


mass of elemental disc

$$dm = \frac{M}{\frac{1}{3}\pi R^2 H} \times \pi r^2 dx$$

$$dm = \frac{3M}{\pi R^2 H} \times \pi\left(\frac{R^2}{H^2}\cdot x^2\right) dx$$

$$dm = \frac{3M}{H^3}\cdot x^2 dx$$

$$\frac{r}{x} = \frac{R}{H}x$$

MI of the elemental disc $dI = \frac{1}{2} dm\, r^2$

$$dI = \frac{1}{2}\left(\frac{3M}{H^3}\cdot x^2 dx\right) \times \frac{R^2}{H^2}\cdot x^2 .$$

$$I = \int dI = \frac{3}{2}\frac{M}{H^5}\cdot R^2 \int_0^H x^4 dx = \frac{3}{2}\frac{M}{H^5}\cdot R^2 \left[\frac{x^5}{5}\right]_0^H$$

$$= \frac{3}{2}\frac{MR^2}{H^5} \times \frac{H^5}{5} = \frac{3}{10}MR^2$$

Web Reference at www.physicsgalaxy.com | Scan for Video Explanation

**Module-9: Moment of Inertia of a Thin Rod**

# Moment of Inertia of a Thin Rod:

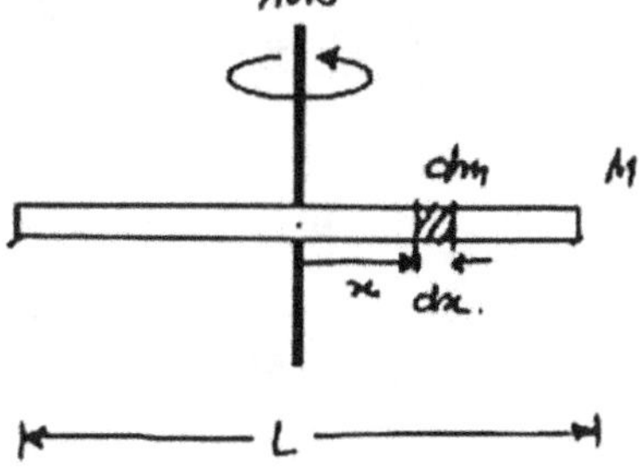


mass of element $dm = \frac{M}{L} dx$

MI of elemental mass is

$$dI = dm\, x^2$$

$$dI = \frac{M}{L} x^2 dx$$

MI of rod $I = \int dI = \int_{-L/2}^{+L/2} \frac{M}{L} x^2 dx = \frac{M}{L}\left[\frac{x^3}{3}\right]_{-L/2}^{+L/2}$

$$= \frac{M}{3L}\left[\frac{L^3}{8} + \frac{L^3}{8}\right]$$

$$= \frac{1}{12} ML^2$$

Web Reference at www.physicsgalaxy.com | Scan for Video Explanation

**Module-9: Moment of Inertia of a Thin Rod**

elemental mass $dm = \frac{M}{L}dx$

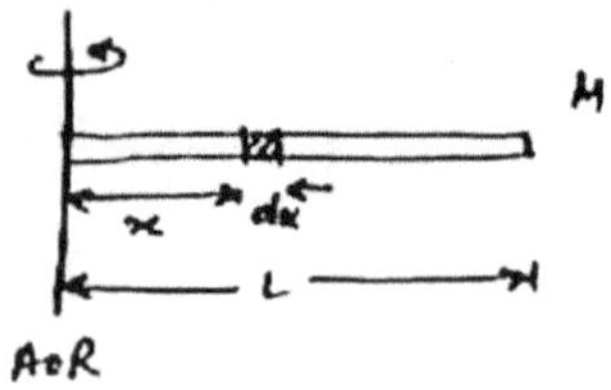


MI of element is

$$dI = dm\,x^2 = \frac{M}{L}x^2 dx$$

MI of rod about side AoR is

$$I = \int dI = \int_0^L \frac{M}{L}x^2 dx$$

$$I = \frac{M}{L}\left[\frac{x^3}{3}\right]_0^L$$

$$\boxed{I = \frac{1}{3}ML^2}$$

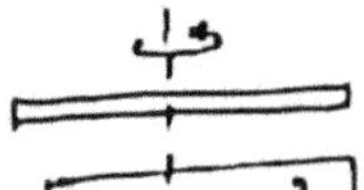

$$\boxed{I = \frac{1}{12}ML^2}$$

Scan for Video Explanation

Web Reference at www.physicsgalaxy.com

**Module-10: Axes Theorem-Perpendicular Axes Theorem**

# Axes Theorems for calculation of Moment of Inertia:

For MI calculation about different AoR, other than standard Central AoR, we use axes theorems.

There are two axes theorems -

(1) Perpendicular Axes Theorem

(2) Parallel Axes Theorem.

Scan for Video Explanation

Web Reference at www.physicsgalaxy.com

**Module-10: Axes Theorem-Perpendicular Axes Theorem**

# Perpendicular Axes Theorem:

This Theorem is applicable only for laminar bodies (2D structure)

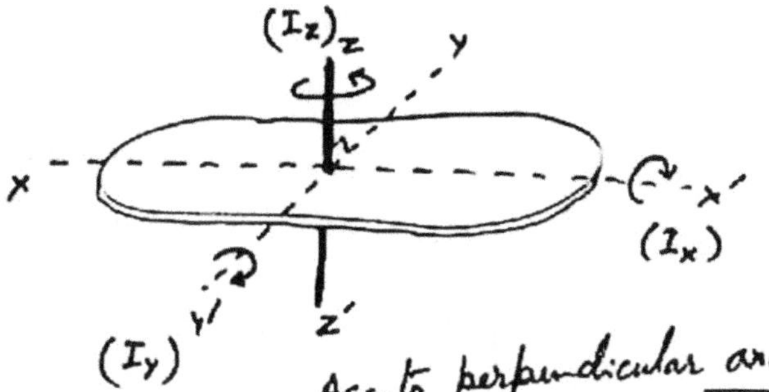


Acc. to perpendicular axes theorem

$$I_z = I_x + I_y$$



**Module-10: Axes Theorem-Perpendicular Axes Theorem**

# Proof of Perpendicular Axes Theorem:

$$I_z = \int dm\, r^2$$

$$I_x = \int dm\, x^2$$

$$I_y = \int dm\, y^2$$

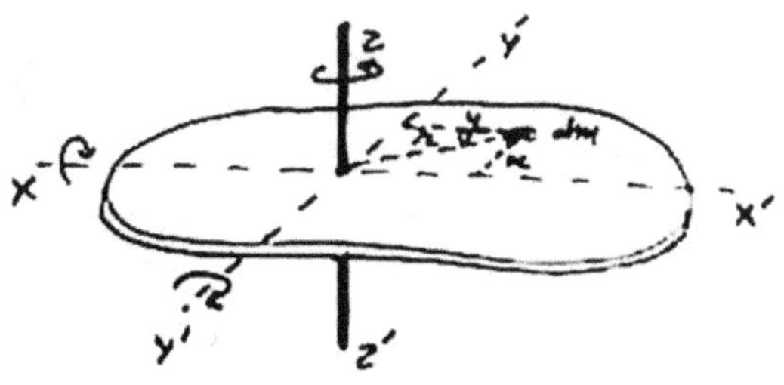


$$x^2 + y^2 = r^2$$

$$I_z = \int dm\, r^2 = \int dm(x^2 + y^2) = \int dm\, x^2 + \int dm\, y^2$$

$$I_z = I_x + I_y$$

**Module-10: Axes Theorem-Perpendicular Axes Theorem**

# Applications of Perpendicular Axes Theorem:

Ring.

$I_c = MR^2$

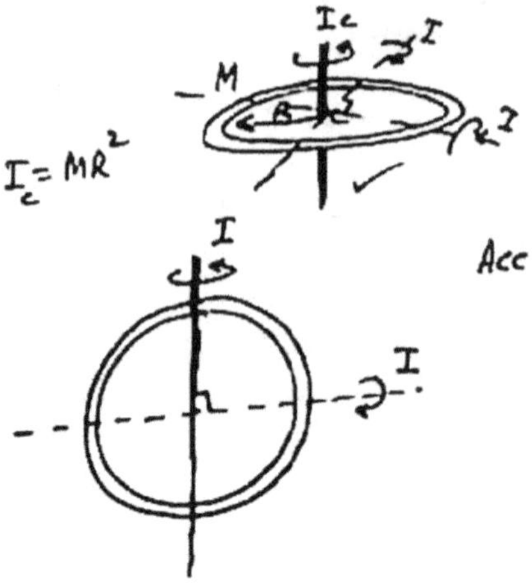


Acc. to ⊥ axes theorem we use

$$I + I = MR^2$$

$$\boxed{I = \frac{1}{2}MR^2}$$

Scan for Video Explanation

Web Reference at www.physicsgalaxy.com

**Module-10: Axes Theorem-Perpendicular Axes Theorem**

Disc.

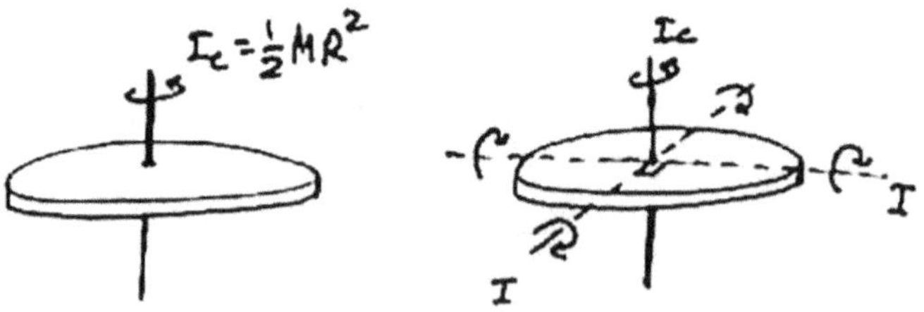


acc. to ⊥ axes theorem

$$I + I = \frac{1}{2}MR^2$$

$$\boxed{I = \frac{1}{4}MR^2}$$

Scan for Video Explanation

Web Reference at www.physicsgalaxy.com

**Module-10: Axes Theorem-Perpendicular Axes Theorem**

Thin rectangular plate:

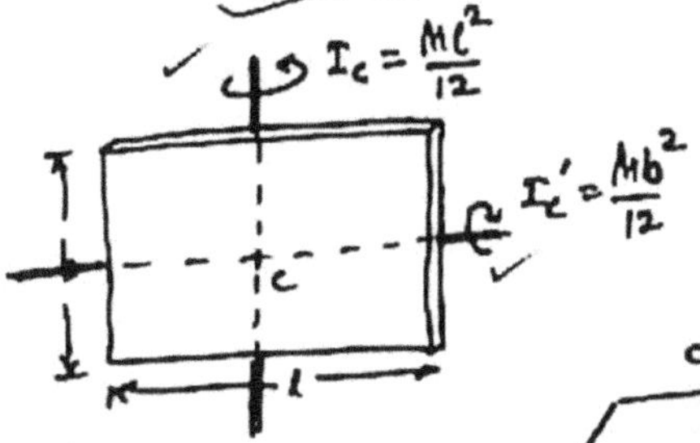


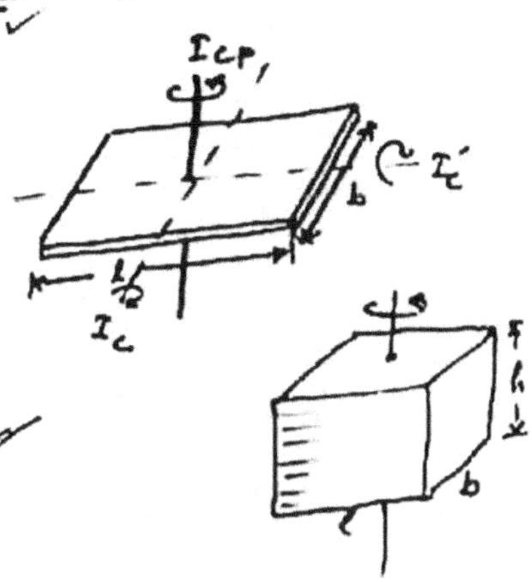


by ⊥ axes th.

$$I_{cp} = I_c + I_c'$$

$$I_{cp} = \frac{M}{12}(l^2 + b^2)$$

Web Reference at www.physicsgalaxy.com | Scan for Video Explanation

**Module-11: Axes Theorem-Parallel Axes Theorem**

# Parallel Axes Theorem:

It is valid for all types of rigid bodies

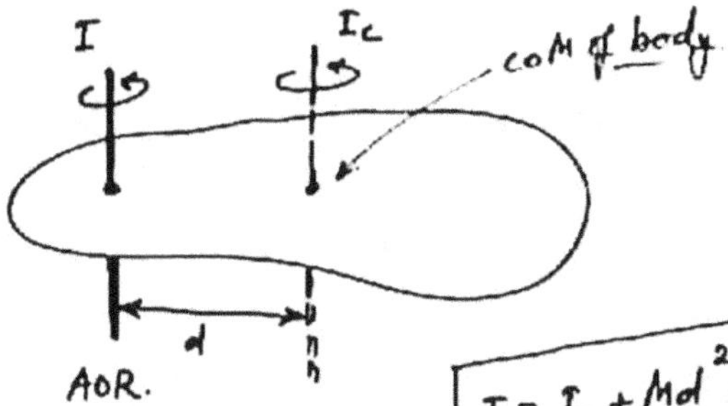


$$I = I_c + Md^2$$

Web Reference at www.physicsgalaxy.com | Scan for Video Explanation

**Module-11: Axes Theorem-Parallel Axes Theorem**

# Proof of Parallel Axes Theorem:

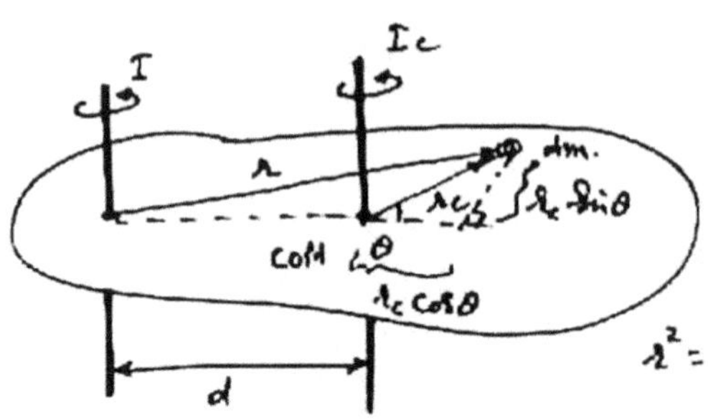


$I_c = \int dm\, r_c^2$

$I = \int dm\, r^2$

$r^2 = (r_c\sin\theta)^2 + (d + r_c\cos\theta)^2$

$r^2 = r_c^2 + d^2 + 2dr_c\cos\theta$

$I = \int dm\, r^2 = \int dm\, r_c^2 + \int dm.d^2 + \int dm(2dr_c\cos\theta)$

$I = I_c + Md^2 + 2d\int dm\, r_c\cos\theta$

$$\boxed{I = I_c + Md^2}$$

Scan for Video Explanation

Web Reference at www.physicsgalaxy.com

**Module-11: Axes Theorem-Parallel Axes Theorem**

# Applications of Parallel Axes Theorem:

Ring:

$I_c = MR^2$

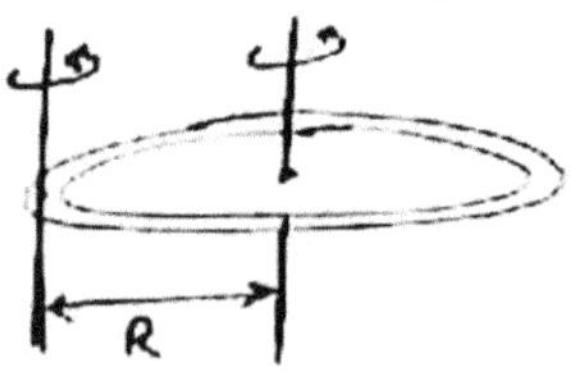


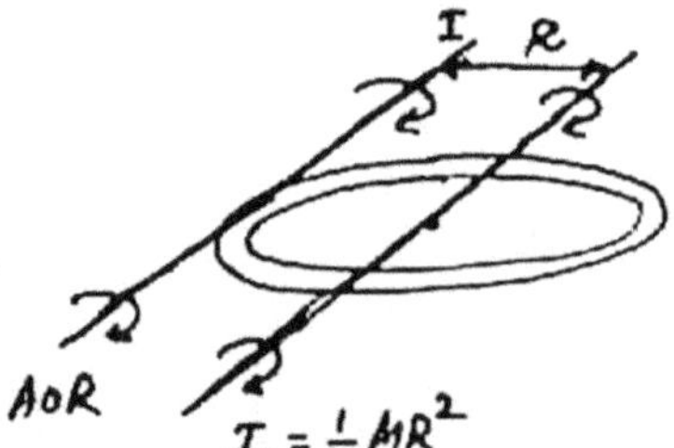


$I_c = \frac{1}{2}MR^2$

$I = I_c + Md^2$

$I = MR^2 + MR^2$

$$\boxed{I = 2MR^2}$$

$I = I_c + Md^2$

$I = \frac{1}{2}MR^2 + MR^2$

$$\boxed{I = \frac{3}{2}MR^2}$$

Scan for Video Explanation

Web Reference at www.physicsgalaxy.com

**Module-11: Axes Theorem-Parallel Axes Theorem**

→ Disc.

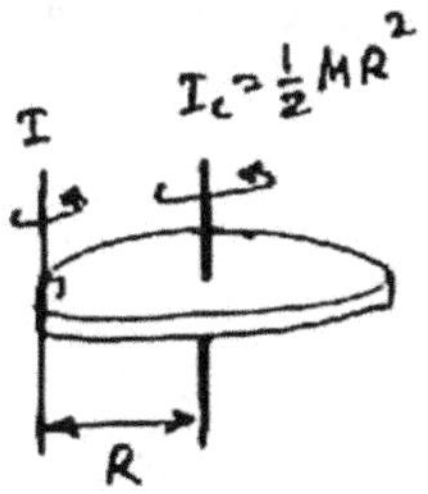


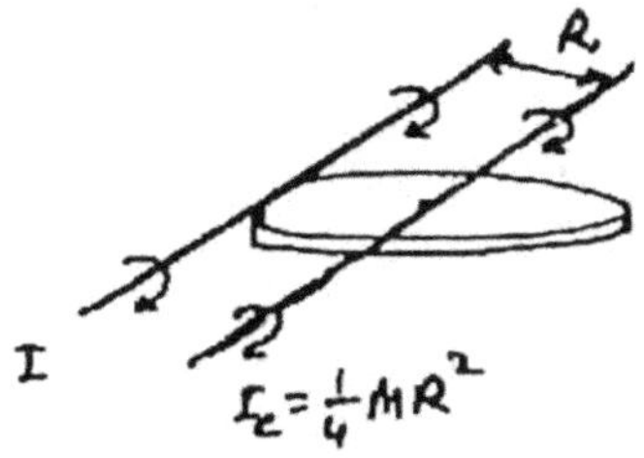


$$I = I_c + Md^2$$
$$I = \frac{1}{2}MR^2 + MR^2$$
$$\boxed{I = \frac{3}{2}MR^2}$$

$$I = I_c + Md^2$$
$$I = \frac{1}{4}MR^2 + MR^2$$
$$\boxed{I = \frac{5}{4}MR^2}$$



**Module-11: Axes Theorem-Parallel Axes Theorem**

→ MI of a hollow & Solid sphere abt tangential AoR.

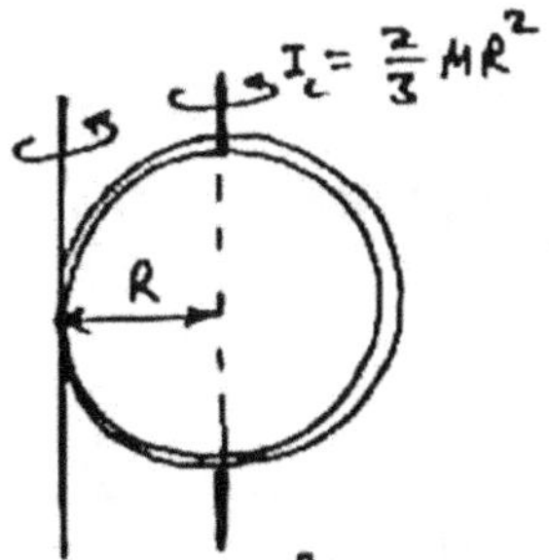


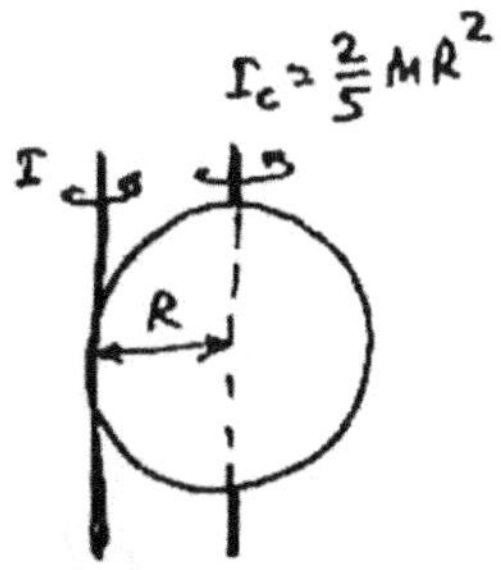


$$I = I_c + Md^2$$
$$I = \frac{2}{3}MR^2 + MR^2$$
$$\boxed{I = \frac{5}{3}MR^2}$$

$$I = I_c + Md^2$$
$$I = \frac{2}{5}MR^2 + MR^2$$
$$\boxed{I = \frac{7}{5}MR^2}$$

**Module-12: Radius of Gyration**

# Radius of Gyration:

It is the eq. radius of cir. motion for a rigid body in rotation.

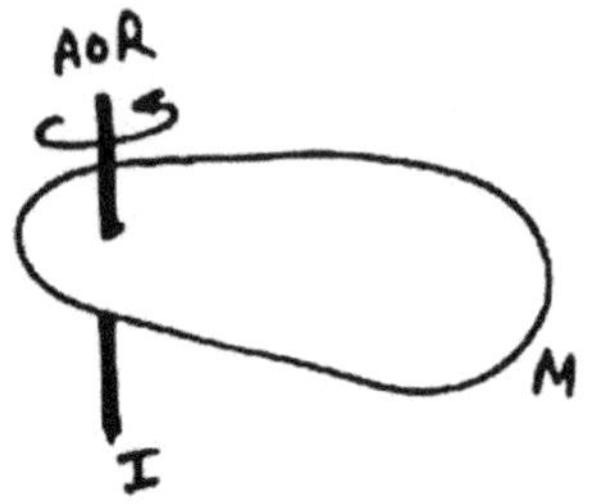


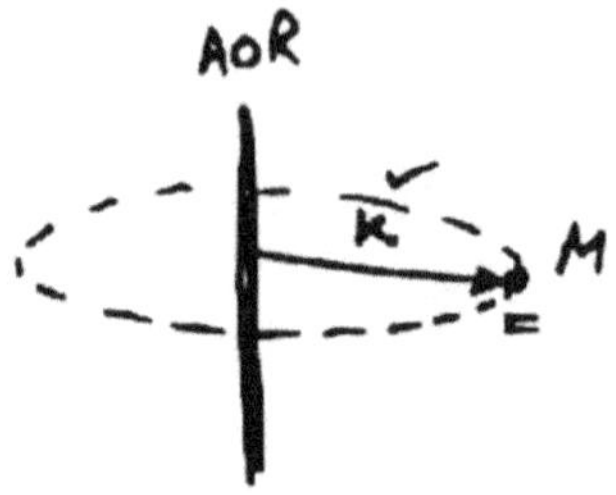


$$I = MK^2$$

$$K = \sqrt{\frac{I}{M}}$$

Scan for Video Explanation

Web Reference at www.physicsgalaxy.com

**Module-12: Radius of Gyration**

for a Ring abt Central AOR $\quad K = R \quad\quad I = MK^2$

abt Diametrical AOR $\quad K = \frac{R}{\sqrt{2}}$

for a sphere abt Central AOR $\quad I = \frac{2}{5}MR^2 = MK^2$

$$K = \sqrt{\frac{2}{5}}R.$$

for a Solid Cone abt Central AOR $\quad I = \frac{3}{10}MR^2 = MK^2$

$$K = \sqrt{\frac{3}{10}}R.$$

Scan for Video Explanation

Web Reference at www.physicsgalaxy.com

**Module-13: Solved Example-1**

Ex : Find moment of inertia of a spherical ball of mass m, radius r with its centre attached at the end of a thin straight rod of mass M and length L about axis passing through the other end of rod is perpendicular to it.

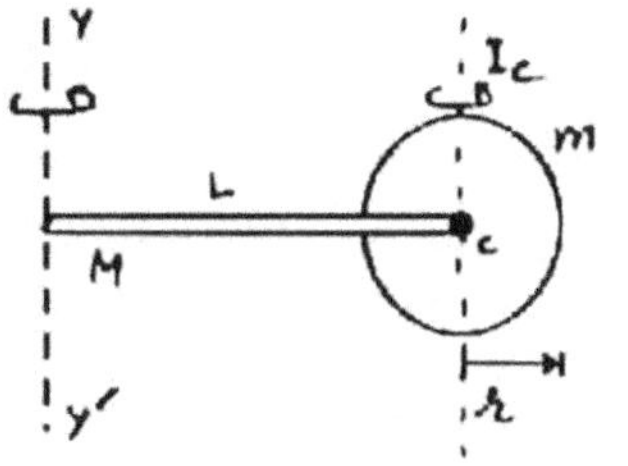


Sol$^n$:

$$I_{Rod} = \frac{ML^2}{3}$$

$$I_{sph_{YY'}} = I_c + mL^2 = \frac{2}{5}mr^2 + mL^2$$

$$I_{Tot_{YY'}} = I_{Rod} + I_{sph_{YY'}} = \frac{ML^2}{3} + \frac{2}{5}mr^2 + mL^2 \quad \text{Ans.}$$

Web Reference at www.physicsgalaxy.com | Scan for Video Explanation

**Module-14: Solved Example-2**

Ex : Find moment of inertia of a half disc of radius $R_2$ from which a smaller half disc of radius $R_1$ is cut as shown in figure, about an axis of rotation passing through centre & perpendicular to plane of disc. Mass of this disc is M.

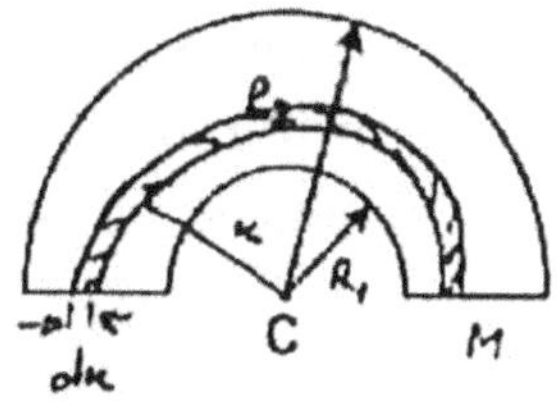


Sol$^n$: mass of elemental half ring

$$dm = \frac{2M}{\pi(R_2^2 - R_1^2)} \times \pi x \times dx$$

MI of element $\quad dI = dm\,x^2 = \frac{2M}{R_2^2 - R_1^2} \times x^3 dx.$

MI of given object $\quad I = \int dI = \frac{2M}{R_2^2 - R_1^2} \cdot \int_{R_1}^{R_2} x^3 dx$

$$= \frac{2M}{R_2^2 - R_1^2} \cdot \frac{1}{4}(R_2^4 - R_1^4)$$

$$= \frac{1}{2}M(R_2^2 + R_1^2) \quad \text{Ans.}$$

Web Reference at www.physicsgalaxy.com | Scan for Video Explanation

**Module-15: Solved Example-3**

**Ex :** Find the moment of inertia of a cylinder of mass M, radius R and length L about an axis passing through its centre and perpendicular to its axis.

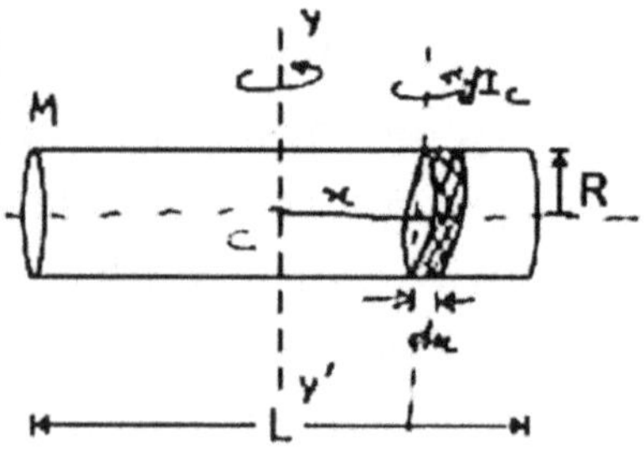


**Solⁿ:** Mass of elemental disc is

$$dm = \frac{M}{L}dx$$

MI of disc about AOR YY'.

$$dI = dI_c + dm\,x^2$$

$$dI = \frac{1}{4}dm R^2 + dm\,x^2$$

Total MI of cylinder

$$I = \int dI = \frac{1}{4}\int_{-L/2}^{+L/2} \frac{M}{L}R^2 dx + \int_{-L/2}^{+L/2} \frac{M}{L}x^2 dx$$

$$I = \frac{1}{4}\frac{MR^2}{L}[x]_{-L/2}^{+L/2} + \frac{M}{L}\left[\frac{x^3}{3}\right]_{-L/2}^{+L/2}$$

$$I = \frac{1}{4}MR^2 + \frac{1}{12}ML^2 \text{ Ans.}$$

Scan for Video Explanation

Web Reference at www.physicsgalaxy.com

**Module-16: Solved Example-4**

**Ex :** Find moment of inertia of the triangular lamina of mass M about the axis of rotation AB as shown in figure.

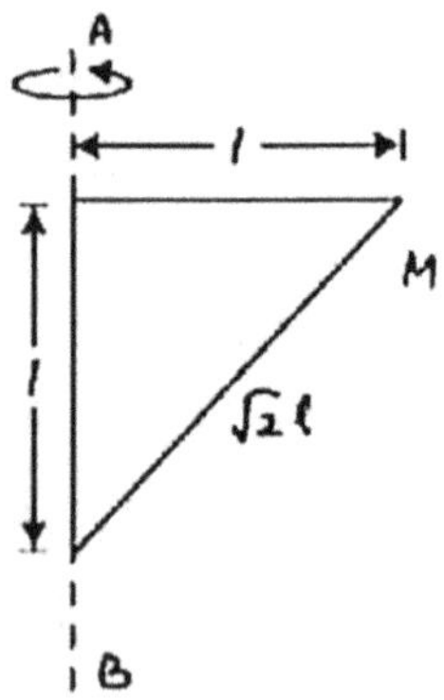


**Solⁿ:**

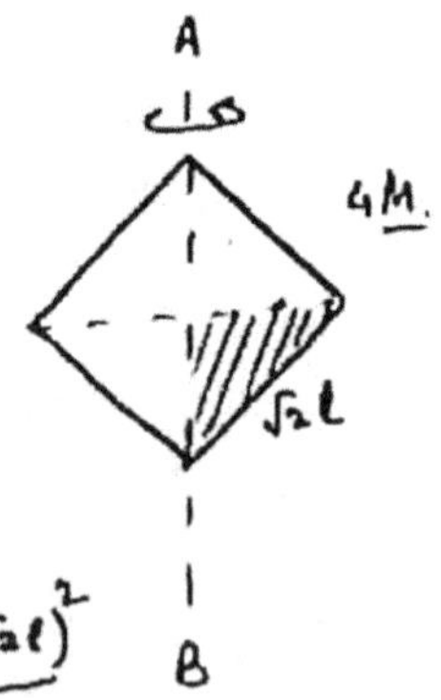


MI of this sq. is –

$$I = \frac{(4M)(\sqrt{2}l)^2}{12} = \frac{2}{3}Ml^2$$

$$I_{each} = \frac{1}{4}I$$

$$I = \frac{\frac{2}{3}Ml^2}{4} = \frac{1}{6}Ml^2 \text{ Ans.}$$

Scan for Video Explanation

Web Reference at www.physicsgalaxy.com

**Module-17: Newton's Second Law in Rotational Motion**

# Newton's Second Law in Rotational Motion:

As we know in Tx motion

$$\Sigma \vec{F} = m\vec{a}$$

Similarly in Rx motion

$$\Sigma \vec{\tau} = I\vec{\alpha}$$

$F_1 > F_2$

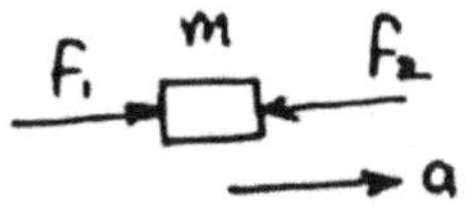


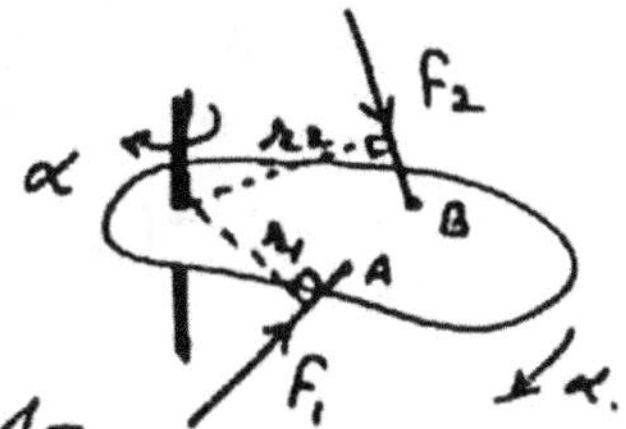


$$F_1 - F_2 = ma.$$

eqn of motion $F_2 r_2 - F_1 r_1 = I\alpha.$

Web Reference at www.physicsgalaxy.com | Scan for Video Explanation

**Module-17: Newton's Second Law in Rotational Motion**

**Ex :** A uniform rod of length L & mass M which is hinged at topmost point is released from a position where it makes an angle θ with vertical as shown. Find the acceleration of the tip of rod just after it is released.

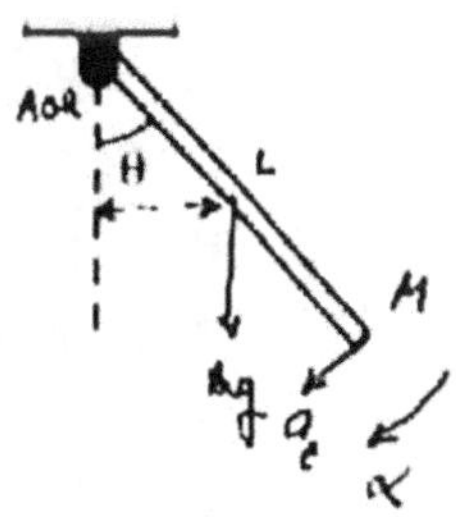


Soln: Torque on rod just after release

$$\tau = Mg \times \frac{L}{2}\sin\theta$$

ang accn $\alpha = \frac{\tau}{I} = \frac{Mg\frac{L}{2}\sin\theta}{\frac{ML^2}{3}} = \frac{3g}{2L}\sin\theta$

$$a_t = L\alpha = \frac{3g}{2}\sin\theta \text{ Ans.}$$

Web Reference at www.physicsgalaxy.com | Scan for Video Explanation

**Module-17: Newton's Second Law in Rotational Motion**

# Analysis of Atwood's Machine with friction between string & pulley:

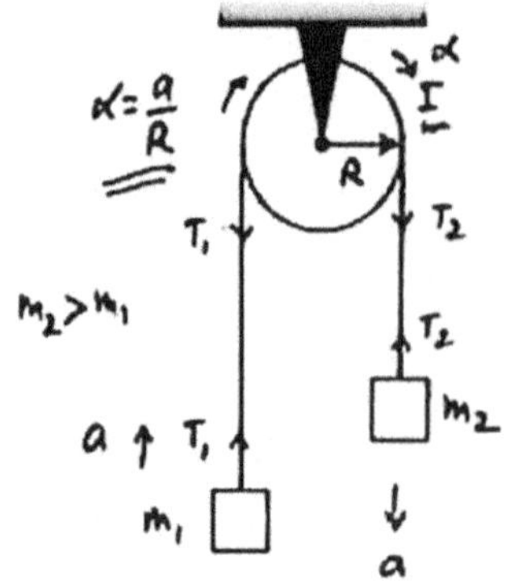


eq$^n$ of motion

$$m_2 g - T_2 = m_2 a \quad -(1)$$

$$T_1 - m_1 g = m_1 a \quad -(2)$$

$$T_2 R - T_1 R = I\left(\frac{a}{R^2}\right) \quad -(3)$$

$$(m_2 - m_1) g = \left(m_1 + m_2 + \frac{I}{R^2}\right) a$$

$$a = \frac{(m_2 - m_1)}{\left(m_1 + m_2 + I/R^2\right)} g.$$

Scan for Video Explanation | Web Reference at www.physicsgalaxy.com

**Module-18: Solved Example-5**

**Ex :** A cylinder of mass m suspended by two strings wrapped around the cylinder, one near each end, the free ends of the strings are attached to ceiling as shown. From rest cylinder is allowed to roll down under gravity. As strings unwind, find acceleration of falling cylinder & tension in strings.

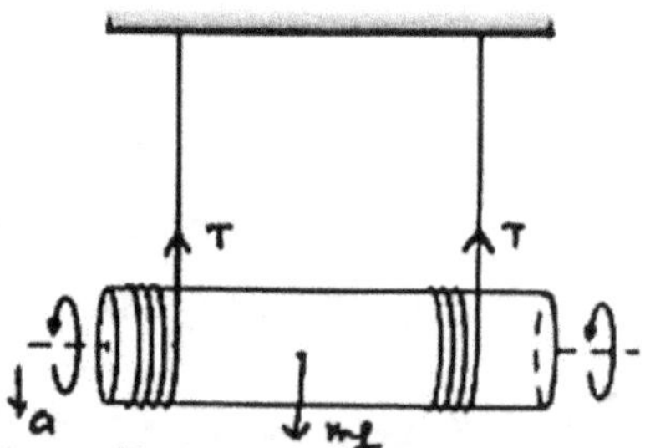


Sol$^n$:

$$mg - 2T = ma \quad -(1)$$

$$2TR = \frac{1}{2} m R^2 \left(\frac{a}{R}\right) \quad -(2)$$

$$mg = \frac{3}{2} m a$$

$$a = \frac{2g}{3} \text{ Ans.}$$

$$mg - 2T = \frac{2mg}{3}$$

$$2T = \frac{mg}{3}$$

$$T = \frac{1}{6} mg \text{ Ans.}$$

Scan for Video Explanation | Web Reference at www.physicsgalaxy.com

**Module-19: Solved Example-6**

**Ex :** A uniform disc of radius R is first spinned about its axis to the angular speed ω and then carefully placed with its flat face on a rough horizontal surface. How long will the disc be rotating on the surface if the friction coefficient is equal to μ ?

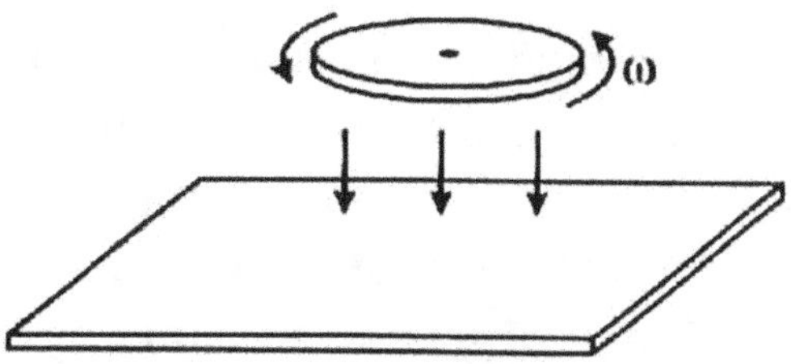


Sol$^n$: mass of elemental ring $dm = \frac{M}{\pi R^2}\times 2\pi x\,dx$

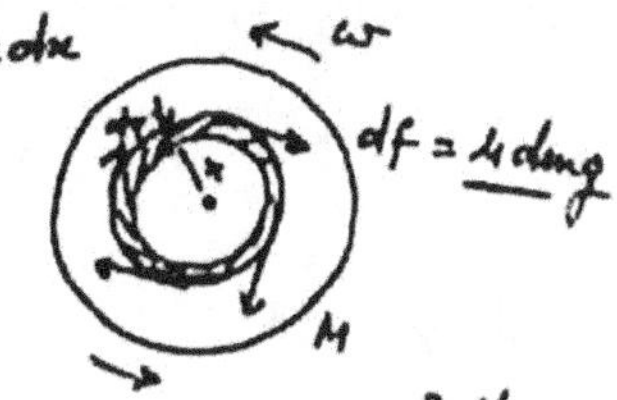


fr. torque due to ring is $d\tau = df\cdot x$

Total torque $\tau = \int d\tau = \int_0^R \frac{M}{R^2}\cdot 2x\,dx\times x\cdot\mu g$

$\tau = \frac{2M\mu g}{R^2}\int_0^R x^2\,dx = \frac{2M\mu g}{R^2}\frac{R^3}{3} = \frac{2}{3}M\mu gR \Rightarrow \alpha = \frac{\tau}{I} = \frac{\frac{2}{3}M\mu gR}{\frac{1}{2}MR^2}$

$\alpha = \frac{4}{3}\frac{\mu g}{R}$.

Speed eq$^n$ $\omega_f = \omega_i + \alpha t \Rightarrow t = \frac{\omega}{\alpha} = \frac{3\omega R}{4\mu g}$ Ans.



**Module-20: Solved Example-7**

**Ex :** A cylinder B of mass m and radius R is mounted on a massless mount A as shown. On cylinder an ideal string is wrapped, other end of which connected to a block C which hangs over an ideal pulley as shown. When the system is set free, find the acceleration of block C.

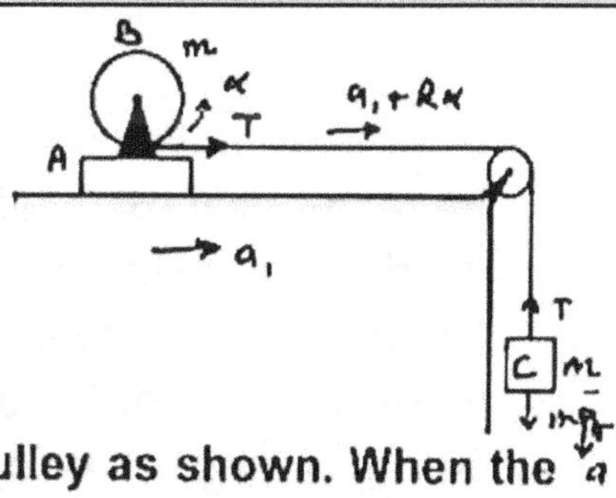


Sol$^n$: $a = a_1 + R\alpha$. —(1)

eq$^n$ of motion for C. $mg - T = ma$. —(2)

for cylinder $T = ma_1$, $TR = \frac{1}{2}mR^2\alpha$. $\rightarrow a_1 = \frac{1}{2}\frac{mR\alpha}{m} = \frac{R\alpha}{2}$

$mg - \frac{mR\alpha}{2} = \frac{3}{2}mR\alpha$

$a = \frac{1}{2}R\alpha + R\alpha = \frac{3}{2}R\alpha$.

$mg = 2mR\alpha$

$a = \frac{3}{2}R\left(\frac{g}{2R}\right) = \frac{3g}{4}$ Ans. $\alpha = \frac{g}{2R}$.

**Module-21: Solved Example-8**

**Ex :** A cylinder of mass M and radius R is hinged from its axis of rotation O from ceiling as shown. An ideal string is wrapped over its surface and other end of which is tied to a block of mass m. If this system is released from rest, find the acceleration of falling mass m.

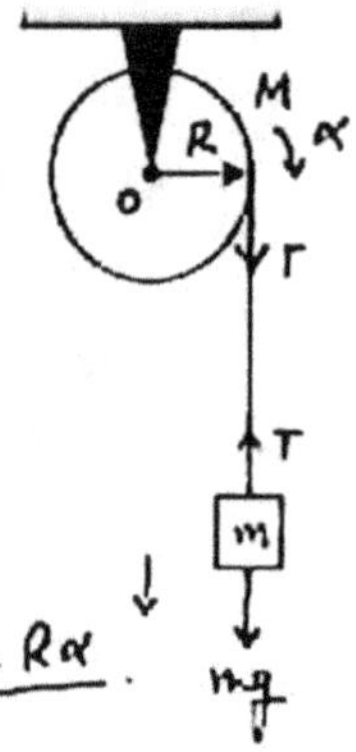


Sol$^n$: eq$^n$ of motion for falling mass $\quad a = R\alpha$

$$mg - T = ma \quad \text{—(1)}$$

for cylinder

$$TR = \left(\tfrac{1}{2}MR^2\right)\left(a/R\right) \quad \text{—(2)}$$

$$mg = \left(m + \frac{M}{2}\right)a$$

$$a = \left(\frac{2m}{2m+M}\right)g \quad \text{Ans.}$$

Scan for Video Explanation

Web Reference at www.physicsgalaxy.com

**Module-22: Solved Example-9**

**Ex :** A solid cylinder A of mass $m_1$ can freely rotate about a horizontal axis fixed to a mount B of mass $m_2$. A constant horizontal force is applied to end P of a light string wound on the cylinder. Neglect friction anywhere, find the acceleration of point P.

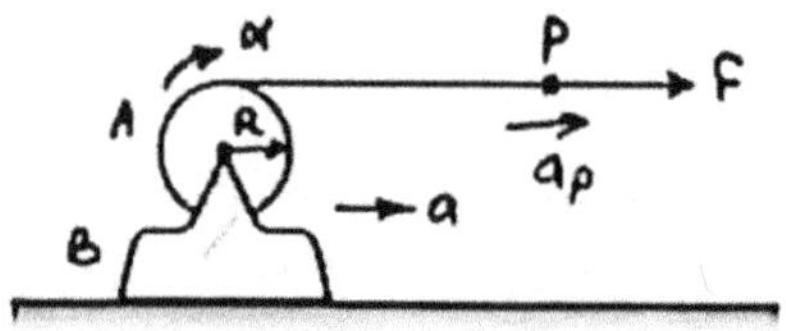


Sol$^n$: $\quad a_p = a + R\alpha$

$$F = (m_1 + m_2)a \longrightarrow a = \frac{F}{m_1 + m_2}$$

$$FR = \left(\tfrac{1}{2}m_1R^2\right)\alpha \longrightarrow \alpha = \frac{2F}{m_1R}$$

$$a_p = a + R\alpha = \frac{F}{m_1+m_2} + \frac{2F}{m_1} = \frac{F(3m_1 + 2m_2)}{m_1(m_1+m_2)} \quad \text{Ans.}$$

Scan for Video Explanation

Web Reference at www.physicsgalaxy.com

**Module-23: Angular Momentum**

# Angular Momentum:

Trans. Motion → linear momentum ($m\vec{v}$)

Rot$^n$ Motion → Angular Momentum

↓

moment of linear momentum.

↓

The product of linear momentum with the shortest distance of line of motion from the pt about which angular momentum is taken.



**Module-23: Angular Momentum**

# Different Cases for calculation of Angular Momentum:

Case-I: Ang. momentum of a particle moving in a st. line

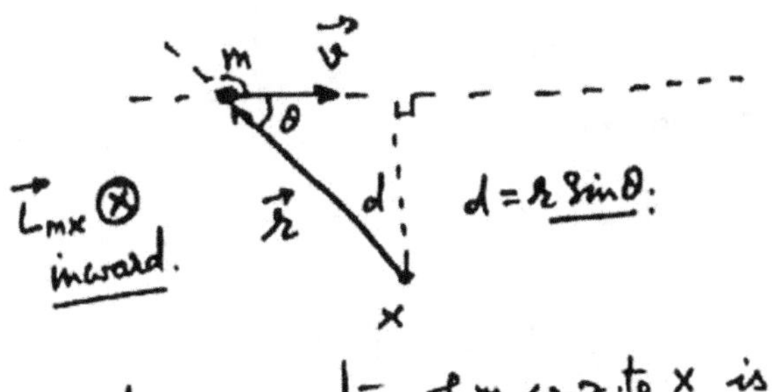


Ang momentum of m w.r.to x is $L_{mx} = m\vec{v}\times d$

$L_{mx} = mv \times r\sin\theta$

$\vec{L}_{mx} = m(\vec{r}\times\vec{v})$

Web Reference at www.physicsgalaxy.com | Scan for Video Explanation

**Module-23: Angular Momentum**

## Case-II: Angular Momentum of a particle in Circular Motion

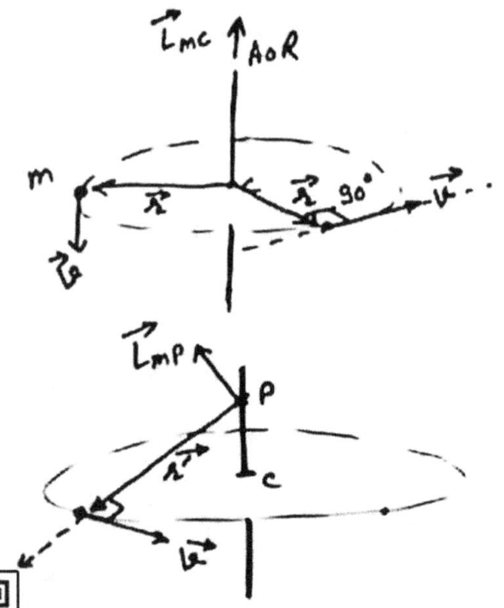


$$\vec{L}_{mc} = m(\vec{r}\times\vec{v})$$

$$L_{mc} = mvr.$$

$$\vec{L}_{mp} = m(\vec{r}'\times\vec{v})$$

$$L_{mp} = mvr'.$$

Scan for Video Explanation

Web Reference at www.physicsgalaxy.com

**Module-23: Angular Momentum**

## Case-III: Angular Momentum of a Rigid Body in Rotation

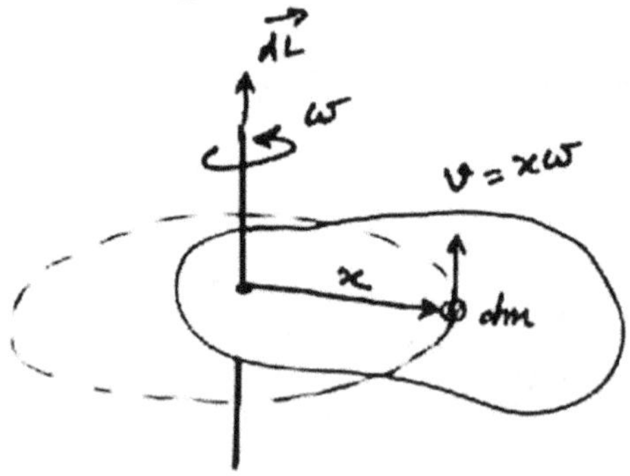

$$L = mvr.$$

Ang momentum of dm about given AoR is

$$dL = dm\, v x = dm(x\omega)x$$

$$dL = dm\, x^2 . \omega$$

for whole body $L = \int dL = \int dm\, x^2 . \omega$

$$\boxed{\vec{L} = I\vec{\omega}}$$

Scan for Video Explanation

Web Reference at www.physicsgalaxy.com

**Module-23: Angular Momentum**

Case-IV: Angular Momentum of a Rotating Body about an AoR passing through CoM & moving translationally.

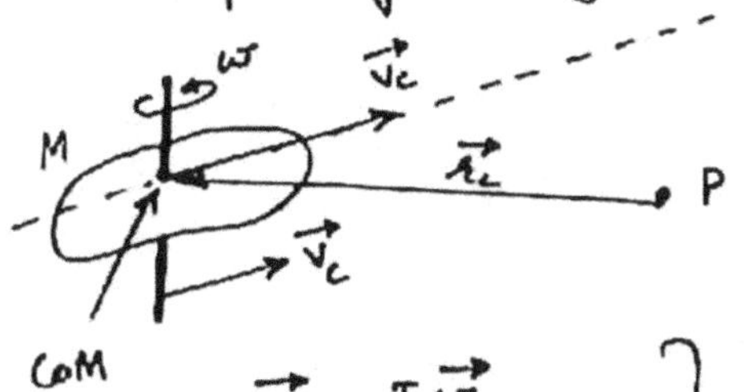


$$\vec{L}_c = I\vec{\omega}$$

$$\vec{L}_T = M(\vec{r}_c \times \vec{v}_c)$$

Ang. momentum of this body w.r.to pt P is

$$\vec{L}_{MP} = \vec{L}_c + \vec{L}_T$$

imp. ✓✓ $$\boxed{\vec{L}_{MP} = I\vec{\omega} + M(\vec{r}_c \times \vec{v}_c)}$$

Web Reference at www.physicsgalaxy.com | Scan for Video Explanation

**Module-24: Conservation of Angular Momentum**

# Law of Conservation of Angular Momentum:

For a rotating rigid body.

$$L = I\omega$$

diff. w.r.to t.

$$\frac{dL}{dt} = I\frac{d\omega}{dt} = I\alpha \leftarrow \tau$$

$$\tau = \frac{dL}{dt} \quad \longleftarrow \quad F = \frac{dP}{dt}$$

Newton's II law for rot^n motion

"Rate of change of angular momentum of a rotating body is equal to net torque acting on it."

Web Reference at www.physicsgalaxy.com | Scan for Video Explanation

**Module-24: Conservation of Angular Momentum**

If for a system of bodies

$$\vec{\tau}_{ext} = \underline{\underline{0}}.$$

$$\Rightarrow \frac{dL}{dt} = 0 \Rightarrow L = \text{Constant}$$

In absence of external torque on a system,
total angular momentum of the system
will remain conserved.

Scan for Video Explanation

Web Reference at www.physicsgalaxy.com

**Module-24: Conservation of Angular Momentum**

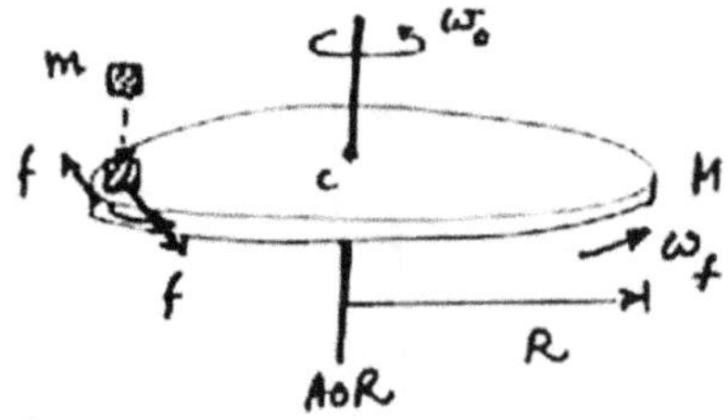


As no ext torque is
present here
$L$ = Conserved.

$$\overset{L_i}{\left(\tfrac{1}{2}MR^2\right)\omega_0} = \overset{L_f}{\left(\tfrac{1}{2}MR^2 + mR^2\right)\omega_f}$$

$$\omega_f = \left(\frac{M\omega_0}{M+2m}\right)$$

Scan for Video Explanation

Web Reference at www.physicsgalaxy.com

**Module-25: Solved Example-10**

**Ex :** **A rod AB of length L and mass M is hinged at the topmost point A as shown & it is free to rotate about this point. If ball of mass M/2, moving at speed u strikes the other end of rod and sticks to it, find the angular speed with which rod will move after collision.**

Sol$^n$:

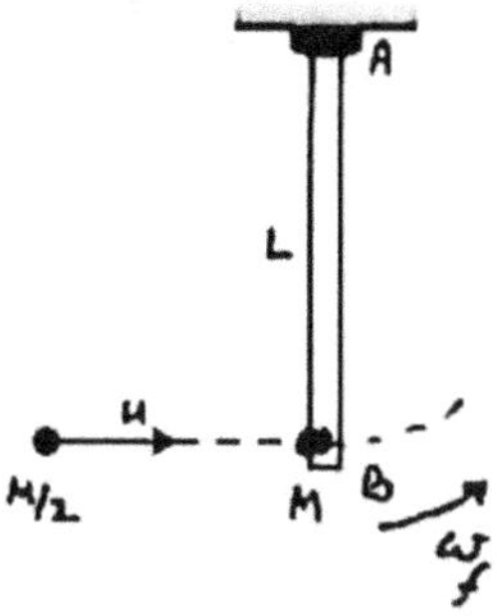


No ext torque is there about A

$$L_i = L_f$$

$$\frac{M}{2} u L = \left(\frac{M}{2}L^2 + \frac{ML^2}{3}\right)\omega_f$$

$$\omega_f = \frac{u/2}{1/2 + 1/3} = \frac{3u}{5L} \text{ Ans.}$$

Web Reference at www.physicsgalaxy.com | Scan for Video Explanation

**Module-26: Solved Example-11**

**Ex :** A child of mass m is standing on periphery of a circular platform of radius R, which can rotate about its central axis. The moment of inertia of platform is I. The child jumps off the platform with a velocity u relative to platform and tangential to it. Find the recoil angular speed of platform after jump.

Sol$^n$: As no ext. torque is present about C

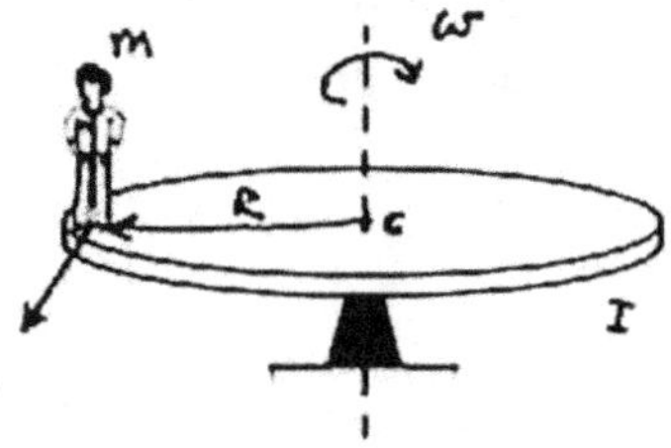


$$L_i = L_f$$

$$0 = m(u - R\omega)R - I\omega$$

$(u - R\omega)$

$$muR = (I + mR^2)\omega$$

$$\omega = \left(\frac{muR}{I + mR^2}\right) \text{ Ans.}$$

Web Reference at www.physicsgalaxy.com | Scan for Video Explanation

**Module-27: Solved Example-12**

**Ex :** **A horizontal circular platform of mass M, radius R and moment of inertia I which is free to rotate about a vertical axis passing through its centre, is rotating at an angular speed $\omega_1$. A boy starts walking from rest long its periphery in opposite direction with speed u relative to platform, what will be the new angular speed of platform.**

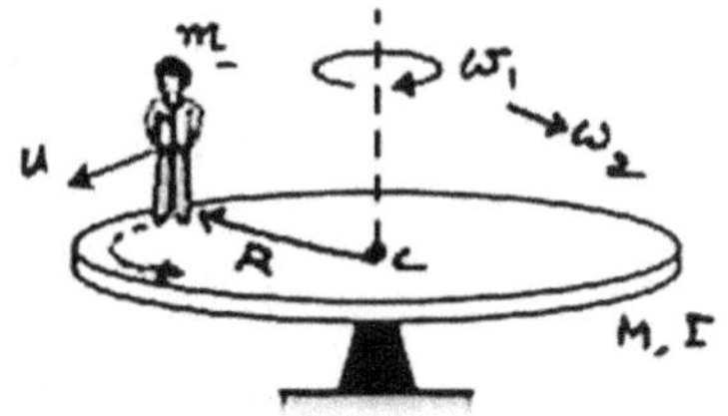


Sol$^n$: As no ext torque is there abt pt C, we use

$$L_i = L_f.$$

$$(I+mR^2)\omega_1 = I\omega_2 - m(u-R\omega_2)R$$

$$\omega_2 = \frac{(I+mR^2)\omega_1}{(I+mR^2)} + \frac{muR}{(I+mR^2)} = \frac{(I+mR^2)\omega_1 + muR}{I+mR^2}$$ Ans.

Scan for Video Explanation

Web Reference at www.physicsgalaxy.com

**Module-28: Solved Example-13**

**Ex :** **A point mass is tied to one end of a cord other end of which passes through a hollow tube, caught in one hand. The point mass is being revolved in a horizontal circle of radius 2m with speed of 4 m/s. The cord is then pulled down so that the radius of circle reduces to 1m. Compute the new linear and angular velocities of the point mass.**

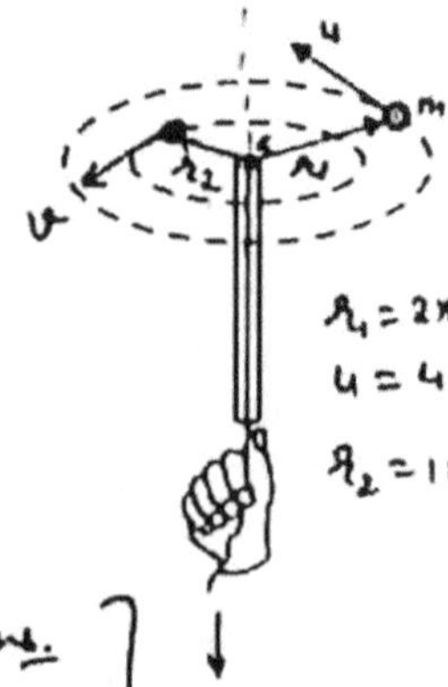


$r_1 = 2m$
$u = 4 m/s$
$r_2 = 1m.$

Sol$^n$: As no ext torque is present abt Centre of circle we can cons. angular momentum of system -

$$L_i = L_f$$

$$mur_1 = mvr_2$$

$$v = \frac{ur_1}{r_2} = \frac{4\times 2}{1} = 8 \text{ m/s} \quad \text{Ans.}$$

$$\omega = \frac{v}{r_2} = \frac{8}{1} = 8 \text{ rad/sec} \quad \text{Ans.}$$

Scan for Video Explanation

Web Reference at www.physicsgalaxy.com

**Module-29: Concept of Angular Impulse**

# Concept of 'Angular Impulse':

for Translational motion $\Delta P = \int_0^t F\,dt$

As for rot$^n$ motion $L = I\omega$.

$$\tau = \frac{dL}{dt}$$

$$dL = \tau\,dt$$

Angular Impulse:
- For a Constant torque for a time $\Delta t$, $\Delta L = \tau\Delta t$
- For a time varying torque for a time 0 to T $\Delta L = \int_0^T \tau\,dt$

Web Reference at www.physicsgalaxy.com | Scan for Video Explanation

**Module-29: Concept of Angular Impulse**

# Impulse - Momentum equation for Rotational Motion:

for $T_r$ motion $P_i \pm J = P_f$

for $R_r$ motion $L_i \pm J_R = L_f$.

$$J_R = \tau\Delta t \quad \text{or} \quad J_R = \int \tau\,dt$$

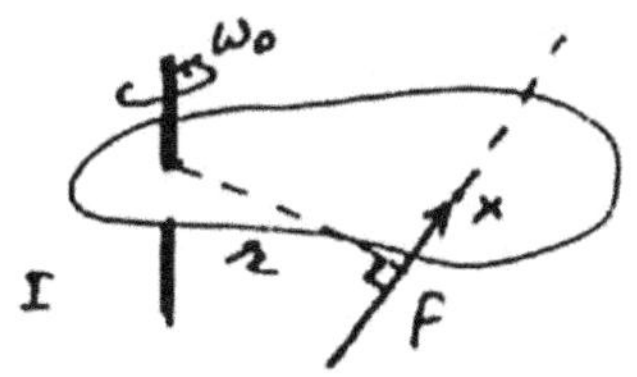


$\tau = Fr$

after time t if ang speed of body changes to $\omega_f$

$$[I\omega_0 + Frt = I\omega_f]$$

Web Reference at www.physicsgalaxy.com | Scan for Video Explanation

**Module-29: Concept of Angular Impulse**

**Ex :** The wheel of radius R and moment of inertia I shown in figure is released from rest with a mass m tied to an ideal string wrapped on wheel. Find the speed of mass m after time t.

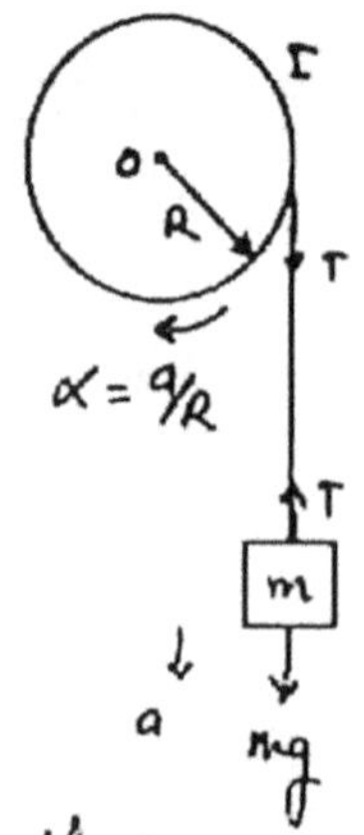


Solⁿ: eqⁿ of motion

$$mg - T = ma.$$

$$T(R) = I\left(a/R^2\right)$$

$$mg = \left(m + I/R^2\right)a$$

$$a = \frac{mg}{m + I/R^2} = \frac{mgR^2}{mR^2 + I}$$

after t

$$v = u + at$$

$$v = at.$$

$$v = \left(\frac{mgR^2}{mR^2 + I}\right)t \quad \text{Ans.}$$

Scan for Video Explanation

Web Reference at www.physicsgalaxy.com

**Module-29: Concept of Angular Impulse**

Alternative Solution:

Solⁿ: we use Impulse eqⁿ as –

$$0 + mgt - Tt = mv \quad -(1)$$

$$0 + TRt = I\left(\frac{v}{R^2}\right) \quad -(2)$$

$$mgt = \left(m + \frac{I}{R^2}\right)v$$

$$v = \frac{mgR^2 t}{I + mR^2} \quad \text{Ans.}$$

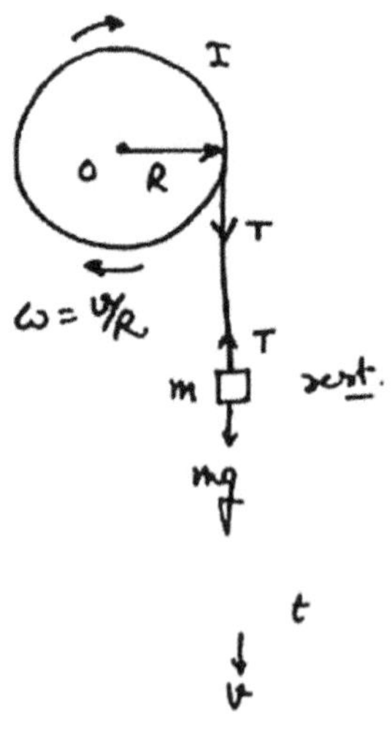


Scan for Video Explanation

Web Reference at www.physicsgalaxy.com

**Module-30: Kinetic Energy of a Rigid body in Rotational Motion**

# Kinetic Energy of a Rigid Body in Rotation:

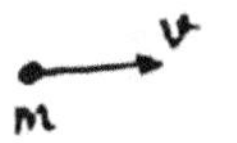


$$K = \frac{1}{2}mv^2$$

KE of dm is –

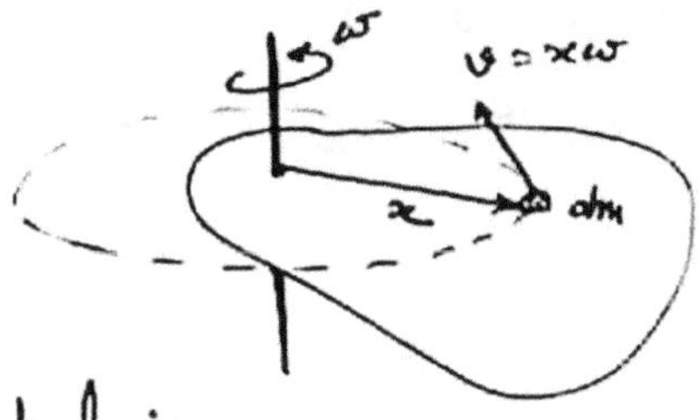


$$dK = \frac{1}{2}dm(x\omega)^2$$
$$= \frac{1}{2}dm\,x^2\omega^2$$

Total KE of rotating body is

$$K = \int dK = \int \frac{1}{2}dm\,x^2\omega^2 = \frac{1}{2}\omega^2\int dm\,x^2$$

$$\boxed{K = \frac{1}{2}I\omega^2}$$

Web Reference at www.physicsgalaxy.com | Scan for Video Explanation

**Module-30: Kinetic Energy of a Rigid body in Rotational Motion**

KE of a rotating body about a moving AOR.

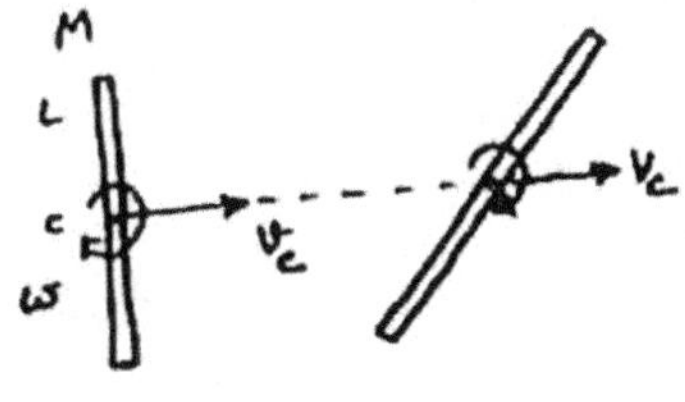


$$K = K_{RC} + \frac{1}{2}Mv_c^2$$

$$\boxed{K = \frac{1}{2}I\omega^2 + \frac{1}{2}Mv_c^2}$$

↓ vel. of COM.

Web Reference at www.physicsgalaxy.com | Scan for Video Explanation

## Module-30: Kinetic Energy of a Rigid body in Rotational Motion

**Ex :** A wheel of moment of inertia I, and radius R is free to rotate about the horizontal axis O and an ideal a string is wrapped on it and attached to a small mass m as shown. If the system is released, find speed of mass m after falling through a distance h.

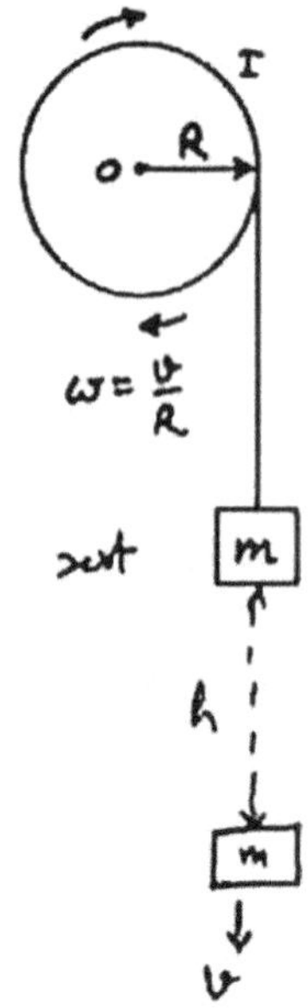


Sol$^n$: using W/E theorem

$$\overset{k_i}{0} + mgh = \overset{k_f}{\frac{1}{2}mv^2 + \frac{1}{2}I\left(\frac{v^2}{R^2}\right)}$$

$$v = \sqrt{\frac{2mghR^2}{(I + mR^2)}} \text{ Ans.}$$

Scan for Video Explanation

Web Reference at www.physicsgalaxy.com

## Module-31: Rolling Motion and Pure Rolling of a body on Inclined Plane

# Rolling Motion:

for a round shaped object

Rolling motion = Rot$^n$ motion + Trans. motion on a Surface.

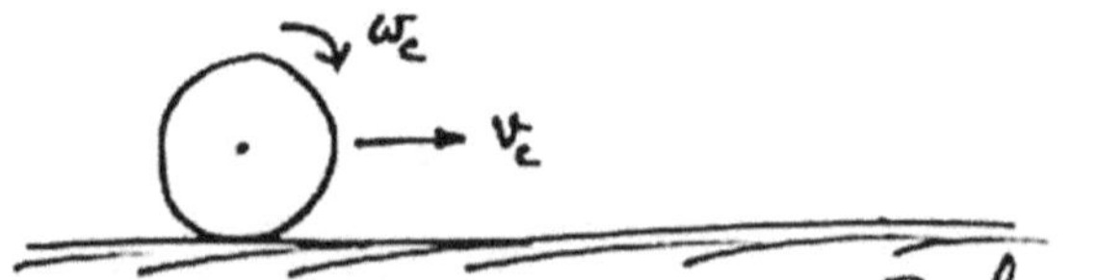


Rolling motion:

Scan for Video Explanation

Web Reference at www.physicsgalaxy.com

**Module-31: Rolling Motion and Pure Rolling of a body on Inclined Plane**

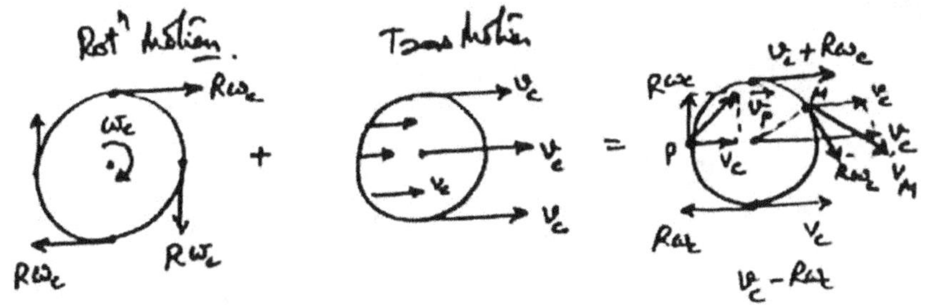


# Types of Rolling:

two ways (i) Pure Rolling – (Rolling without Sliding)

(ii) Rolling with Sliding



**Module-31: Rolling Motion and Pure Rolling of a body on Inclined Plane**

# Pure Rolling:

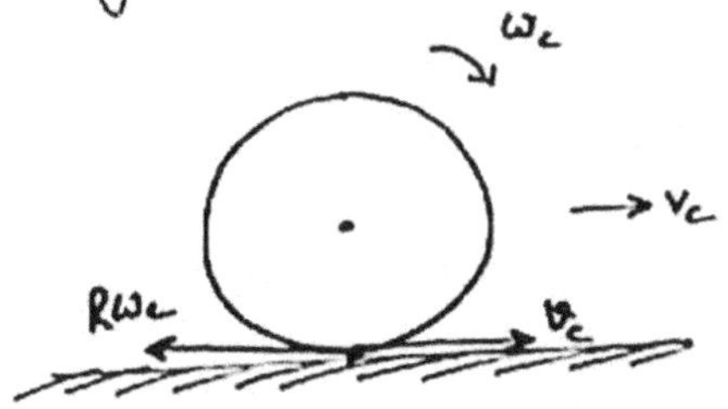


if $v_c = R\omega_c$ ⇒ the Contact pt will be at rest.

if $F_{ext}$ is present ⇒ No Dynamic friction will act on body.

f → static

if No $F_{ext}$ is present & friction if exist, will be static friction only.

$f = 0$. (No friction)

**Module-31: Rolling Motion and Pure Rolling of a body on Inclined Plane**

# Rolling with Sliding: ($v_c \neq R\omega_c$)

Case-I: $v_c > R\omega_c$.

$\mu N = f_k$

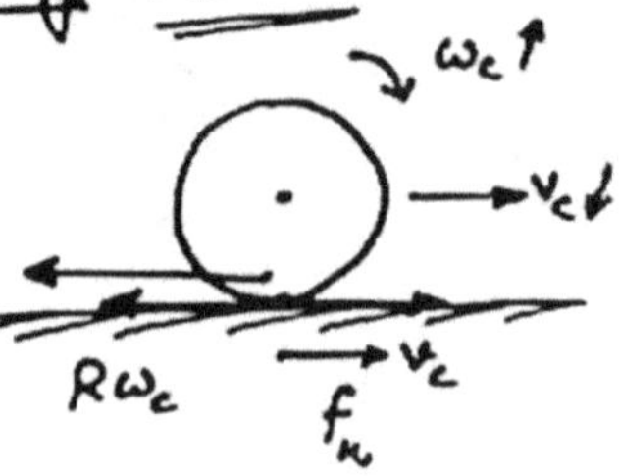


after some time

$v_f = R\omega_f$

pure rolling will start ($f = 0$)

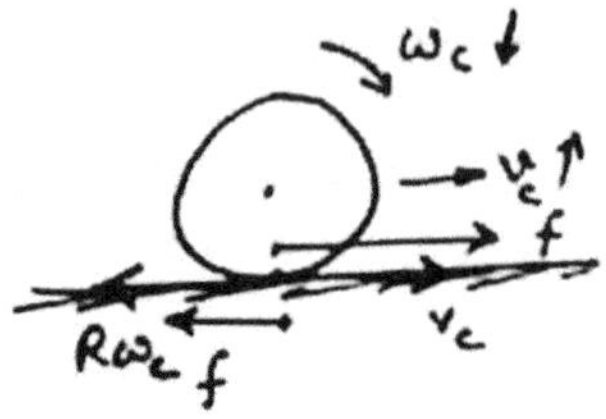


Case-II: $v_c < R\omega_c$

after some time $v_f = R\omega_f$

pure rolling will start ($f = 0$)

Scan for Video Explanation

Web Reference at www.physicsgalaxy.com

**Module-31: Rolling Motion and Pure Rolling of a body on Inclined Plane**

# Pure Rolling of a body on Inclined Plane:

for pure rolling $v = R\omega$.

using W-E Theorem.

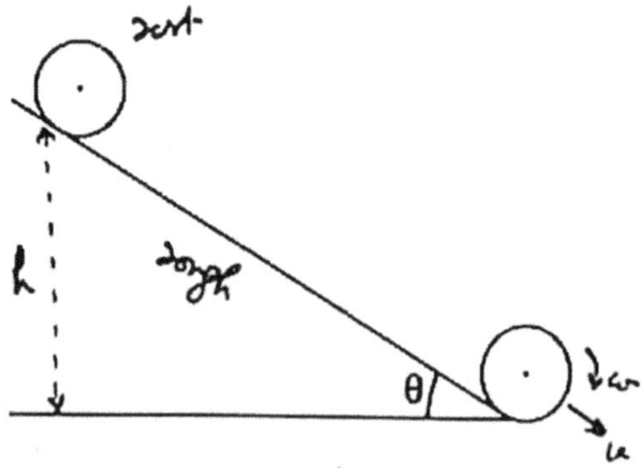


$$\overset{K_i}{0} + Mgh = \overset{K_f}{\frac{1}{2}Mv^2 + \frac{1}{2}(MK^2)\omega^2}$$

$I = MK^2$

$$2gh = v^2 + K^2\left(\frac{v^2}{R^2}\right)$$

$$\boxed{v = \sqrt{\frac{2gh}{1 + K^2/R^2}}}$$

Scan for Video Explanation

Web Reference at www.physicsgalaxy.com

**Module-31: Rolling Motion and Pure Rolling of a body on Inclined Plane**

# Acceleration of a body in pure rolling on inclined Plane:

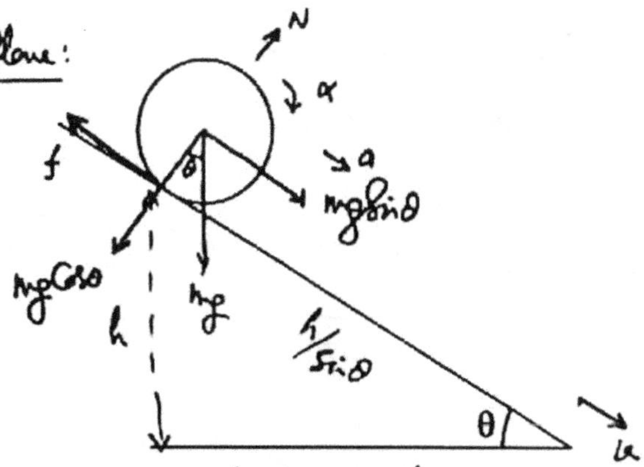


(T) $mg\sin\theta - f = ma$

$fR = I\left(\frac{a}{R^2}\right)$

$mg\sin\theta = \left(m+\frac{I}{R^2}\right)a.$

$a = \frac{mg\sin\theta}{m + I/R^2}$

for pure rolling $a = R\alpha$.

abt Central axis $I = Mk^2$

$a = \frac{Mg\sin\theta}{M + \frac{Mk^2}{R^2}} \rightarrow \boxed{a = \frac{g\sin\theta}{1 + k^2/R^2}}$

$v = \sqrt{\frac{2ah}{\sin\theta}} = \sqrt{\frac{2gh}{1+k^2/R^2}}$

Web Reference at www.physicsgalaxy.com | Scan for Video Explanation

**Module-32: Solved Example-14**

**Ex :** A force F is applied tangential at the topmost point of a sphere of mass M and radius R. If the ground is rough enough to prevent sliding, find the linear acceleration of sphere.

Sol$^n$:

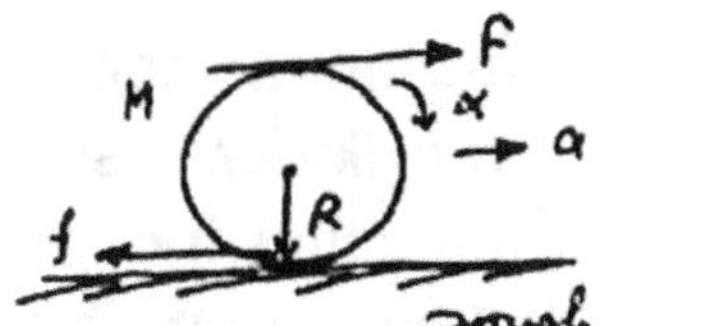


eq$^n$ of motions

$F - f = Ma$ — (1)

$FR + fR = \left(\frac{2}{5}MR^2\right)\left(a/R\right)$ — (2)

for pure rolling $\alpha = a/R$.

$2F = \frac{7}{5}Ma$

$a = \frac{10F}{7M}$ Ans.

friction

$f = F - Ma = F - \frac{10F}{7}$

$f = -\frac{3F}{7}$ ↓ opposite dir.

Web Reference at www.physicsgalaxy.com | Scan for Video Explanation

**Module-32: Solved Example-14**

**Alternative Solution:**

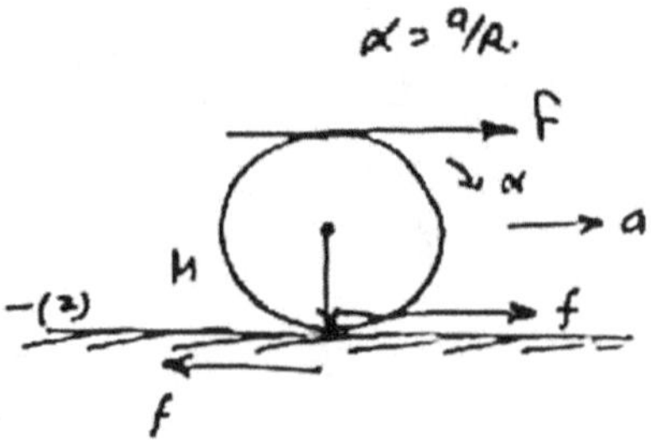


eqⁿ of body -

$$(T) \rightarrow F + f = Ma \quad \text{---(1)}$$

$$(R) \rightarrow FR - fR = \left(\frac{2}{5}MR^2\right)\left(\frac{a}{R}\right) \quad \text{---(2)}$$

$$2F = \frac{7}{5}Ma$$

$$a = \frac{10F}{7M} \text{ Ans.}$$

$$F + f = M\left(\frac{10F}{7M}\right)$$

$$f = \frac{10}{7}F - F$$

$$f = \frac{3F}{7}$$

NOTE: In cases of pure rolling in a prob, solution will not be affected by direction of friction on body.

Scan for Video Explanation

Web Reference at www.physicsgalaxy.com

**Module-33: Solved Example-15**

**Ex :** A cylinder of mass M and radius R rolls without slipping down an inclined plane of length $l$ as shown. Find the linear acceleration of cylinder. Also find the minimum friction coefficient required on the inclined plane for which the cylinder does not slip.

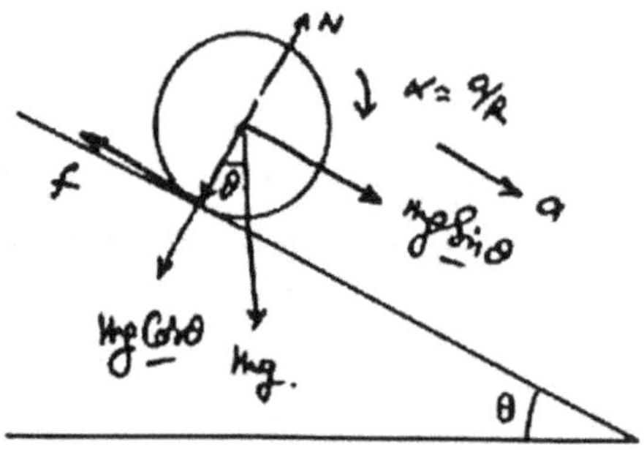


Solⁿ: As we know $a = \frac{g\sin\theta}{1 + k^2/R^2} = \frac{g\sin\theta}{1 + 1/2}$ $\quad k = R/\sqrt{2}$

$$a = \frac{2}{3}g\sin\theta \text{ Ans.}$$

for pure rolling $f < f_L(\mu N)$

$$mg\sin\theta - f = m\left(\frac{2}{3}g\sin\theta\right)$$

$$f = \frac{1}{3}mg\sin\theta$$

$$\frac{1}{3}mg\sin\theta < \mu\, mg\cos\theta$$

$$\mu > \frac{1}{3}\tan\theta$$

$$\text{min } \mu = \frac{1}{3}\tan\theta \text{ Ans.}$$

Scan for Video Explanation

Web Reference at www.physicsgalaxy.com

**Module-34: Solved Example-16**

**Ex :** Figure shows a cylinder on which a string is wrapped and other end of which is connected to a hanging mass m. If cylinder moves in pure rolling, find the acceleration of falling mass.

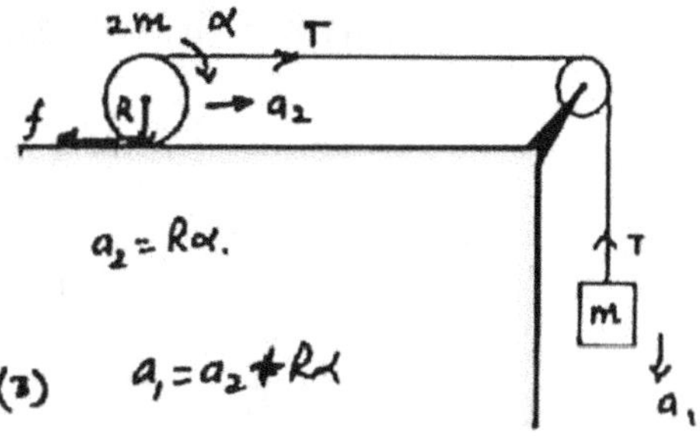


**Sol$^n$:** eq$^n$ of motion

$$mg - T = ma_1 \quad —(1)$$

$$T - f = 2m\left(\frac{a_1}{2}\right) \quad —(2)$$

$$TR + fR = \frac{1}{2}(2m)R^2\left(\frac{a_1}{2R}\right) \quad —(3)$$

$$2T = \frac{3}{2}ma_1$$

$$4T = 3ma_1$$

$a_2 = R\alpha$.

$$a_1 = a_2 + R\alpha$$

$$a_1 = 2a_2$$

from eq$^n$ (1)

$$mg - \frac{3ma_1}{4} = ma_1$$

$$\frac{7}{4}ma_1 = mg \Rightarrow a_1 = \frac{4g}{7} \text{ Ans.}$$

Web Reference at www.physicsgalaxy.com

Scan for Video Explanation

**Module-35: Solved Example-17**

**Ex :** Figure shows a ball of mass m placed on rough surface of a plank of mass 4 m which is placed on a smooth ground. If plank is pulled by a horizontal force F. Find linear acceleration of plank and cylinder if cylinder is not slipping on plank.

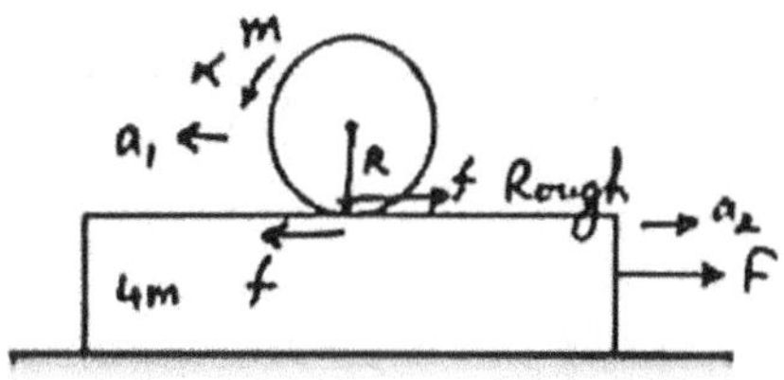


**Sol$^n$:** eq$^n$ of motion

$$F - f = 4ma_2$$

$$f = m(a_2 - a_1)$$

$$F = 5ma_2 - ma_1$$

$$fR = \frac{2}{5}mR^2\left(\frac{a_1}{R}\right).$$

for pure rolling of ball.

$$a_1 = R\alpha.$$

$$\frac{2}{5}ma_1 = ma_2 - ma_1$$

$$\frac{7}{5}a_1 = a_2$$

$$F = 5ma_2 - m\left(\frac{5}{7}a_2\right)$$

$$F = \frac{30}{7}ma_2$$

$$a_2 = \frac{7F}{30m} \text{ Ans.}$$

Web Reference at www.physicsgalaxy.com

Scan for Video Explanation

**Module-36: Solved Example-18**

**Ex :** A plank of mass m is placed over a cylinder of mass m as shown. A horizontal force F is applied on plank, if slipping is not there at any surface find linear acceleration of plank.

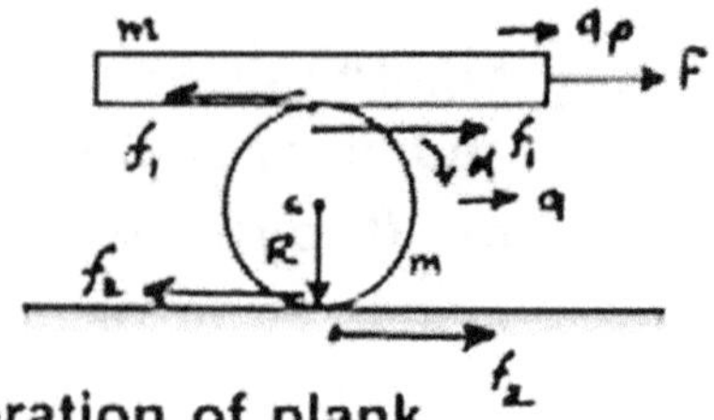


Sol$^n$: $f_1, f_2 \rightarrow$ static friction.

for pure rolling

$a = R\alpha$

$a_p = a + R\alpha = 2a$

eq$^n$ of motion. $F - f_1 = m(2a)$ —(1)

$f_1 - f_2 = ma$ —(2)

$f_1 R + f_2 R = \left(\frac{1}{2}mR^2\right)\left(\frac{a}{R}\right)$ —(3)

(2) + (3) $\quad 2f_1 = \frac{3}{2}ma.$

from (1) $F - \frac{3}{4}ma = 2ma$

$F = \frac{11}{4}ma.$

$a = \frac{4F}{11m}.$

acc$^n$ of Plank $a_p = 2a = \frac{8F}{11m}$ Ans.

Scan for Video Explanation

Web Reference at www.physicsgalaxy.com

**Module-37: Analysis of Rolling with Sliding**

# Analysis of Rolling with Sliding:

↳ Kinetic friction acts

$v \neq R\omega$ : $v > R\omega$, $v < R\omega$

When $F_{ext} = 0$.

Case-I $v > R\omega$.

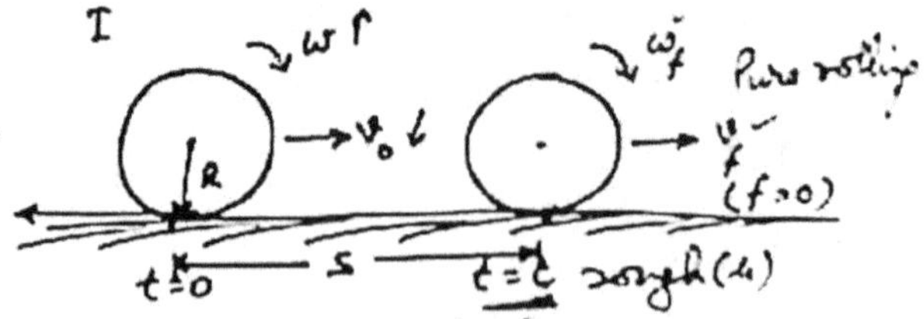


Impulse eq$^n$

(T) $\quad Mv_0 - \mu Mgt = Mv_f$ —(1)

(R) $\quad 0 + \mu MgRt = I\left(\frac{v_f}{R^2}\right)$ —(2)

$v_f = R\omega_f$

$S = v_0 t - \frac{1}{2}\mu g t^2$

$Mv_0 = \left(M + \frac{I}{R^2}\right)v_f \Rightarrow v_f = \frac{Mv_0}{\left(M + \frac{I}{R^2}\right)}$

Scan for Video Explanation

Web Reference at www.physicsgalaxy.com

**Module-37: Analysis of Rolling with Sliding**

Case-II: $v < R\omega$

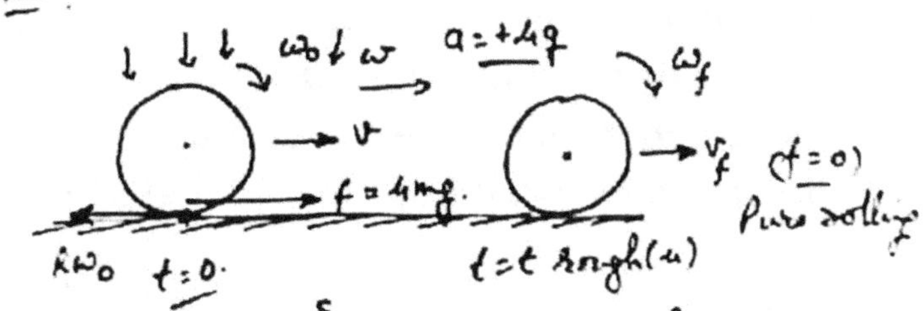


Impulse eqn:

(T) $0 + \mu mgt = Mv_f$ — (1)

(R) $\frac{I\omega_0}{R} - \mu mgt = I\left(\frac{v_f}{R^2}\right)$ — (2)

$$\frac{I\omega_0}{R} = \left(M + \frac{I}{R^2}\right)v_f \rightarrow v_f = \frac{I\omega_0}{R\left(M + \frac{I}{R^2}\right)}$$

$$S = \frac{1}{2}\mu g t^2 = \text{- - -}$$

Web Reference at www.physicsgalaxy.com | Scan for Video Explanation

**Module-38: Solved Example-19**

**Ex :** A thin spherical shell of mass m and radius R lying on a rough surface is hit sharply by a cue horizontally as shown. Find h so that shell does not slip over the surface.

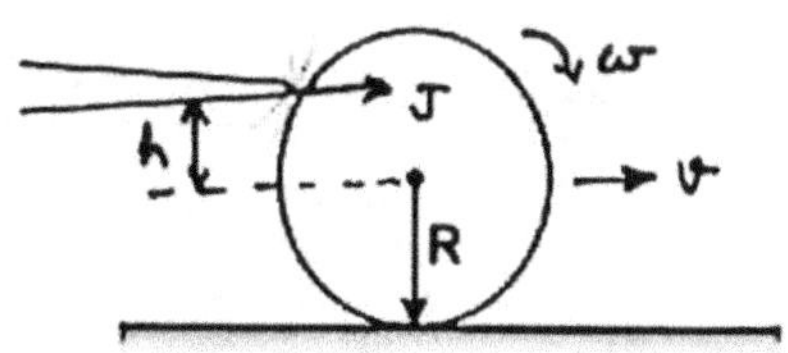


Soln. Shell does not slip if from beginning it starts pure rolling.

$$v = R\omega.$$

Impulse eqn:

$$J = Mv$$

$$Jh = \left(\frac{2}{3}MR^2\right)\left(\frac{v}{R}\right)$$

$$Mvh = \frac{2}{3}MRv \rightarrow h = \frac{2}{3}R \text{ Ans.}$$

Web Reference at www.physicsgalaxy.com | Scan for Video Explanation

**Module-39: Solved Example-20**

**Ex :** A solid cylinder of radius R is set in rotation about its axis with an angular speed $\omega_0$ then lowered with its lateral surface onto a horizontal plane and released. Friction coefficient between cylinder and plane is $\mu$, find the time after which cylinder will start pure rolling.

Sol$^n$: Impulse eq$^n$:

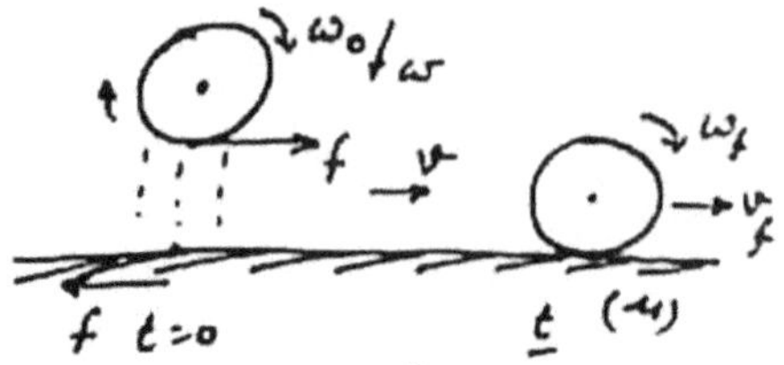


(T) $0 + \mu Mg t = M v_f$ — (1)

(R) $\left(\frac{1}{2}MR^2\right)\omega_0 - \mu MgRt = \left(\frac{1}{2}MR^2\right)\left(\frac{v_f}{R}\right)$ — (2)

$f = \mu mg$

Pure rolling $v_f = R\omega_f$

$\frac{1}{2}R\omega_0 = \frac{3}{2}v_f$

$v_f = \frac{1}{3}R\omega_0$

from eq$^n$ (1) $t = \frac{v_f}{\mu g} = \frac{R\omega_0}{3\mu g}$ Ans.

Scan for Video Explanation

Web Reference at www.physicsgalaxy.com

**Module-40: Work & Power in Rotational Motion**

# Work & Power in Rotational Motion:

for Trans motion

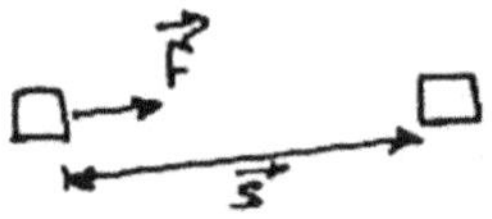


$W = \vec{F}\cdot\vec{s}$

$W = \int_0^s \vec{F}\cdot\vec{dx}$

Torque $= FR$.

W/D in disp by $\theta$ is

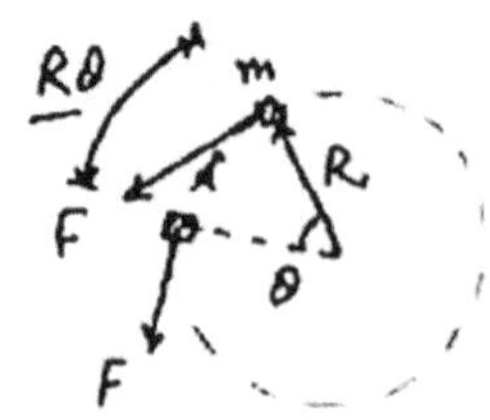


$W = F(R\theta)$

$W = \tau\theta$

$\boxed{W = \vec{\tau}\cdot\vec{\Delta\theta}}$

$W = \int_0^{\Delta\theta} \tau\, d\theta$

Scan for Video Explanation

Web Reference at www.physicsgalaxy.com

**Module-40: Work & Power in Rotational Motion**

Power → Rate of doing Work.

$$P = \frac{dW}{dt} = \frac{\tau d\theta}{dt} = \tau\omega$$

vector rep. $\boxed{P = \vec{\tau}\cdot\vec{\omega}}$ ✓

- Avg Power → $\Delta W/\Delta t$
- * Instantaneous Power → $dW/dt$

Web Reference at www.physicsgalaxy.com | Scan for Video Explanation

**Module-40: Work & Power in Rotational Motion**

**Ex :** An electric motor of power P is connected with a pulley & a string to which a mass m is hanging. Find the maximum speed at which mass can be raised up.

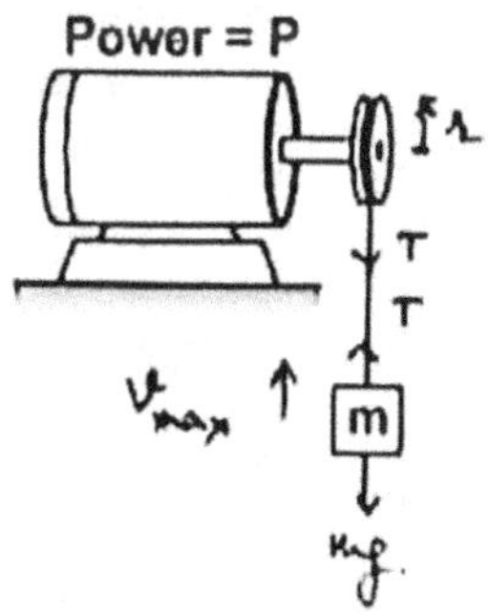


**Sol^n:**

Torque on motor due to wt of block is

$$\tau = mg \cdot r$$

max power of motor $\quad P = mgr \cdot \omega \quad (\tau)\cdot(\omega)$

$$\omega = \frac{P}{mgr}$$

$$v_{max} = r\omega = \frac{P}{mg} \quad \text{Ans.}$$

Web Reference at www.physicsgalaxy.com | Scan for Video Explanation

**Module-41: Solved Example-21**

**Ex :** A uniform rod of mass 3 m and length L is hanging from a hinge H as shown. A small ball of mass m moving horizontally at speed u strikes the end of rod elastically. Find the speed of ball and angular speed of rod just after collision.

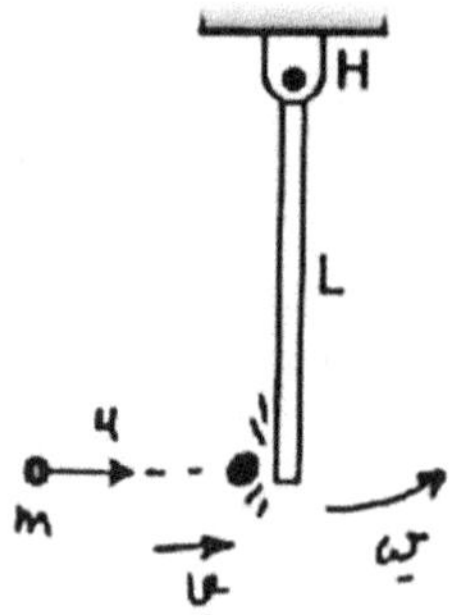


Sol$^n$: by Cons of ang. momentum abt H.

$$muL = mvL + \frac{1}{3}(3m)L^2 \cdot \omega$$

$$e = 1 = \frac{v_{sep}}{v_{app}}$$

$$u = v + L\omega \quad —(1)$$

by $e = 1$: $$u = L\omega - v \quad —(2)$$

$$2u = 2L\omega \Rightarrow \omega = u/L$$

from eq$^n$ (1) $u = v + u \Rightarrow v = 0$ Ans.

Scan for Video Explanation

Web Reference at www.physicsgalaxy.com

**Module-42: Solved Example-22**

**Ex :** A small ball of mass m, moving at speed u strikes normally on a rod of mass M and length L placed on a smooth surface as shown. Find M for which after collision ball will come to rest,coefficient of restitution is e.

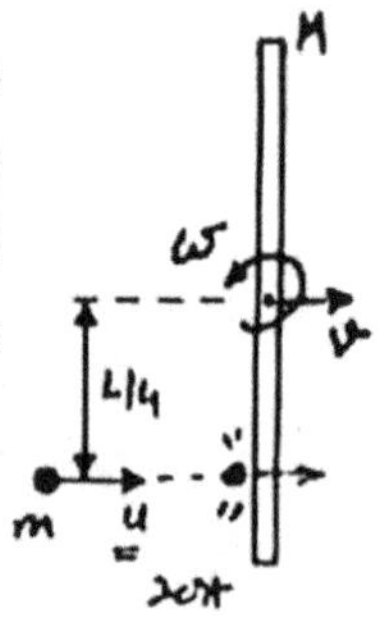


Sol$^n$: by linear momentum Cons.

$$mu = Mv \quad —(1) \rightarrow v = \frac{mu}{M}$$

$$v_{sep} = e\, v_{app}$$

by ang momentum Cons.

$$mu\left(\frac{L}{4}\right) = \frac{ML^2}{12}\cdot\omega \quad —(2) \rightarrow \omega = \frac{3mu}{ML}$$

by CoR. $$e = \frac{v_{sep}}{v_{app}} = \frac{v + \frac{L}{4}\omega}{u} \Rightarrow v + \frac{L}{4}\omega = ue$$

$$\frac{mu}{M} + \frac{L}{4}\left(\frac{3mu}{ML}\right) = ue \quad M = \frac{7}{4}\frac{m}{e}$$

$$7m = 4eM \quad \text{Ans.}$$

Scan for Video Explanation

Web Reference at www.physicsgalaxy.com

# Chapter 12
# Fluid Statics

## Lecture Notes Modules

**Module-1:** Introduction to Fluids
**Module-2:** Pressure in a Fluid
**Module-3:** Floatation of a Body due to Fluid Pressure
**Module-4:** Fluid Pressure due to Gravity
**Module-5:** Average Pressure on Side Walls of a Vessel
**Module-6:** Force on Side Walls of a Random Shaped Vessel
**Module-7:** Archimede's Principle
**Module-8:** Buoyant Force on a Body at the Interface of Two Liquids
**Module-9:** Apparent Weight of a Body Submerged in a Liquid
**Module-10:** Solved Example-1
**Module-11:** Solved Example-2
**Module-12:** Solved Example-3
**Module-13:** Solved Example-4
**Module-14:** Solved Example-5
**Module-15:** Pascal's Principle of Pressure Distribution
**Module-16:** Pressure at Different Levels in a Liquid
**Module-17:** Mechanism of a Hydraulic Lift
**Module-18:** Concept of Open tube Manometer
**Module-19:** Pressure Distribution in an Accelerated Frame
**Module-20:** Pressure Distribution in a Closed Accelerated Container
**Module-21:** Pressure Distribution in a Rotating tube Filled with a Liquid
**Module-22:** Solved Example-6
**Module-23:** Solved Example-7
**Module-24:** Solved Example-8
**Module-25:** Solved Example-9

**Module-1: Introduction to Fluids**

# Fluids:

– Substances which do not have definite shape but have definite volume are called fluids (liq/gas) [almost constant density]

– A given amount of fluid is always separated by an interface from other solids or gases.

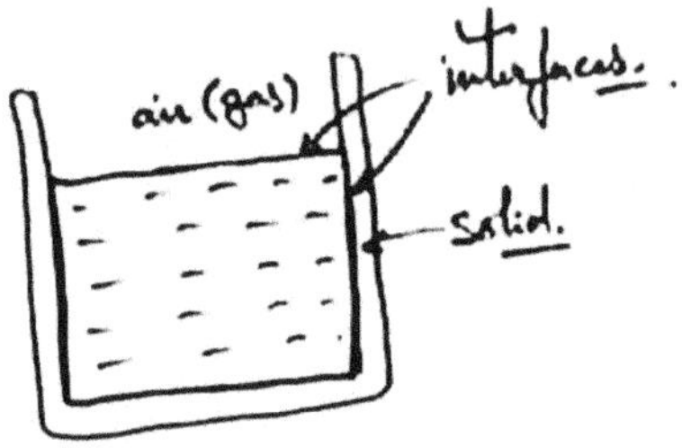


– The fluid interface bounded by a gas is called free surface of the fluid.

Scan for Video Explanation

Web Reference at www.physicsgalaxy.com

**Module-2: Pressure in a Fluid**

# Pressure in a Fluid:

When an external force is applied on a fluid body, it develops an internally distributed pressure at all pts of fluid body. This pressure is called fluid pressure.

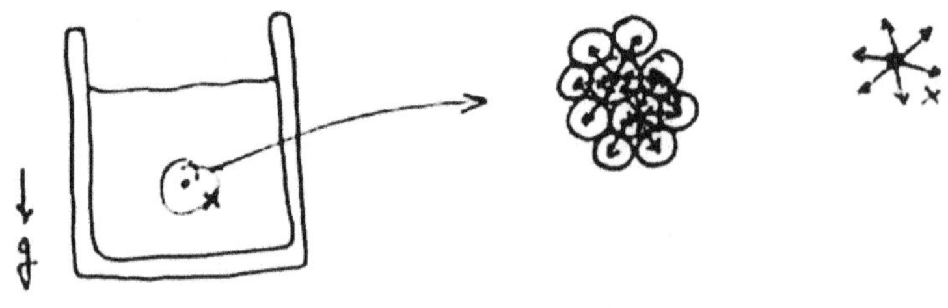


Scan for Video Explanation

Web Reference at www.physicsgalaxy.com

**Module-2: Pressure in a Fluid**

# Some Important points about Fluid Pressure:

(1) It acts at all pts of fluid body. (distributed)

(2) At one point in fluid body it acts uniformly in all direction

(3) If fluid is in contact with any solid surface, fluid pressure will exert a force normal to that surface.

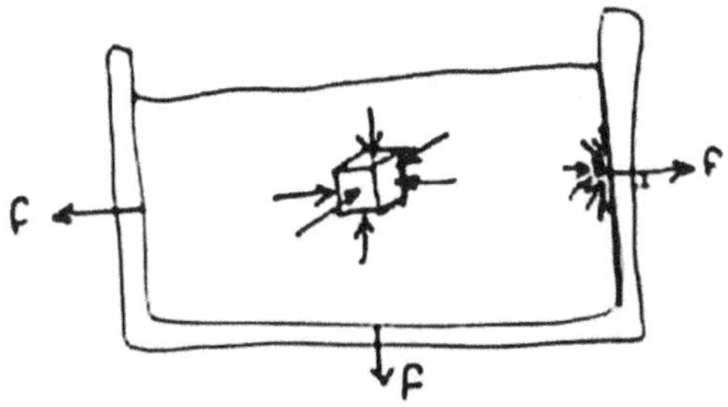


Web Reference at www.physicsgalaxy.com

Scan for Video Explanation

**Module-3: Floatation of a Body due to Fluid Pressure**

# Floatation of a body due to Fluid Pressure:

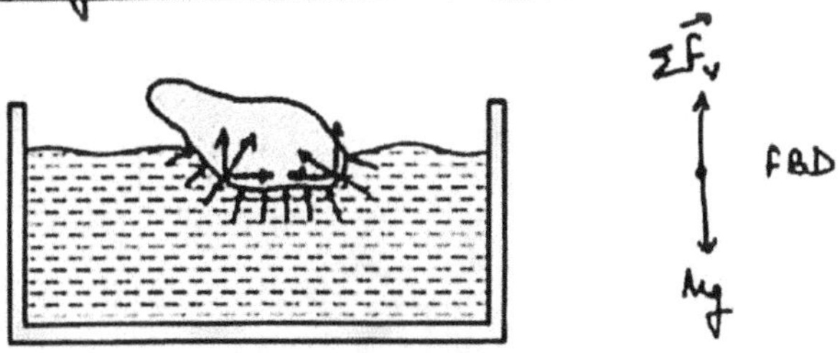


All the forces on body surface due to fluid pressure have their horizontal components balanced.

all vertical components of forces due to fluid pressure balance the wt. of body under floatation.

Web Reference at www.physicsgalaxy.com

Scan for Video Explanation

**Module-4: Fluid Pressure due to Gravity**

# Fluid Pressure due to Gravity:

Fluid Pressure at pt A is due to the whole wt of liquid

Pressure at pt A is

$$P_A = \frac{Mg}{S} = \frac{(hS\rho)g}{S}$$

← Gauge pressure.

$$\boxed{P_A = h\rho g}$$

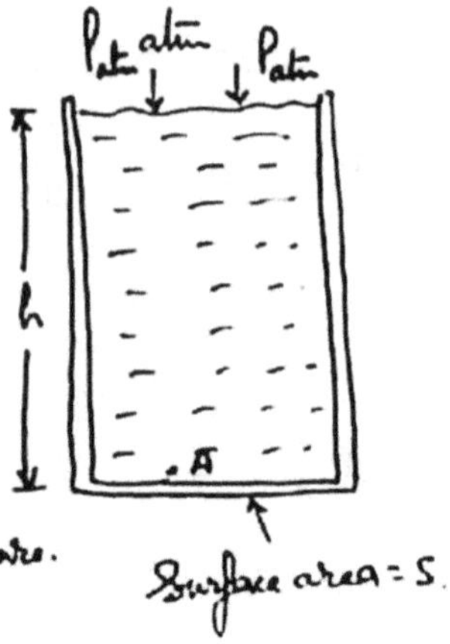


h → height of fluid column above pt A.

net pressure at pt A $\quad P_{AT} = P_{atm} + h\rho g$.

Scan for Video Explanation | Web Reference at www.physicsgalaxy.com

**Module-4: Fluid Pressure due to Gravity**

$$P_x = P_{atm} + h\rho g$$

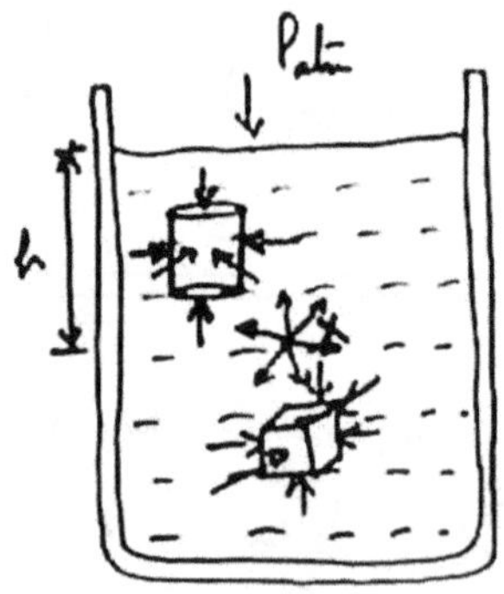


Scan for Video Explanation | Web Reference at www.physicsgalaxy.com

**Module-4: Fluid Pressure due to Gravity**

# Fluid pressure does not depend on shape of Vessel:

$P_x = P_{atm} + h\rho g$

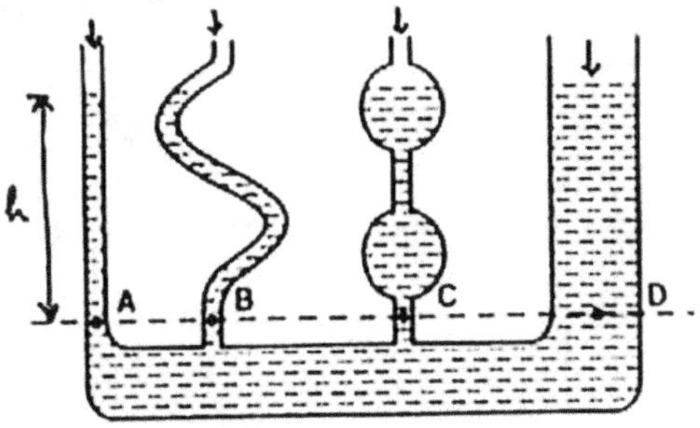


$P_A = P_B = P_C = P_D = P_{atm} + h\rho g$

$P_x = P_{atm} + h\rho g$

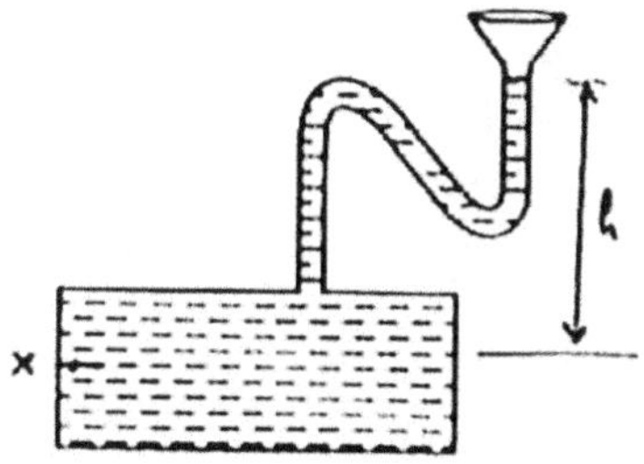


Web Reference at www.physicsgalaxy.com | Scan for Video Explanation

**Module-5: Average Pressure on Side Walls of a Vessel**

# Average Pressure on side walls of a vessel:

Pressure due to fluid at a pt behind the elemental strip is

$P = y\rho g$

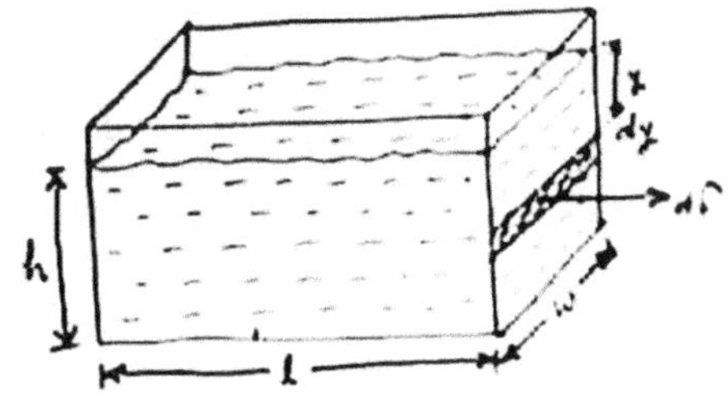


$P_p = h\rho g$

outward force on elemental strip is

$dF = y\rho g \times w\,dy$

Total force on side wall of vessel due to fluid pressure -

$F = \int dF = \int_0^h y\rho g w\,dy = \rho g w \left[\frac{y^2}{2}\right]_0^h = \frac{1}{2}\rho g w h^2$

avg pressure on side wall is $P_{av} = \frac{F}{hw} = \frac{\frac{1}{2}\rho g w h^2}{hw} = \boxed{\frac{1}{2}h\rho g}$

half of the pressure which fluid exert on bottom

Web Reference at www.physicsgalaxy.com | Scan for Video Explanation

**Module-6: Force on Side Walls of a Random Shaped Vessel**

# Force on Side walls of a random shaped vessel:

Pressure at bottom

$$P = h\rho g$$

force acting on the bottom of Cont is $F = h\rho g \cdot A = N_1$

FBD of fluid body

for eq^n of fluid body

$$N_1 + N_V = Mg$$

$$\boxed{N_V = Mg - h\rho g A}$$

Scan for Video Explanation

Web Reference at www.physicsgalaxy.com

**Module-6: Force on Side Walls of a Random Shaped Vessel**

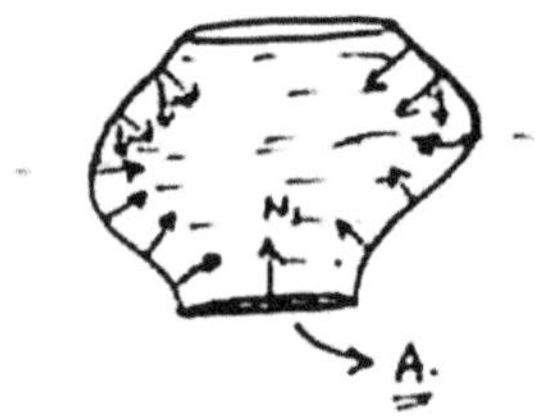


FBD.

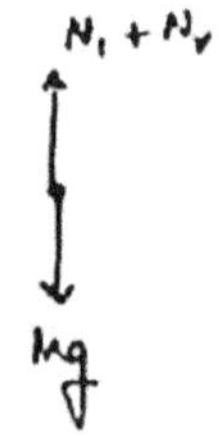


for eq^n

$$N_1 + N_V = Mg$$

$$N_V = Mg - N_1$$

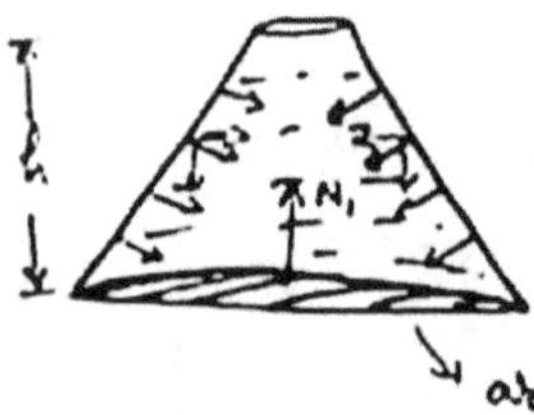


$$N_1 = h\rho g . A .$$

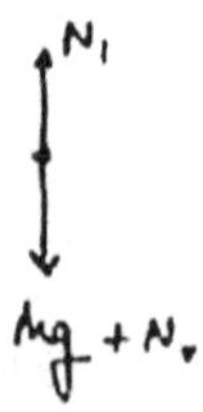


$$N_1 = Mg + N_V$$

$$N_V = N_1 - Mg$$

Scan for Video Explanation

Web Reference at www.physicsgalaxy.com

**Module-7: Archimede's Principle**

# Archimede's Principle :

"A body whether partially or completely submerged in a fluid is buoyed upward by a force due to fluid pressure that is equal to the weight of displaced liquid."

Buoyant force on object

$F_B = F_2 - F_1$

$F_1 = (P_{atm} + l\rho g) \cdot S$

$F_2 = [P_{atm} + (l+h)\rho g] \cdot S$

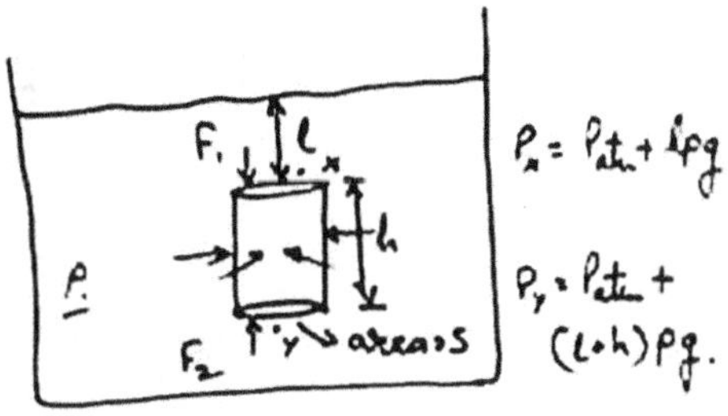


$\rightarrow F_B = hS\rho g = V\rho g$ [wt of liq displaced by the object]

Web Reference at www.physicsgalaxy.com

Scan for Video Explanation

**Module-7: Archimede's Principle**

# Law of Floatation :

$F_B = V\rho g$

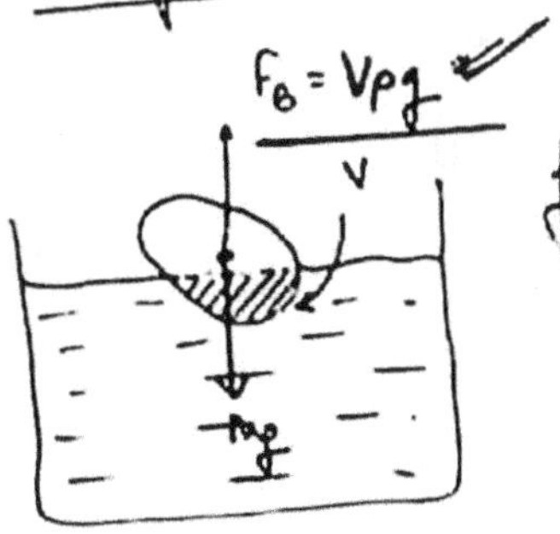


for eq$^n$ of floating object

$$V\rho g = mg.$$

Web Reference at www.physicsgalaxy.com

Scan for Video Explanation

**Module-8: Buoyant Force on a Body at the Interface of Two Liquids**

# Buoyant Force on a body at the interface of two liquids:

$F_1 = P_A . S$
$= (P_{atm} + l\rho_1 g) S$

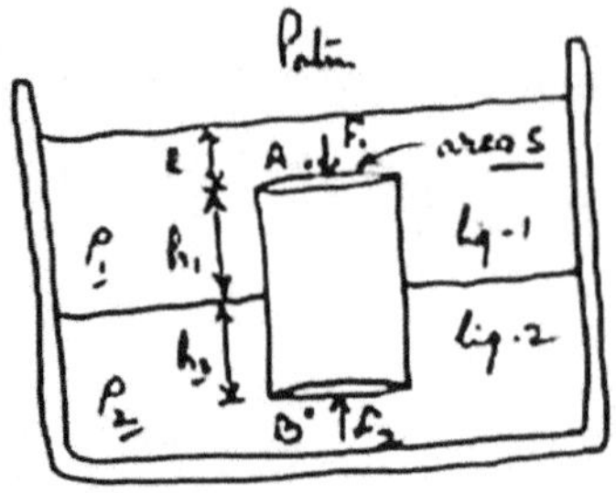


$F_2 = P_B . S$
$= (P_{atm} + (l+h_1)\rho_1 g + h_2 \rho_2 g) . S$

$F_{net} = F_2 - F_1$
$=$

Buoyant force on object is

$$F_B = h_1 S \rho_1 g + h_2 S \rho g.$$

Scan for Video Explanation

Web Reference at www.physicsgalaxy.com

**Module-9: Apparent Weight of a Body Submerged in a Liquid**

# Apparent weight of a body submerged in a liquid:

If mass M is hanging through Sp.

its reading $= \frac{T}{g} = M$ kgf.

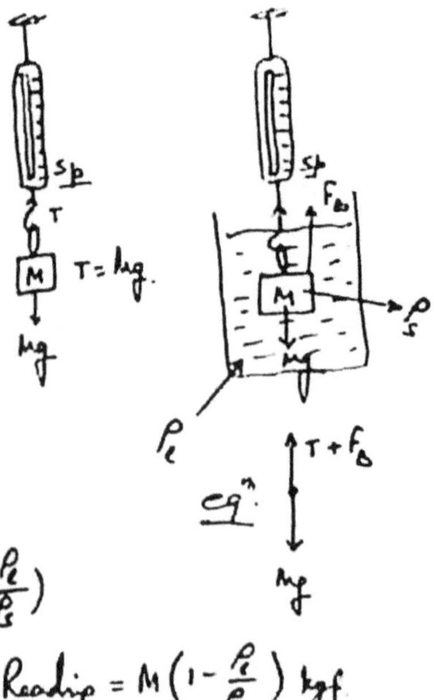


Volume of weight $V = \left(\frac{M}{\rho_s}\right)$

Buoyant force on block $F_B = V\rho_l g$

for eqm of weight →

$T + F_B = Mg$

$T = Mg - F_B = Mg - Mg\left(\frac{\rho_l}{\rho_s}\right)$

apparent wt of object ←

$$T = Mg\left(1 - \frac{\rho_l}{\rho_s}\right) \rightarrow \text{Reading} = M\left(1 - \frac{\rho_l}{\rho_s}\right) \text{kgf}$$

Scan for Video Explanation

Web Reference at www.physicsgalaxy.com

**Module-10: Solved Example-1**

**Ex :** A tank containing water of mass M is placed on a weighing machine. A stone of weight W is hung and lowered in water without touching the sides. If volume of stone is V and density of water is ρ, what will be the reading of weighing machine now.

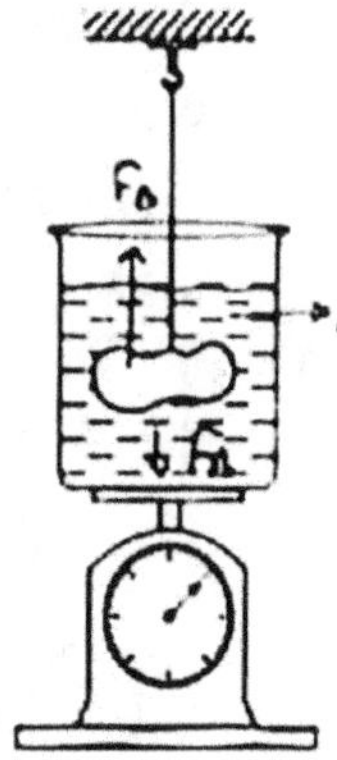


Sol: Total force on weighing machine is

$$F = Mg + F_B \qquad F_B = V\rho g$$

Reading of machine in kgf

$$R = \left(\frac{Mg + V\rho g}{g}\right) = (M + V\rho) \quad \text{Ans.}$$

Web Reference at www.physicsgalaxy.com | Scan for Video Explanation

**Module-11: Solved Example-2**

**Ex :** A rubber ball of mass m and radius r is submerged in water to a depth h. On releasing the ball, find to what height above the water surface ball will jump. Neglect the resistance of water. Take density of water $\rho_\omega$.

Sol: from pts (i) to (f) using W/E th we get

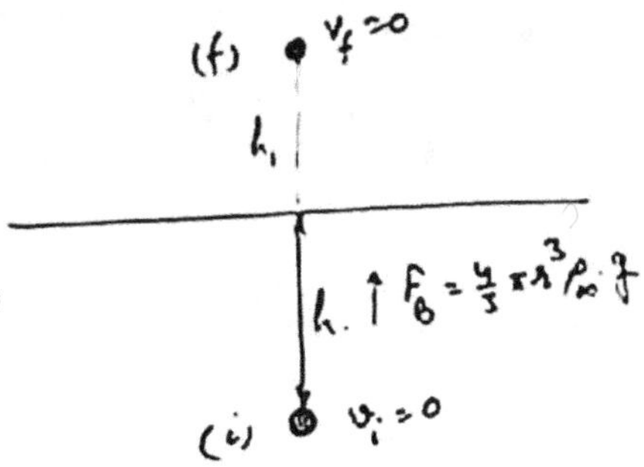


$$\overset{k_i}{0} + F_B \cdot h - mg(h + h_1) = 0$$

$$h_1 = \left(\frac{F_B - mg}{mg}\right) h.$$

$$h_1 = \left(\frac{\frac{4}{3}\pi r^3 \rho_w - m}{m}\right) h. \quad \text{Ans.}$$

Web Reference at www.physicsgalaxy.com | Scan for Video Explanation

**Module-12: Solved Example-3**

**Ex :** A cube of wood supporting 200 gm mass just floats in water in fully submerged state. When the mass is removed cube rises by 2cm above water level. Find the size of cube ? (Take $\rho_{water}$ = 1000 kg/m$^3$)

**Sol$^n$:** for equilibrium of this system.

we use

$$\left(l^3 \rho_{wd} g + 0.2 g\right) = l^3 \rho g$$

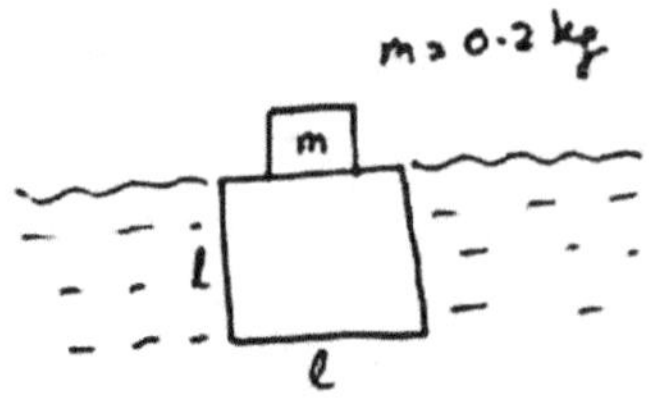


after removal of block, we use

$$l^3 \rho_{wd} g = (l-h) l^2 \rho g$$

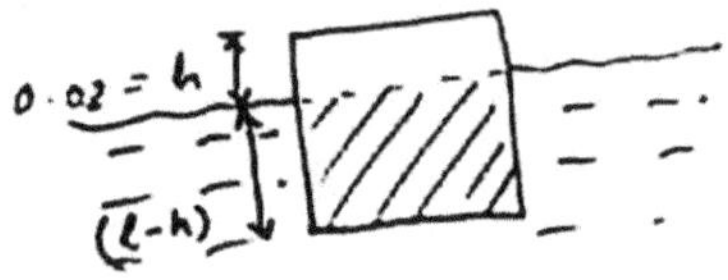


$$l^3 \rho + 0.2 = l^3 \rho + h l^2 \rho$$

$$l^2 = \frac{0.2}{h\rho} = \frac{0.2}{0.02 \times 1000} = 0.01$$

$$l = 0.1\ m = 10\ cm. \quad \text{Ans.}$$

Scan for Video Explanation

Web Reference at www.physicsgalaxy.com

**Module-13: Solved Example-4**

**Ex :** A piece of alloy of mass 96gm is composed of two metals whose specific gravity are 11.4 and 7.4 respectively. If the weight of alloy is 86gm in water, find mass of each metal in alloy.

**Sol$^n$:** let mass of first metal in alloy is (x) (say) gm

then that of second metal is (96 − x) gm

Weight in water = Actual weight − Force of Buoyancy

$$86 = 96 - \frac{x}{11.4} \times 1 - \frac{96-x}{7.4} \times 1$$

Solving we get:

$$x = 62.7\ gm. \ (\text{metal-1})$$

$$96 - x = 33.3\ gm \ (\text{metal-2})$$

Ans.

Scan for Video Explanation

Web Reference at www.physicsgalaxy.com

**Module-14: Solved Example-5**

**Ex :** **A piece of copper having an internal cavity weighs 264gm in air and 221gm in water. Find the volume of cavity. Take density of copper 8.8 gm/cm³.**

Sol$^n$:

apparent weight = actual weight − Buoyant force

$$W_{ap} = W - V\rho g$$

$$221g = 264g - \left[\frac{264}{8.8} + V_c\right] \cdot 1 \cdot g$$

$$221 = 264 - V_c - \frac{264}{8.8}$$

$$V_c = 13\ cm^3 \text{ Ans.}$$

Web Reference at www.physicsgalaxy.com

Scan for Video Explanation

**Module-15: Pascal's Principle of Pressure Distribution**

# Pascal's Principle of Pressure Distribution:

"The pressure applied at one pt in an enclosed fluid body is transmitted uniformly to every part of fluid."

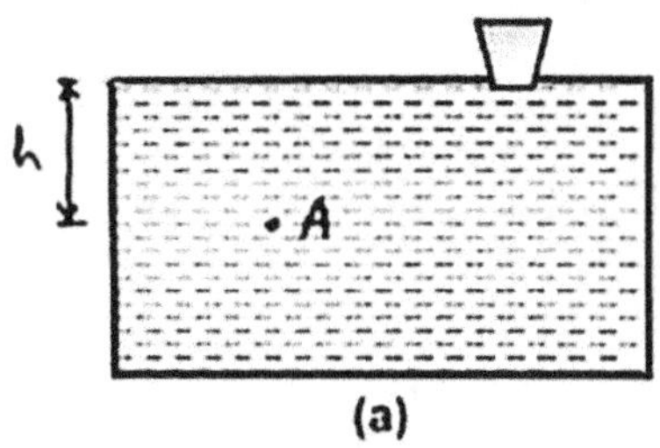


(a)

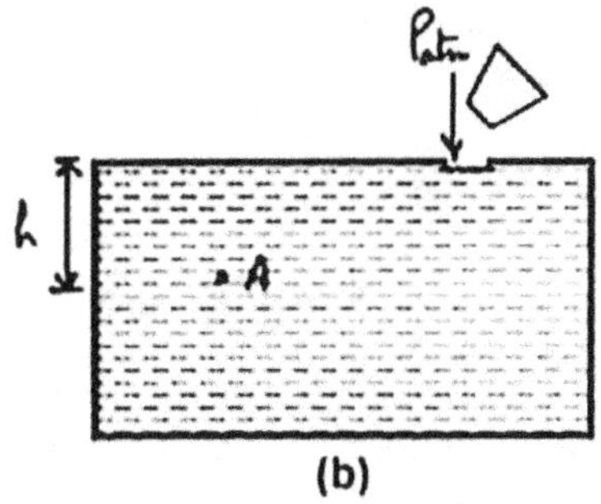


(b)

$$P_A = h\rho g$$

$$P_A = h\rho g + P_{atm}$$

Web Reference at www.physicsgalaxy.com

Scan for Video Explanation

**Module-15: Pascal's Principle of Pressure Distribution**

# Applications of Hydraulic Forces:

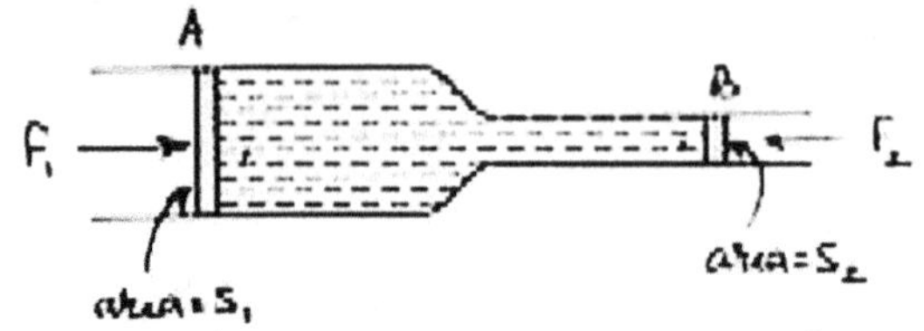


$$P_2 = \frac{F_2}{S_2}$$

$$P_1 = \frac{F_1}{S_1}$$

$$P_1 = P_2$$

$$\frac{F_1}{S_1} = \frac{F_2}{S_2} \Rightarrow F_1 = F_2\left(\frac{S_1}{S_2}\right)$$

$$\left[\text{if } S_1 > S_2 \Rightarrow F_1 > F_2\right]$$

Scan for Video Explanation

Web Reference at www.physicsgalaxy.com

**Module-16: Pressure at Different Levels in a Liquid**

# Pressure at different levels in a liquid:

- In a static fluid, its free surface is always normal to the effective acceleration acting on the fluid.

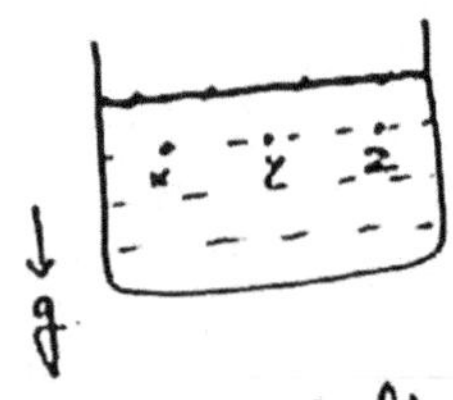


$$P_x = P_y = P_z$$

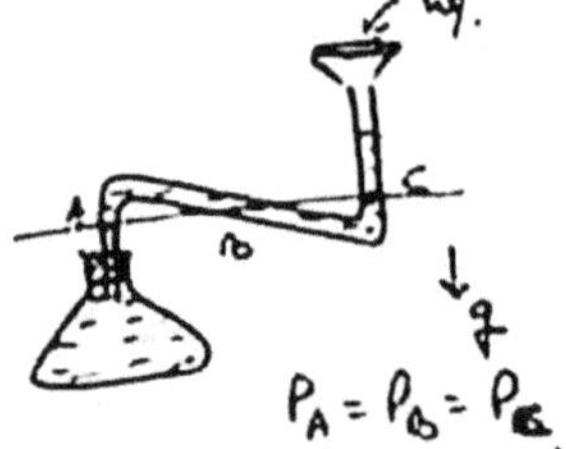


$$P_A = P_B = P_C$$

- In a layer of fluid parallel to free surface of fluid, pressure remains same for all pts.

Scan for Video Explanation

Web Reference at www.physicsgalaxy.com

**Module-16: Pressure at Different Levels in a Liquid**

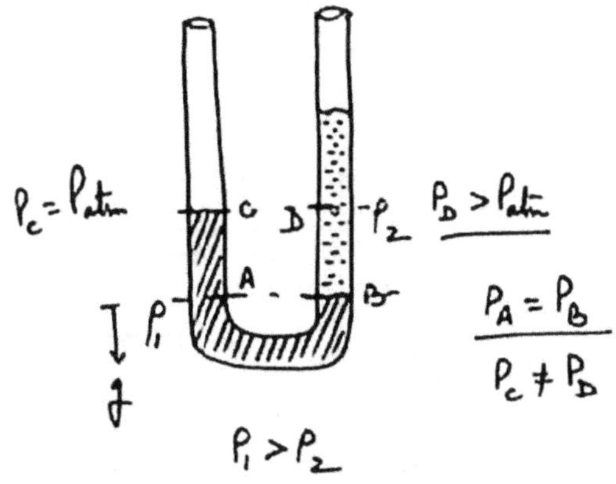


**Web Reference at www.physicsgalaxy.com** | **Scan for Video Explanation**

**Module-17: Mechanism of a Hydraulic Lift**

# Mechanism of a Hydraulic Lift:

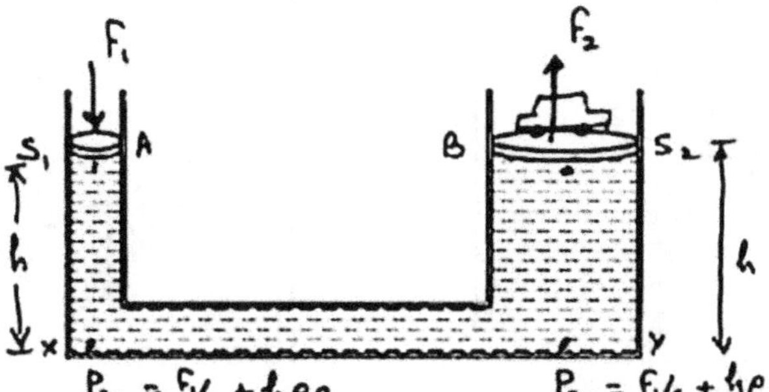


$P_{Bx} = F_1/S_1 + h\rho g$

$P_{By} = F_1/S_1 + h\rho g$

$P_A = F_1/S_1$

$P_B = \left(\frac{F_1}{S_1} + h\rho g\right) - h\rho g$

force by fluid on piston B

$F_2 = P_B \cdot S_2 = \frac{F_1}{S_1} \cdot S_2$ if $S_2 >> S_1$

$\Rightarrow F_2 >> F_1$.

**Web Reference at www.physicsgalaxy.com** | **Scan for Video Explanation**

## Module-18: Concept of Open tube Manometer

# Concept of Open Tube Manometer:

used to measure pressure gauge.

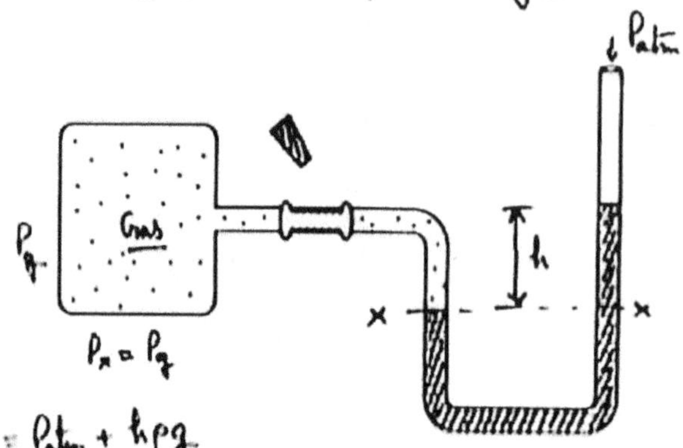


$$P_x = P_{atm} + h\rho g$$

$$\boxed{P_g - P_{atm} = h\rho g}$$



## Module-19: Pressure Distribution in an Accelerated Frame

# Pressure distribution in an accelerated frame:

In a static fluid free surface of fluid is always normal to effective acceleration acting on it.

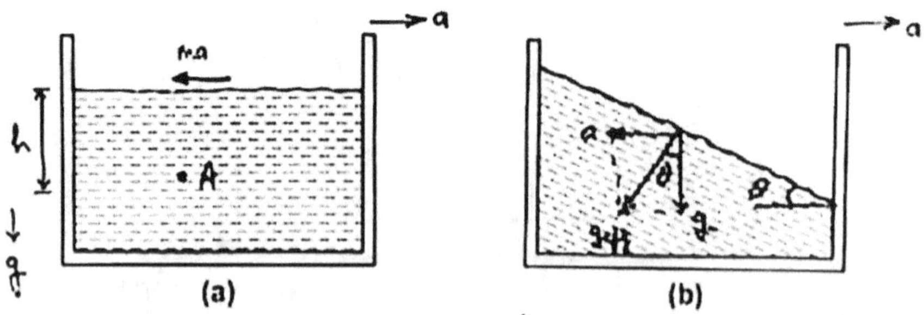


$$P_A = P_{atm} + h\rho g$$

$$g_{eff} = \sqrt{a^2 + g^2}$$

$$\theta = \tan^{-1}(a/g)$$

**Module-19: Pressure Distribution in an Accelerated Frame**

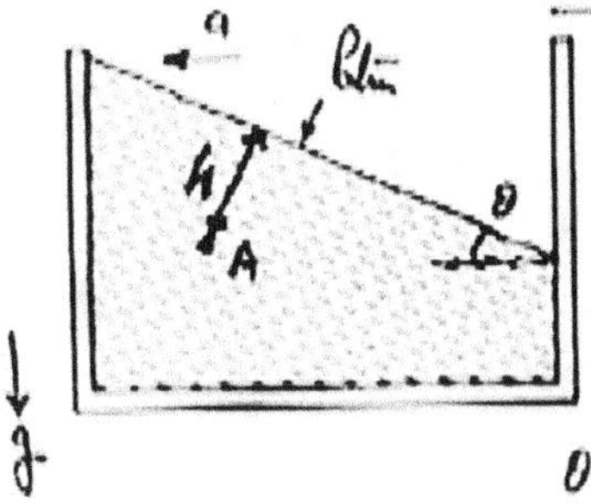


$\theta = \tan^{-1}(a/g)$

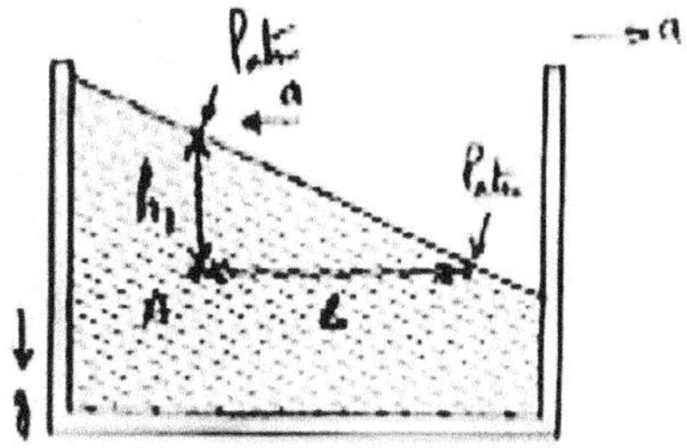


$$P_A = P_{atm} + h\rho\sqrt{g^2+a^2}$$

$$P_A = P_{atm} + h_1\rho g$$

$$P_A = P_{atm} + l\rho a$$

Web Reference at www.physicsgalaxy.com | Scan for Video Explanation

**Module-20: Pressure Distribution in a Closed Accelerated Container**

\# Pressure distribution in a closed Accelerated container:

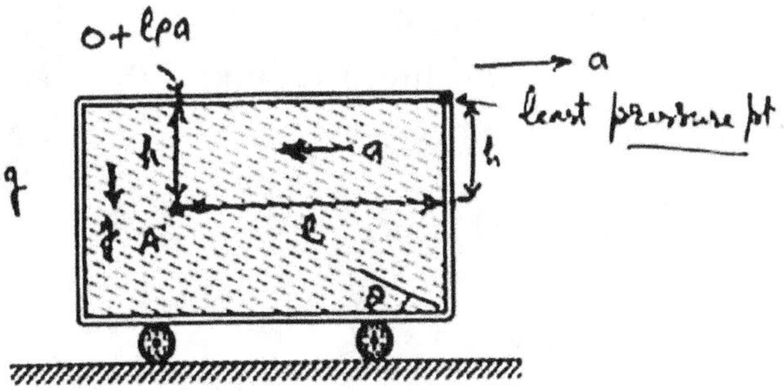


$P_A = 0 + l\rho a + h\rho g$

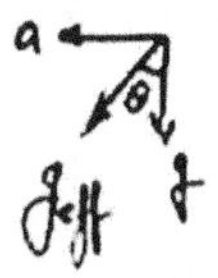


$\theta = \tan^{-1}(a/g)$

$$0 + h\rho g + l\rho a = P_A$$

$$\boxed{P_A = h\rho g + l\rho a}$$

Web Reference at www.physicsgalaxy.com | Scan for Video Explanation

**Module-21: Pressure Distribution in a Rotating tube Filled with a Liquid**

# Pressure distribution in a rotating tube filled with a liquid:

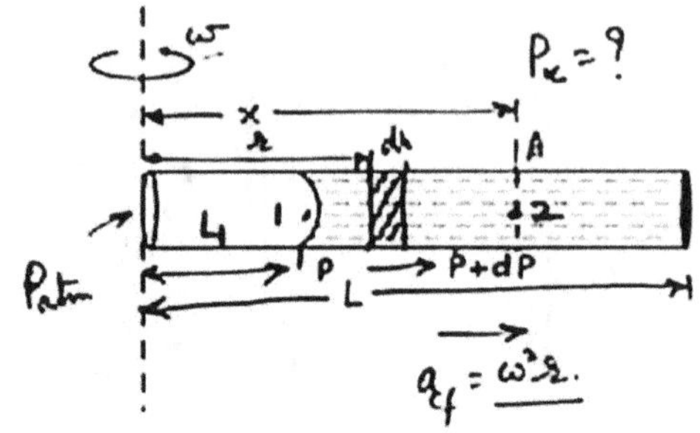


$$dP = dr \cdot \rho \cdot \omega^2 r$$

$$P_2 - P_1 = \int dP = \int_{L_1}^{x} \rho\omega^2 r\, dr = \rho\omega^2 \left[\frac{r^2}{2}\right]_{L_1}^{x}$$

$$P_2 - P_1 = \frac{1}{2}\rho\omega^2 [x^2 - L_1^2]$$

$$\boxed{P_A = P_{atm} + \frac{1}{2}\rho\omega^2 (x^2 - L_1^2)}$$

Scan for Video Explanation

Web Reference at www.physicsgalaxy.com

**Module-22: Solved Example-6**

Ex : The liquids shown in figure in the two arms are mercury (specific gravity = 13.6) and water. If the difference of height of the mercury column is 2 cm, find the height of the water column.

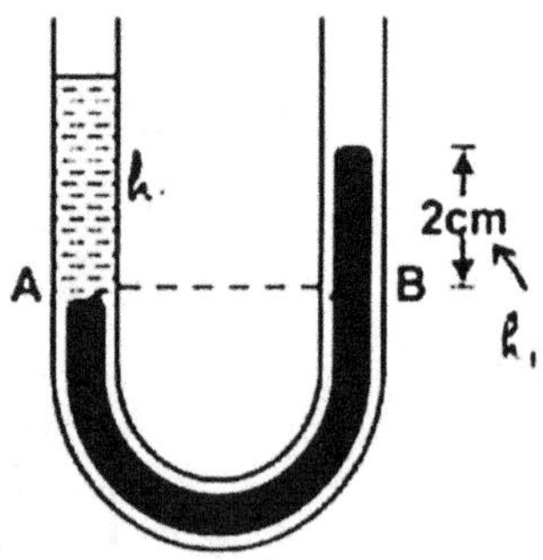


Sol: Pressure at pt A $P_A = P_{atm} + h\rho_w g$

Pressure at pt B $P_B = P_{atm} + h_1 \rho_{Hg} g$

we can use $P_A = P_B$

$$h\rho_w g = h_1 \rho_{Hg} g$$

$h = h_1 \times 13.6 = 2 \times 13.6 = 27.2$ cm Ans.

Scan for Video Explanation

Web Reference at www.physicsgalaxy.com

**Module-23: Solved Example-7**

**Ex :** **An open U-tube of uniform cross section contains some mercury. When 40.8 cm length of water is poured in one arm of U-tube, find how high does the mercury rise in other arm from its initial level ? Take $\rho_{\omega} = 1 gm/cm^3$ and $\rho_{Hg} = 13.6 gm/cm^3$.**

Solⁿ: In final state, we have -

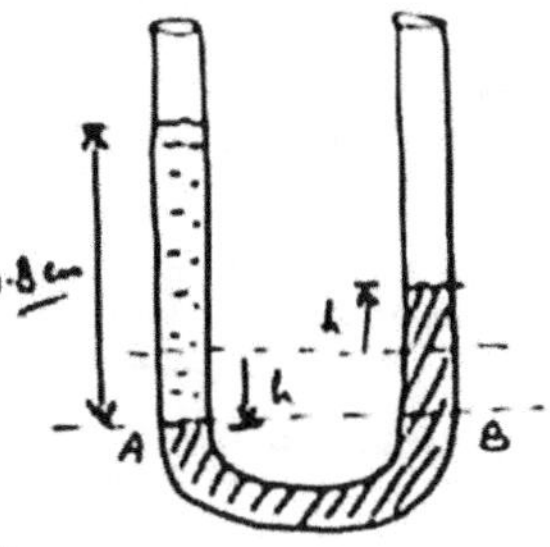


$$P_A = P_B$$

$$P_{atm} + 40.8 \times 1 \times g = P_{atm} + 2h \times 13.6 \times g$$

$$h = \frac{40.8}{2 \times 13.6} = 1.5 \text{ cm Ans.}$$

Web Reference at www.physicsgalaxy.com | Scan for Video Explanation

**Module-24: Solved Example-8**

**Ex :** **In the figure shown the liquid in manometer tube is of specific gravity 0.9. If the difference in levels of the liquid in tube is 40 cm, find the height difference of water level in the two tanks.**

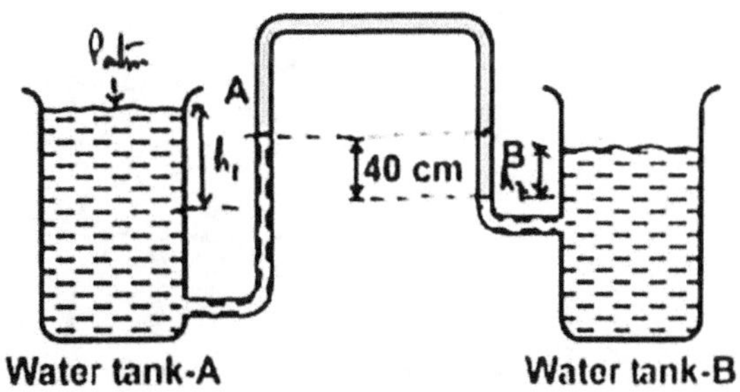


Solⁿ:

$$P_A + 40.\rho_l.g = P_B$$

$$P_{atm} + (h_1 - 40) \times 1 \times g + 40 \times 0.9.g = P_{atm} + h_2 \times 1 \times g$$

$$h_1 - 40 + 40 \times 0.9 = h_2$$

$$h_1 - h_2 = 40 \times 0.1 = 4 \text{ cm Ans}$$

Web Reference at www.physicsgalaxy.com | Scan for Video Explanation

**Module-25: Solved Example-9**

**Ex :** Figure shows a U-tube of base length L in which a liquid of density ρ is filled such that it completely fills the base length only. If the tube is now rotated at angular speed ω as shown, find the level rise of liquid in outer arm of tube.

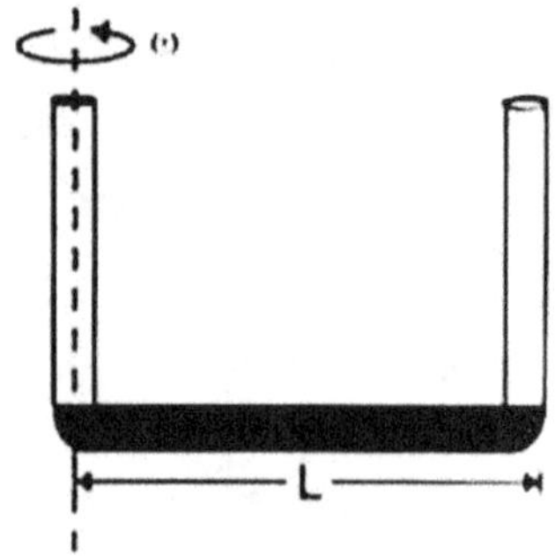


Sol: Pressure at x $P_x = P_{atm} + y\rho g$

on the base length of U-tube, we have

$$dP = dx.\rho.\omega^2 x$$

$$P_x - P_{atm} = \int_y^L dP = \int \rho\omega^2 x\,dx$$

$$P_x = P_{atm} + \frac{\rho\omega^2}{2}[L^2 - y^2] = P_{atm} + y\rho g$$

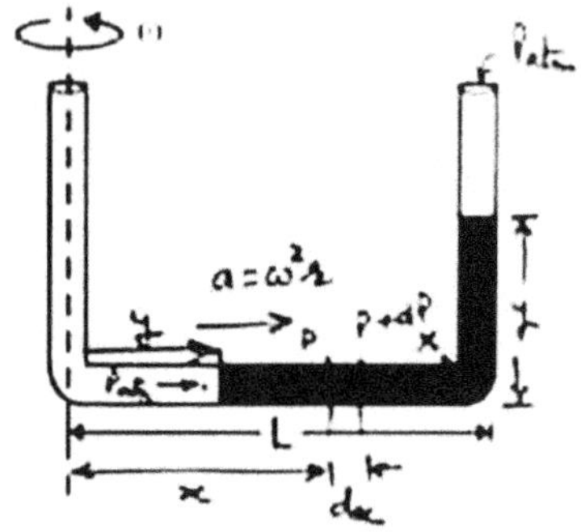


Scan for Video Explanation

Web Reference at www.physicsgalaxy.com

**Module-25: Solved Example-9**

$$\frac{\rho\omega^2}{2}[L^2 - y^2] = y\rho g$$

$$y^2 + \frac{2g}{\omega^2}y - L^2 = 0$$

on Solving

$$y = \frac{-\frac{2g}{\omega^2} + \sqrt{4g^2/\omega^4 + 4L^2}}{2} \quad \text{Ans.}$$

Scan for Video Explanation

Web Reference at www.physicsgalaxy.com

# Chapter 13
# Fluid Dynamics

## Lecture Notes Modules


**Module-1:** Introduction to Fluid Dynamics
**Module-2:** Representation of Streamlines
**Module-3:** Velocity Profile of a Flowing Fluid
**Module-4:** Turbulent Flow
**Module-5:** Equation of Continuity
**Module-6:** Solved Example-1
**Module-7:** Bernoulli's Theorem
**Module-8:** Pressure-Velocity Trade-off
**Module-9:** Solved Example-2
**Module-10:** Solved Example-3
**Module-11:** Solved Example-4
**Module-12:** Solved Example-5
**Module-13:** Working and Application of Pitot Tube
**Module-14:** Working and Application of Venturimeter
**Module-15:** Solved Example-6
**Module-16:** Solved Example-7
**Module-17:** Torecelli's Theorem
**Module-18:** Solved Example-8
**Module-19:** Solved Example-9
**Module-20:** Cross-section of a Freely Falling Liquid
**Module-21:** Reaction due to Ejection of a Liquid
**Module-22:** Solved Example-10
**Module-23:** Solved Example-11

**Module-1: Introduction to Fluid Dynamics**

# Fluid Dynamics: study & analyze motion of fluids.

For basic analysis we use ideal fluids instead of real fluids.

For an ideal fluid we make some assumptions –

(1) Streamline flow: Also termed as laminar flow.
at a point vel of fluid part does not change in mag/dir.

(2) Incompressible flow: we assume that during flow fluid density remain Constant

(3) Non viscous flow: For ideal fluids we neglect viscous forces between fluid layers.

(4) Irrotational flow: we assume that fluid particles do not impart any rotational motion to the uniform bodies in path of fluid which are moving along with fluid.

Scan for Video Explanation

Web Reference at www.physicsgalaxy.com

**Module-2: Representation of Streamlines**

# Representation of Streamlines: hypothetical lines which represent fluid flow.

- Stream lines in a flowing fluid are the representation of fluid particles motion
- For a flowing fluid dir of flow vel of fluid part is given by the dir of stream lines & speed of fluid flow is meas by the density of streamlines
- In a region where flow vel is more ⇒ s. lines are denser & in a region where flow vel is less ⇒ s. lines are rarer.

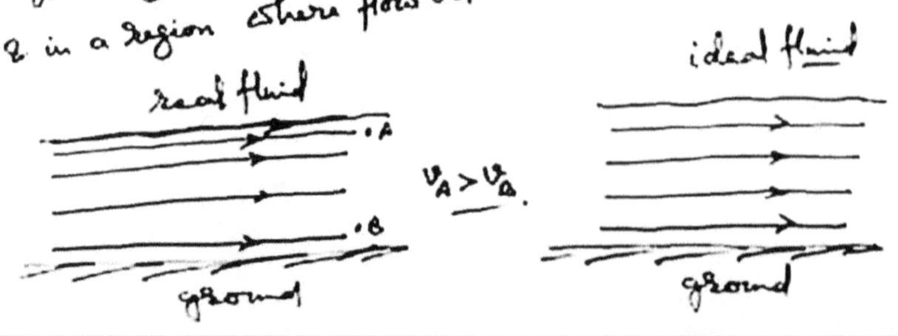


Scan for Video Explanation

Web Reference at www.physicsgalaxy.com

**Module-3: Velocity Profile of a Flowing Fluid**

# Velocity Profile of a flowing fluid:

Due to friction between layers of a fluid, the particles which are close to walls/surface of a pipe tend to move slowly (due to fr) and those which are away from walls/surfaces move relatively fast.

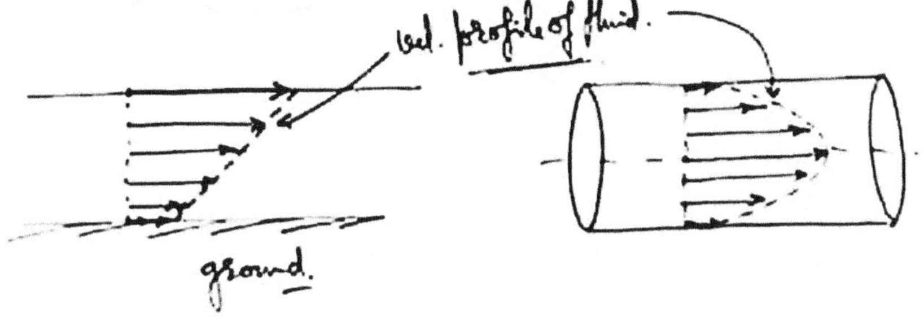


Web Reference at www.physicsgalaxy.com

Scan for Video Explanation

**Module-4: Turbulent Flow**

# Turbulent Flow:

When in a streamline flow, flow vel of fluid increases, at a certain point flow particles become eratic, no streamlines exist after this & this flow is called turbulent flow.

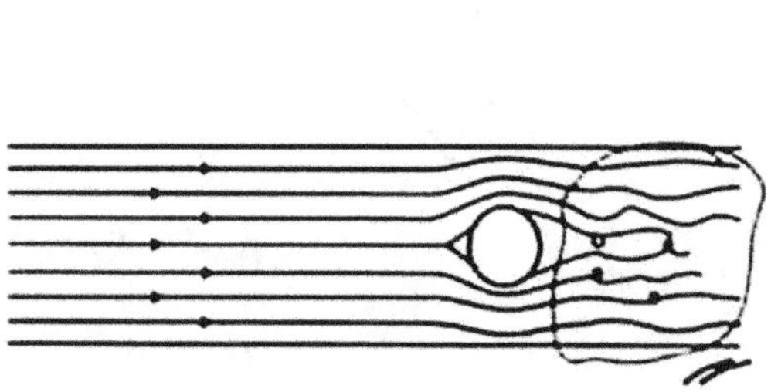

Turbulent flow

Web Reference at www.physicsgalaxy.com

Scan for Video Explanation

**Module-5: Equation of Continuity**

# Equation of Continuity:

This eqⁿ defines that for streamline flow for a fluid the mass entring per second at one end is equal the mass leaving/sec at the other end.

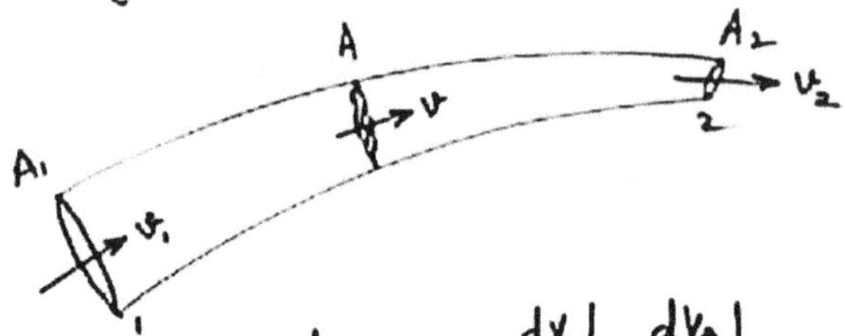


Acc. to eqⁿ of Continuity

$$\left.\frac{dV_1}{dt}\right|_1 = \left.\frac{dV_2}{dt}\right|_2$$

$$\boxed{A_1 v_1 = A_2 v_2}$$

$$Av = \text{Constant}$$

Scan for Video Explanation

Web Reference at www.physicsgalaxy.com

**Module-6: Solved Example-1**

**Ex :** A river of width 12 m flowing at 20 kph mixes with another river of width 8 m flowing at 16 kph and spreads in another stream of width 16 m. Find the flow velocity of water after mixing assuming the depth of rivers are same.

Solⁿ:

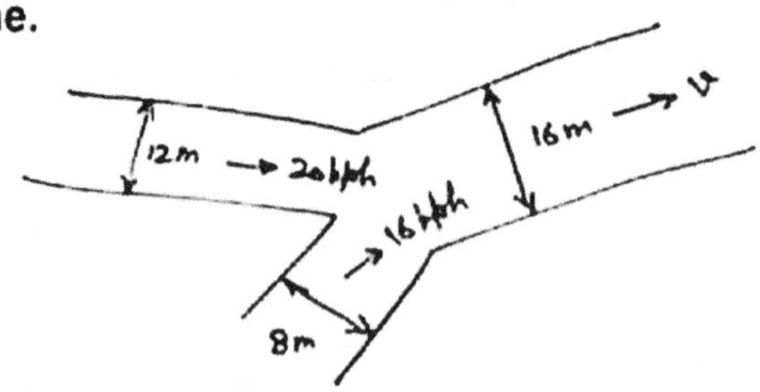


Acc. to Eqⁿ of Continuity

$$A_1 v_1 + A_2 v_2 = A_3 v_3$$

$$12 \times 20 + 8 \times 16 = 16 \times v$$

$$v = \frac{\overset{3\times 5}{\cancel{12\times 20}} + 8 \times \cancel{16}}{\cancel{16}}$$

$$= 15 + 8 = 23 \text{ kph} \quad \text{Ans.}$$

Scan for Video Explanation

Web Reference at www.physicsgalaxy.com

**Module-7: Bernoulli's Theorem**

# Bernoulli's Theorem:

This is a theorem about conservation of energy for a flowing fluid from one end to another in streamline flow.

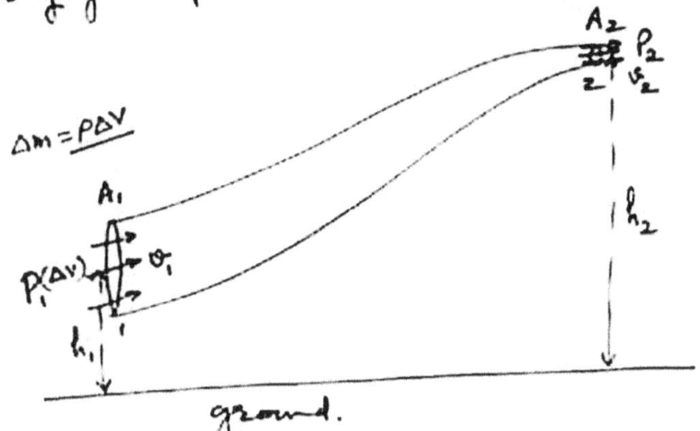


use W/E theorem for a volume $\Delta V$ of fluid from end 1 to end 2,

$$\frac{1}{2}[\rho \Delta V]v_1^2 + P_1\Delta V - [\rho\Delta V]g(h_2 - h_1) - P_2\Delta V = \frac{1}{2}[\rho\Delta V]v_2^2$$

($K_i$ on the left, $K_f$ on the right)



**Module-7: Bernoulli's Theorem**

$$\frac{1}{2}[\rho \Delta V]v_1^2 + P_1\Delta V - [\rho\Delta V]g(h_2 - h_1) - P_2\Delta V = \frac{1}{2}[\rho\Delta V]v_2^2$$

$$\frac{1}{2}\rho v_1^2 + P_1 - \rho g h_2 + \rho g h_1 - P_2 = \frac{1}{2}\rho v_2^2$$

$$\frac{1}{2}\rho v_1^2 + P_1 + \rho g h_1 = \frac{1}{2}\rho v_2^2 + P_2 + \rho g h_2$$

at all sections of the flowing fluid body, we can use -

Bernoulli's Eq$^n$ → $\boxed{\frac{1}{2}\rho v^2 + P + \rho g h = \text{Constant}}$

- $\frac{1}{2}\rho v^2$ → KE per unit vol of fluid
- $P$ → Pressure Energy per unit volume.
- $\rho g h$ → grav. PE per unit volume w.r.t. given Ref.

**Module-8: Pressure-Velocity Trade-off**

# Pressure-Velocity Trade-off:

For a fluid flowing at a horizontal level acc. to Bernoulli's theorem, we use -

$$P + \frac{1}{2}\rho v^2 = \text{Constant}$$

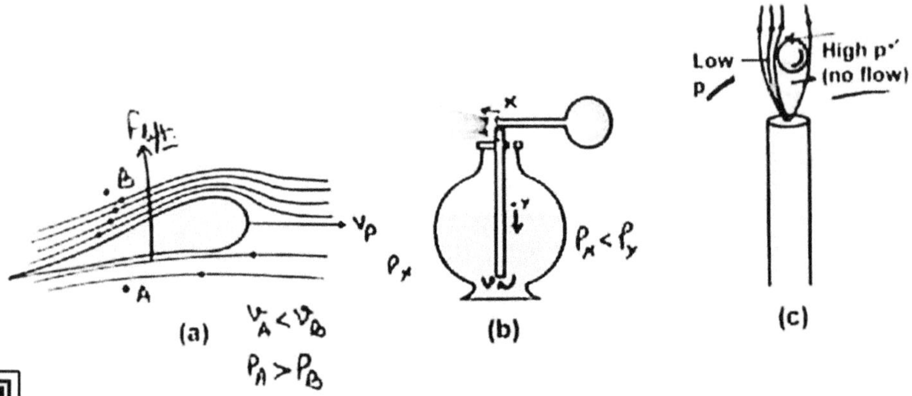


Scan for Video Explanation

Web Reference at www.physicsgalaxy.com

**Module-9: Solved Example-2**

**Ex :** In a horizontal pipe line, the flowing oil pressure falls by 8 N/m$^2$ between two points separated by 1 km. Find the change in kinetic energy per kg of oil at these points. Take density of oil 800 kg/m$^3$.

**Sol$^n$:** By using Bernoulli's theorem, we have

$$P_1 + \frac{1}{2}\rho v_1^2 = P_2 + \frac{1}{2}\rho v_2^2$$

$$\frac{1}{2}\rho(v_2^2 - v_1^2) = P_1 - P_2 = 8 \text{ N/m}^2$$

$$\text{change in KE/kg} = \frac{1}{2}(v_2^2 - v_1^2) = \frac{P_1 - P_2}{\rho} = \frac{8}{800} = 0.01 \text{ J/kg} \quad \text{Ans.}$$

Scan for Video Explanation

Web Reference at www.physicsgalaxy.com

**Module-10: Solved Example-3**

**Ex :** **Calculate the gross lift force on an airplane wing of length 10 m and width 2 m. The air speed above the wing is 120 m/s and below the wing is 90 m/s. Take air density 1.3 kg/m$^3$.**

Sol$^n$:

lift force $F = (P_B - P_T) \times A$

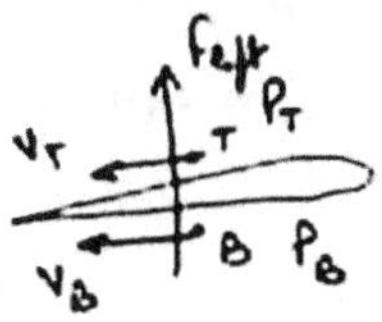


$P_B > P_T$

we use Bernoulli's theorem at pts T & B –

$$P_T + \frac{1}{2}\rho v_T^2 = P_B + \frac{1}{2}\rho v_B^2$$

$$P_B - P_T = \frac{1}{2}\rho (v_T^2 - v_B^2)$$

$$= \frac{1}{2} \times 1.3 \times [(120)^2 - (90)^2] = 4.1 \times 10^3 \text{ N/m}^2$$

Gross lift force $F = (P_B - P_T) \times A$

$$= 4.1 \times 10^3 \times 20 = 8.2 \times 10^4 \text{ N} \quad \text{Ans.}$$

Web Reference at www.physicsgalaxy.com

Scan for Video Explanation

**Module-11: Solved Example-4**

**Ex :** **Water flows through a tube which has one end A at a height 20 cm above ground and cross-sectional area 1 cm$^2$ and other end B at a height 15 cm above ground and cross-sectional area 0.5 cm$^2$. The speed of water at A is 10 cm/s find the speed at end B and pressure difference at A and B.**

Sol$^n$: Using eq$^n$ of Continuity

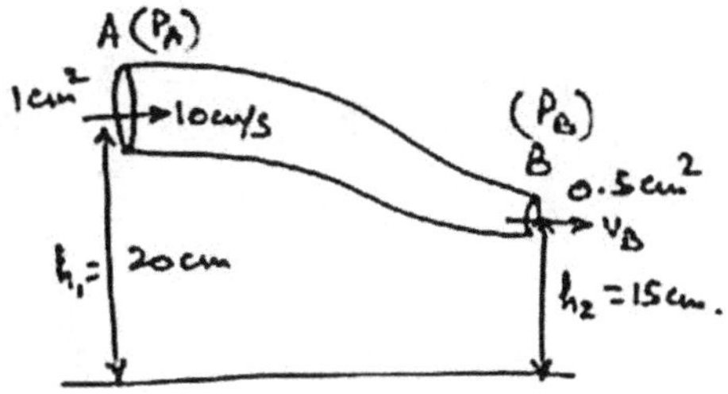


$$A_1 v_1 = A_2 v_2$$

$$1 \times 10 = 0.5 \times v_B$$

$$v_B = 20 \text{ cm/s} \quad \text{Ans.}$$

using Bernoulli's Eq$^n$ at ends A & B.

$v_A = 0.1$

$v_B = 0.2$

$$P_A + \frac{1}{2}\rho v_A^2 + \rho g h_1 = P_B + \frac{1}{2}\rho v_B^2 + \rho g h_2$$

$$P_B - P_A = \rho g (h_1 - h_2) + \frac{1}{2}\rho (v_A^2 - v_B^2)$$

$$= 1000 \times 10 \times 0.05 + \frac{1}{2} \times 1000 [0.01 - 0.04] = 500 - 15$$

$$= 485 \text{ N/m}^2 \quad \text{Ans}$$

Web Reference at www.physicsgalaxy.com

Scan for Video Explanation

**Module-12: Solved Example-5**

**Ex :** **In a bucket shown in figure a liquid of density $1.25 \times 10^3$ kg/m$^3$ is kept. Find speed at which liquid will be coming out if the bottom surface is removed and its top level is maintained by pouring the liquid through the tap.**

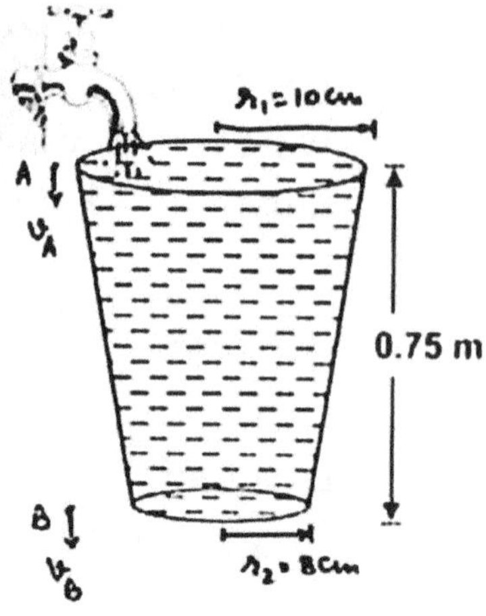


Solⁿ: using Bernoulli's eqⁿ at A and B.

$$P_{atm} + \frac{1}{2}\rho v_A^2 + \rho g h = P_{atm} + \frac{1}{2}\rho v_B^2$$

$$A_A v_A = A_B v_B$$

$$v_B = v_A \cdot \frac{r_1^2}{r_2^2}$$

$$v_A = v_B \cdot \frac{r_2^2}{r_1^2}$$

$$v_B^2\left[1 - \frac{r_2^4}{r_1^4}\right] = 2gh$$

$$v_B = \sqrt{\frac{2gh}{1 - r_2^4/r_1^4}} = \sqrt{\frac{2 \times 10 \times 0.75}{1 - (0.8)^4}} = 6.45 \text{ m/s}$$

Ans.

Scan for Video Explanation

Web Reference at www.physicsgalaxy.com

**Module-13: Working and Application of Pitot Tube**

# Working and Application of a 'Pitot Tube':

↓ A device used to meas flow vel of a fluid.

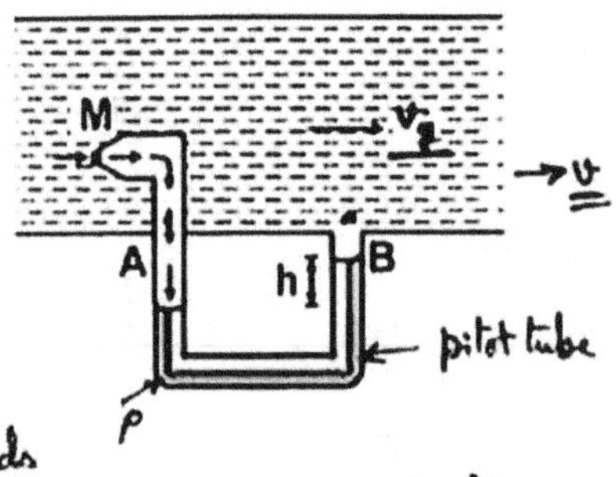


if flowing fluid is a gas.

$$P_A - P_B = h\rho g$$

$$P_A - P_B = h(\rho - \rho_l)g$$

if flowing fluid is a liquids

if we use Bernoulli's th at pts A & B in the flowing fluid,

$$P_A + 0 = P_B + \frac{1}{2}\rho v^2 \Rightarrow \boxed{v = \sqrt{\frac{2(P_A - P_B)}{\rho}}}$$

Scan for Video Explanation

Web Reference at www.physicsgalaxy.com

**Module-14: Working and Application of Venturimeter**

# Working and Applications of 'Venturimeter':

A device used to meas flow vel of liquids in pipelines.

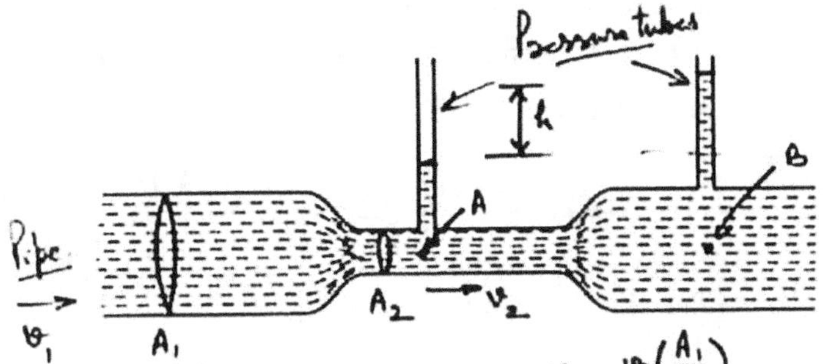


by Cont eqn $A_1 v_1 = A_2 v_2 \longrightarrow v_2 = v_1\left(\frac{A_1}{A_2}\right)$

using Bernoulli's th at pts A & B - $\frac{1}{2}\rho v_2^2 + P_A = \frac{1}{2}\rho v_1^2 + P_B$

$$\frac{1}{2}\rho(v_2^2 - v_1^2) = P_B - P_A = h\rho g$$

$$\frac{1}{2}v_1^2\left[\frac{A_1^2}{A_2^2} - 1\right] = hg$$

$$\boxed{v_1 = \sqrt{2gh\Big/\left(\frac{A_1^2}{A_2^2} - 1\right)}}$$



**Module-15: Solved Example-6**

**Ex :** In a venturi meter diameter of the pipe is 4 cm and that of constriction is 3 cm. When water flows the heights of water level in pressure tubes are 15 cm and 20 cm. Find the discharge rate of pipe. (Take g = 10 m/s$^2$)

**Soln:**

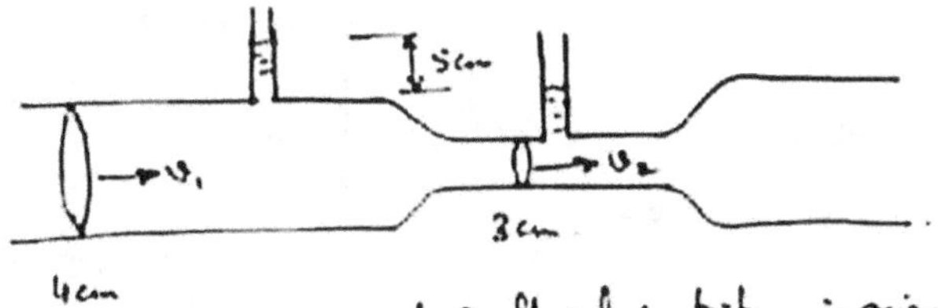


in venturimeter the flow vel of fluid in pipe is given as-

$$v_1 = \sqrt{\frac{2gh}{\left(\frac{A_1^2}{A_2^2} - 1\right)}} = \sqrt{\frac{2 \times 10 \times 0.05}{\left(\frac{4}{3}\right)^4 - 1}} = \sqrt{\frac{1}{\frac{256}{81} - 1}} = \frac{9}{5\sqrt{7}} \text{ m/s}$$

$$\text{Discharge Rate of pipe} = A_1 v_1 = \pi(2)^2 \times \frac{900}{5\sqrt{7}}$$

$$= \frac{3600\pi}{5\sqrt{7}} \text{ cm}^3/\text{sec} \quad \text{Ans.}$$

**Module-16: Solved Example-7**

**Ex :** **Water flows through the tube shown in figure. The areas of cross-section of wide and narrow sections are 5 cm² and 2 cm² respectively. Rate of water flow in tube is 500 cm³/sec. Find the difference in mercury levels in the U-tube**

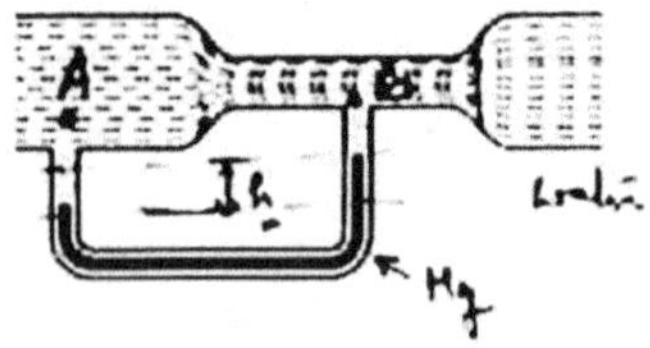


Sol$^n$: Acc. to Bernoulli's theorem at pts A & B.

$$P_A + \frac{1}{2}\rho v_A^2 = P_B + \frac{1}{2}\rho v_B^2$$

$$P_A - P_B = \frac{1}{2}\rho(v_B^2 - v_A^2)$$

Continuity eq$^n$ –

$$A_A v_A = A_B v_B = 500$$

$$v_A = 100 \text{ cm/s}$$

$$v_B = 250 \text{ cm/s}.$$

$$P_A - P_B = h(\rho_{Hg} - \rho)g$$

$$h(13600 - 1000) \times 10 = \frac{1}{2}\rho\left[(2.5)^2 - (1)^2\right]$$

$$h = \frac{5.25}{2 \times 126} = 0.0208 \text{ m}$$

$$= 2.08 \text{ cm} \text{ Ans.}$$

Scan for Video Explanation

Web Reference at www.physicsgalaxy.com

**Module-17: Torecelli's Theorem**

# Torricelli's Theorem:

This theorem is abt calculation of efflux vel of a fluid flowing out through a hole in a container.

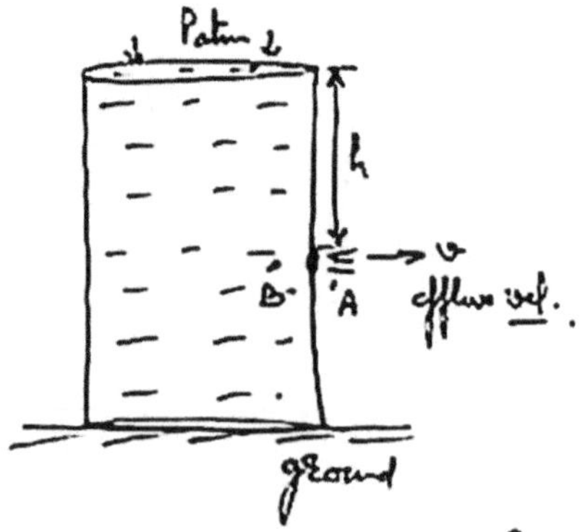


If we use Bernoulli's theorem at just inside & outsides pts of hole –

$$P_{atm} + \frac{1}{2}\rho v^2 = P_{atm} + h\rho g + 0$$

$$\boxed{v = \sqrt{2gh}}$$

Torricelli's eq$^n$ of efflux vel.

area of hole << area of Cont.

Scan for Video Explanation

Web Reference at www.physicsgalaxy.com

**Module-18: Solved Example-8**

**Ex :** Water stands upto a height H in a tank as shown. A hole is made at a depth h below the open end of tank from which water ejects out. Find h for which the range of ejecting water on ground will be maximum.

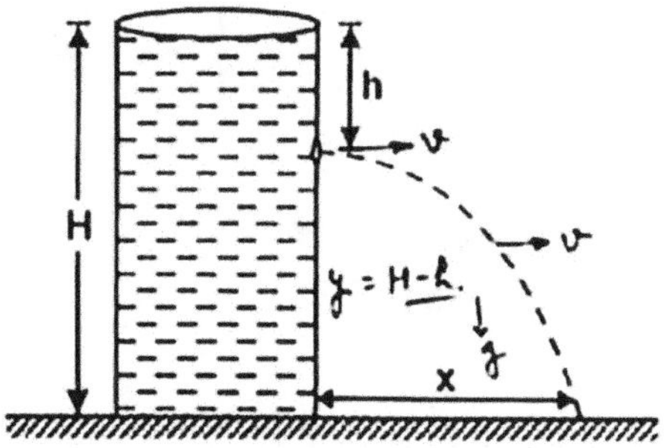


Soln: As we know, the efflux vel of water is $v=\sqrt{2gh}$.

time to reach ground level by water is $t=\sqrt{\frac{2y}{g}}=\sqrt{\frac{2(H-h)}{g}}$

horizontal range on ground $x = vt = \sqrt{2gh}\cdot\sqrt{\frac{2(H-h)}{g}} = 2\sqrt{h(H-h)}$

for x to be max $\frac{dx}{dh}=0 \Rightarrow \frac{2}{2\sqrt{h(H-h)}}\times(H-2h)=0$

$h=H/2$ Ans.

Web Reference at www.physicsgalaxy.com | Scan for Video Explanation

**Module-19: Solved Example-9**

**Ex :** Figure shows a cylindrical tank of cross-sectional area A filled with water upto its full height H. A small hole of area a is made at its bottom through which water starts flowing. Find the time taken in which the tank will get empty to half volume.

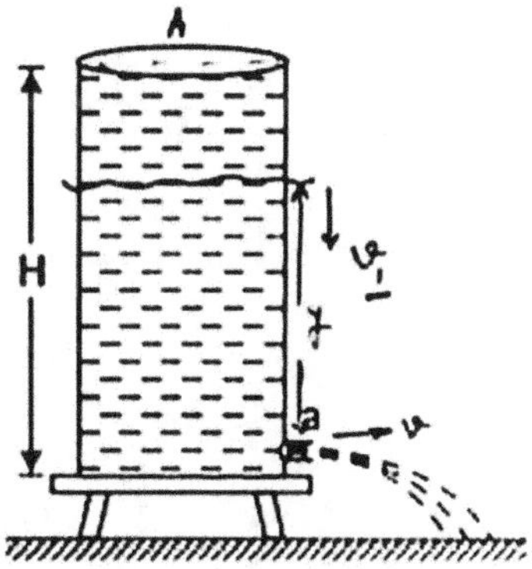


Soln: efflux vel $v=\sqrt{2gy}$

Acc to Cont eqn

$Av_1 = a.v$

$v_1 = \frac{a}{A}.v = -\frac{dy}{dt}$

$\frac{dy}{dt} = -\frac{a}{A}\sqrt{2gy}$

$\int_H^{H/2}\frac{1}{\sqrt{y}}\,dy = -\frac{a}{A}\sqrt{2g}\int_0^t dt$

$\left[2\sqrt{y}\right]_H^{H/2} = -\frac{a}{A}\sqrt{2g}\,t$

$t=\frac{A}{a}\frac{1}{\sqrt{2g}}\left[2\sqrt{H}-\sqrt{H/2}\right]$

$t=\frac{A}{a}\sqrt{\frac{H}{g}}(\sqrt{2}-1)$ Ans.

Web Reference at www.physicsgalaxy.com | Scan for Video Explanation

**Module-20: Cross-section of a Freely Falling Liquid**

# Cross-Section of a freely falling liquid:

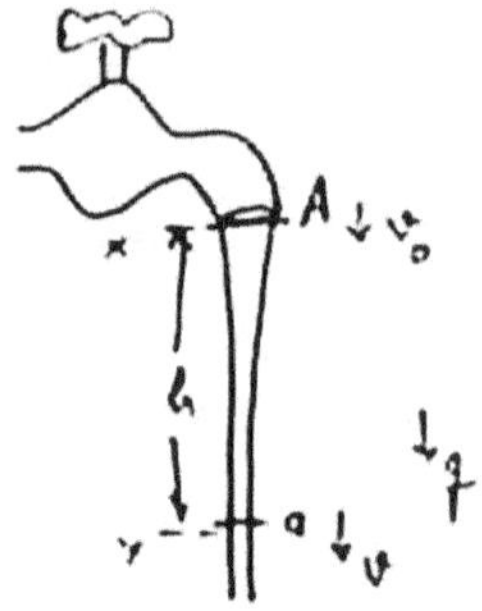


we can use Bernoulli's Th at pts x & Y –

$$\cancel{P_{atm}} + \frac{1}{2}\cancel{\rho} v_0^2 + \cancel{\rho} gh = \cancel{P_{atm}} + \frac{1}{2}\cancel{\rho} v^2$$

$$v^2 = v_0^2 + 2gh$$

using eqn of Continuity

$$v_0 A = v a$$

$$a = \frac{v_0}{v} A$$

$$\left[ a = \frac{v_0 A}{\sqrt{v_0^2 + 2gh}} \right]$$

Scan for Video Explanation

Web Reference at www.physicsgalaxy.com

**Module-21: Reaction due to Ejection of a Liquid**

# Reaction due to ejection of a liquid:

Flow rate of liquid ejecting out is

$$\frac{dV}{dt} = Av$$

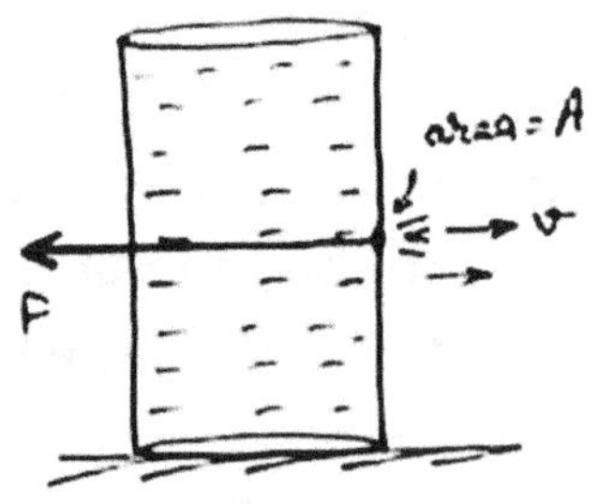


mass rate of ejecting liq is

$$\frac{dm}{dt} = \rho A v$$

momentum ejecting per ~~secto~~ second in fluid is

$$F = (\rho A v) v = \rho A v^2$$

force of reaction on Container due to ej of liq is $\boxed{F_R = \rho A v^2}$

Scan for Video Explanation

Web Reference at www.physicsgalaxy.com

**Module-22: Solved Example-10**

Ex : On the opposite sides of a wide vertical vessel filled with water, two identical holes are opened, each of area S. The height difference between them is Δh. Find the net force on vessel due to water flowing out of the vessel.

Sol$^n$: net force on vessel due to ejection of water

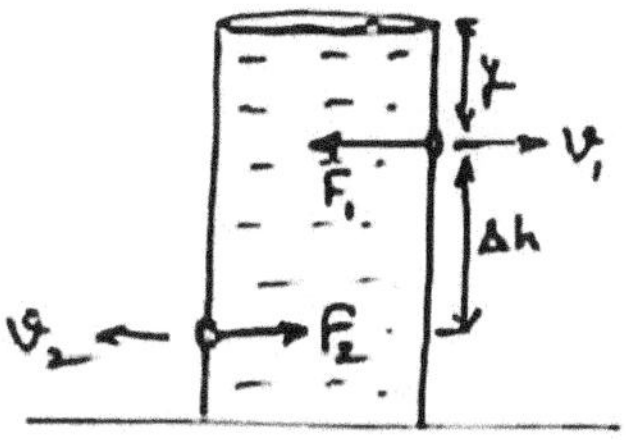


$$F_{net} = F_2 - F_1$$
$$= \rho S(v_2^2 - v_1^2)$$
$$= \rho S[2g\Delta h]$$
$$F_{net} = 2\rho Sg\Delta h \quad \text{Ans.}$$

$$F = \rho A v^2$$
$$v_1 = \sqrt{2gy}$$
$$v_2 = \sqrt{2g(y+\Delta h)}$$

Web Reference at www.physicsgalaxy.com | Scan for Video Explanation

**Module-23: Solved Example-11**

Ex : A cylindrical vessel filled with water upto a height of 2m stand on a rough horizontal plane. At its bottom in its side wall a small hole is opened and it is found that the vessel slides on plane as water comes out. Find the diameter of hole if friction coefficient between vessel and plane is 0.4 and total mass of vessel with water is 100 kg.

Sol$^n$: Reaction force due to ejection of water

$$F = \rho A v^2$$

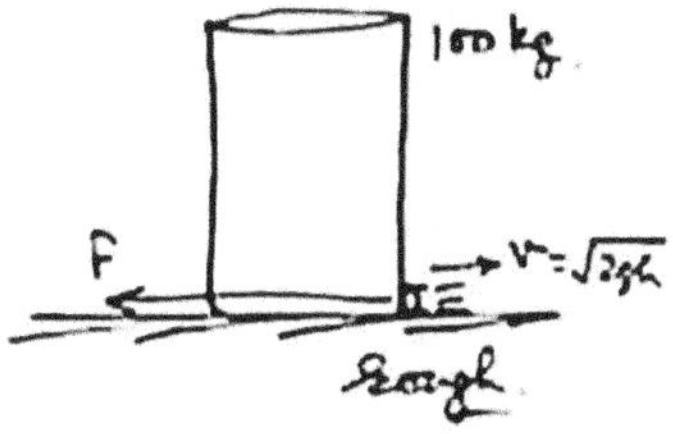


to slide the vessel

$$f_L = \mu Mg = 0.4 \times 100 \times 10 = 400\text{N}.$$

$$\rho A v^2 = 400$$
$$\rho\left(\frac{\pi d^2}{4}\right)v^2 = 400$$
$$d = \sqrt{\frac{4 \times 400}{\rho \pi \times 2g \times 2}} = \sqrt{\frac{4 \times 400}{1000 \times 3.14 \times 20 \times 2}} = 0.114\text{ m. Ans.}$$

Web Reference at www.physicsgalaxy.com | Scan for Video Explanation

# Chapter 14
# Surface Tension

## Lecture Notes Modules


**Module-1:** Surface Tension
**Module-2:** Significance of Surface Tension
**Module-3:** Some Facts About Surface Tension
**Module-4:** Solved Example-1
**Module-5:** Solved Example-2
**Module-6:** Concept of Surface Energy
**Module-7:** Solved Example-3
**Module-8:** Solved Example-4
**Module-9:** Excess Pressure Inside a Liquid Drop
**Module-10:** Excess Pressure Inside a Soap Bubble
**Module-11:** Excess Pressure Inside a Surface Having Two Curvatures
**Module-12:** Connecting Soap Bubbles of Different Radii
**Module-13:** Angle of Contact
**Module-14:** Shape of Liquid Meniscus in a Tube
**Module-15:** Capillary Action
**Module-16:** Force Between Two Parallel Plates Due to Liquid Between Them
**Module-17:** Capillary Action with a Capillary of Insufficient Height
**Module-18:** Solved Example-5
**Module-19:** Solved Example-6
**Module-20:** Solved Example-7
**Module-21:** Solved Example-8
**Module-22:** Solved Example-9

**Module-1: Surface Tension**

# Surface Tension:

The prop of a liquid due to which its free surface tries to have min surface area and it behaves if it is under a tension like a stretched membrane is called Surface Tension.

- It is due to intermolecular force among liq. particles.

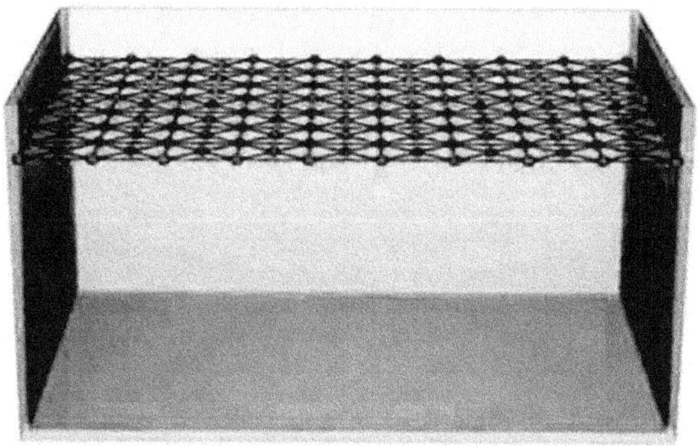

Scan for Video Explanation

Web Reference at www.physicsgalaxy.com

**Module-2: Significance of Surface Tension**

# Significance of Surface Tension:

Surface tension is numerically defined as the force acting per unit length of a hypothetical line drawn on a liq. surface normal to it in either direction.

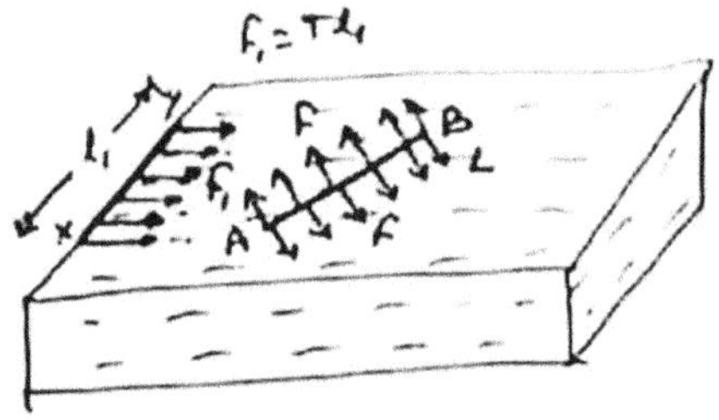


Surface tension

$T = \frac{F}{L}$ Nt/m.

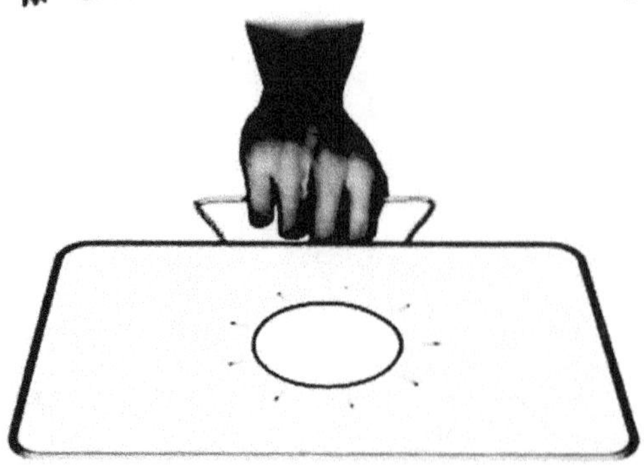

Scan for Video Explanation

Web Reference at www.physicsgalaxy.com

**Module-3 Some Facts About Surface Tension**

# Some Facts about Surface Tension:

- It is a Scalar.
- It depends only on nature of liquid & its intermolecular forces.
- ST decreases with rise in temp & becomes 0 at critical temperature.
- It depends on impurities added to a liquid.
- For insoluble impurities ST decreases & for Soluble impurities it may increase or decrease.

Web Reference at www.physicsgalaxy.com

Scan for Video Explanation

**Module-4 Solved Example-1**

**Ex :** A thin wooden rod of mass M and length *l* is floating in water. Find the force required in vertically upward direction to lift the rod from water surface. Surface tension of water is T.

Sol$^n$: To lift the rod

$$F = Mg + 2Tl.$$ Ans.

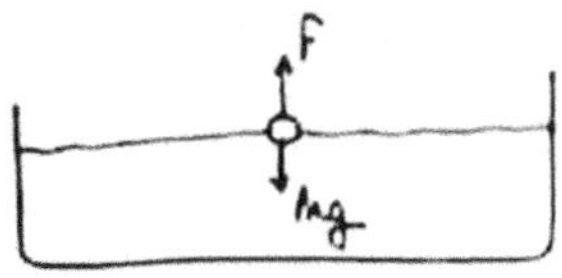


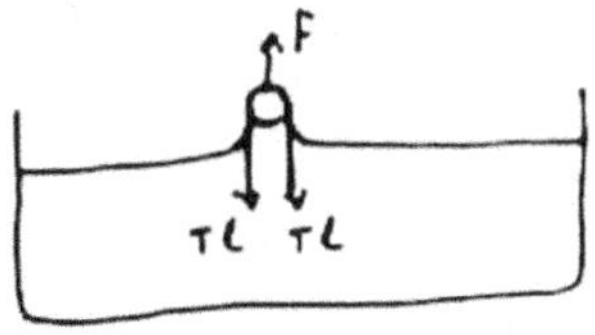


Web Reference at www.physicsgalaxy.com

Scan for Video Explanation

**Module-5 Solved Example-2**

**Ex :** A ring of radius R and mass m is kept at rest on the surface of a liquid of surface tension T by a string tied as shown in figure. Find the force required to lift the ring above the liquid.

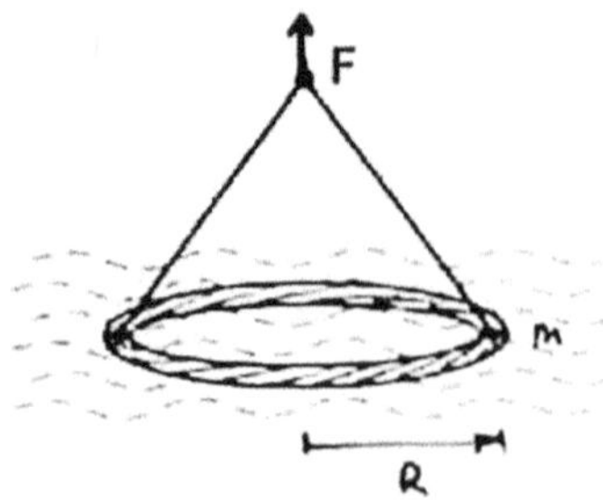


Sol$^n$: Force reqd to lift the ring

= wt of ring + force of ST

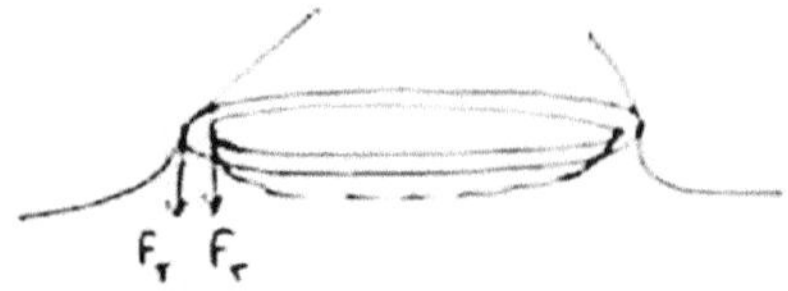


$F_T = T.2\pi R$

$F = mg + 2T(2\pi R)$

$F = mg + 4\pi RT$ Ans.

Scan for Video Explanation

Web Reference at www.physicsgalaxy.com

**Module-6 Concept of Surface Energy**

# Concept of Surface Energy:

Stretched spring → stores some p.E in it

free surface of a liquid → some PE is stored in it
↓
Surface energy

force on slider due to ST is $F = 2Tl$

to slowly move the slider out $F_{ext} = 2Tl$

W/D in displacing slider by $\Delta x$ is

$\Delta W = F_{ext}\cdot \Delta x = 2Tl\Delta x$

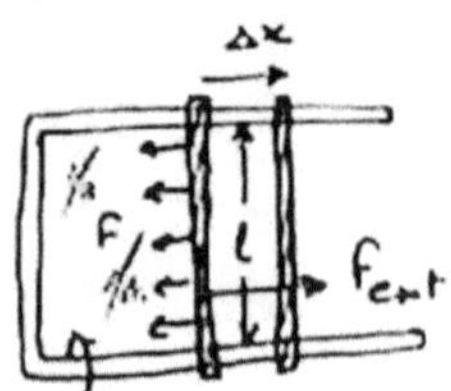


Soap film (ST = T)

inc in surface area of film

$\Delta S = 2l\Delta x$

inc in Surface energy of film ⇒ $\Delta U = \Delta W = T\Delta S$

$\boxed{T = \frac{\Delta U}{\Delta S}}$ ← Surface energy per unit area.

Scan for Video Explanation

Web Reference at www.physicsgalaxy.com

**Module-7 Solved Example-3**

Ex : A soap bubble of radius R and film thickness t (t << R) coalesce to form a liquid drop. Find the rise in temperature of drop. Surface tension of soap solution is T, density of soap solution is ρ and its specific heat is S.

Sol. Initial surface area of bubble is $A_i = 2\times 4\pi R^2 = 8\pi R^2$

If radius of liq drop formed is r we use –

$$\frac{4}{3}\pi r^3 = 4\pi R^2 t \Rightarrow r = (3R^2t)^{1/3}$$

Final surface area $A_f = 4\pi r^2 = 4\pi(3R^2t)^{2/3}$

If temp of drop rises by $\Delta\theta$, we use

$$T(A_i - A_f) = mS\Delta\theta = \frac{4}{3}\pi r^3 \rho S\Delta\theta$$

$$\Delta\theta = \frac{T(8\pi R^2 - 4\pi(3R^2t)^{2/3})}{\frac{4}{3}\pi r^3 \rho S} = \frac{T[2R^2 - (3R^2t)^{2/3}]}{\rho S R^2 t}$$ Ans.

Web Reference at www.physicsgalaxy.com | Scan for Video Explanation

**Module-8 Solved Example-4**

Ex : A film of water is formed between two straight parallel wires each 10 cm long at a separation of 0.5 cm. Calculate the work required to increase their separation by 1 mm. Surface tension of water is $72 \times 10^{-3}$ N/m.

Sol. Increase in surface area of film is $\Delta S = 2l\Delta x$

$$= 2\times 0.1\times 1\times 10^{-3}$$
$$= 2\times 10^{-4}\ m^2$$

W/D in the process is equal to increase in surface energy

$$W = T\Delta S = 72\times 10^{-3}\times 2\times 10^{-4}$$
$$= 144\times 10^{-7}$$
$$= 14.4\ \mu J$$ Ans.

Web Reference at www.physicsgalaxy.com | Scan for Video Explanation

## Module-9 Excess Pressure Inside a Liquid Drop

# Excess Pressure inside a liquid drop:

Due to ST, the surface of a liq drop tend to contract & compresses the enclosed matter due to which pressure inside increase

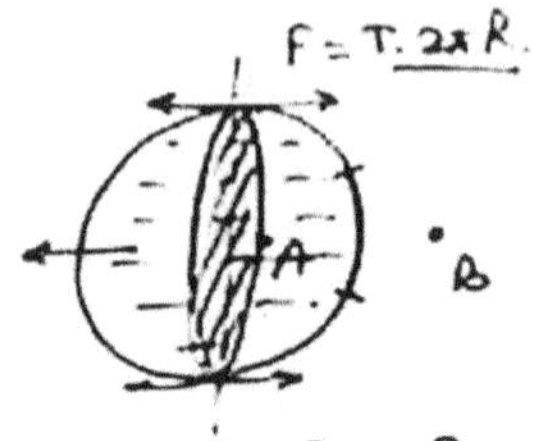


$P_A > P_B$

In eq$^n$ of a drop, the diff in inside & outside pressure is called excess pressure $(P_{ex})$

for eq$^n$ of the hemispherical section of drop.

$$T.(2\pi R) = P_{ex}.\pi R^2$$

$$\boxed{P_{ex} = \frac{2T}{R}}$$

Scan for Video Explanation

Web Reference at www.physicsgalaxy.com

## Module-10 Excess Pressure Inside a Soap Bubble

# Excess Pressure inside a Bubble:

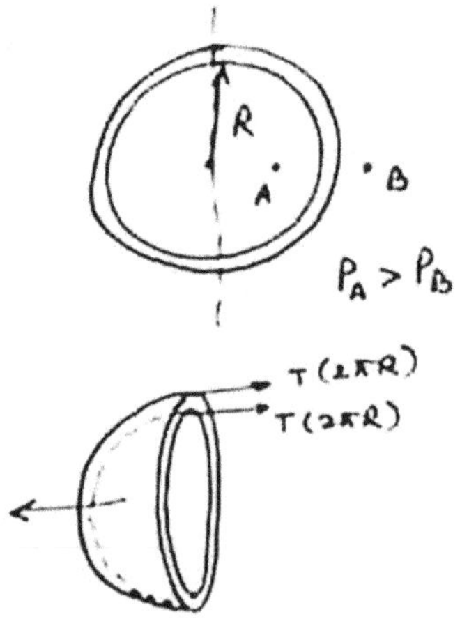


$P_A > P_B$

For eq$^n$ of hemispherical section of the bubble, we can use

$$P_{ex}.\pi R^2 = 2(T.2\pi R)$$

$$\boxed{P_{ex} = \frac{4T}{R}}$$

$F_{out} = P_{ex}.\pi R^2$

Scan for Video Explanation

Web Reference at www.physicsgalaxy.com

**Module-11 Excess Pressure Inside a Surface Having Two Curvatures**

# Excess Pressure inside a Surface having Two curvature:

net outward force on liq Surface is $F_{out} = (P_A - P_B) R_1 R_2 \theta \phi$

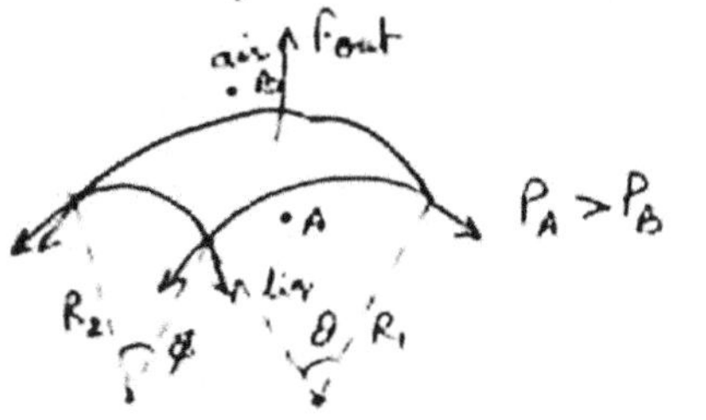


$\theta, \phi \to$ Small.

net inward force on liq surface due to ST is

$F_{in} = T(R_2\phi)\sin\theta/2 \times 2 + T(R_1\theta)\sin\phi/2 \times 2$

$= T(R_2\phi)\cdot\theta + T(R_1\theta)\phi$

$F_{in} = T\phi\theta(R_1 + R_2)$

Web Reference at www.physicsgalaxy.com | Scan for Video Explanation

**Module-11 Excess Pressure Inside a Surface Having Two Curvatures**

for eqm of liq Surface

$$F_{out} = F_{in}$$

$$(P_A - P_B) R_1 R_2 \theta\phi = T\theta\phi(R_1 + R_2)$$

$$P_{ex} = T\left[\frac{1}{R_1} + \frac{1}{R_2}\right]$$

if this Surface is a film $P_{ex} = 2T\left[\frac{1}{R_1} + \frac{1}{R_2}\right]$

# Excess Pressure inside a cylindrical Surface of liq.

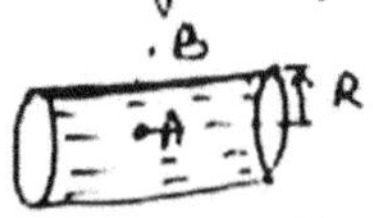


$R_1 = R$

$R_2 = \infty$

$P_A - P_B = T/R$

$P_{ex} = T\left[\frac{1}{R} + \frac{1}{\infty}\right] = T/R$ for a cyl film $P_{ex} = 2T/R$.

Web Reference at www.physicsgalaxy.com | Scan for Video Explanation

**Module-12 Connecting Soap Bubbles of Different Radii**

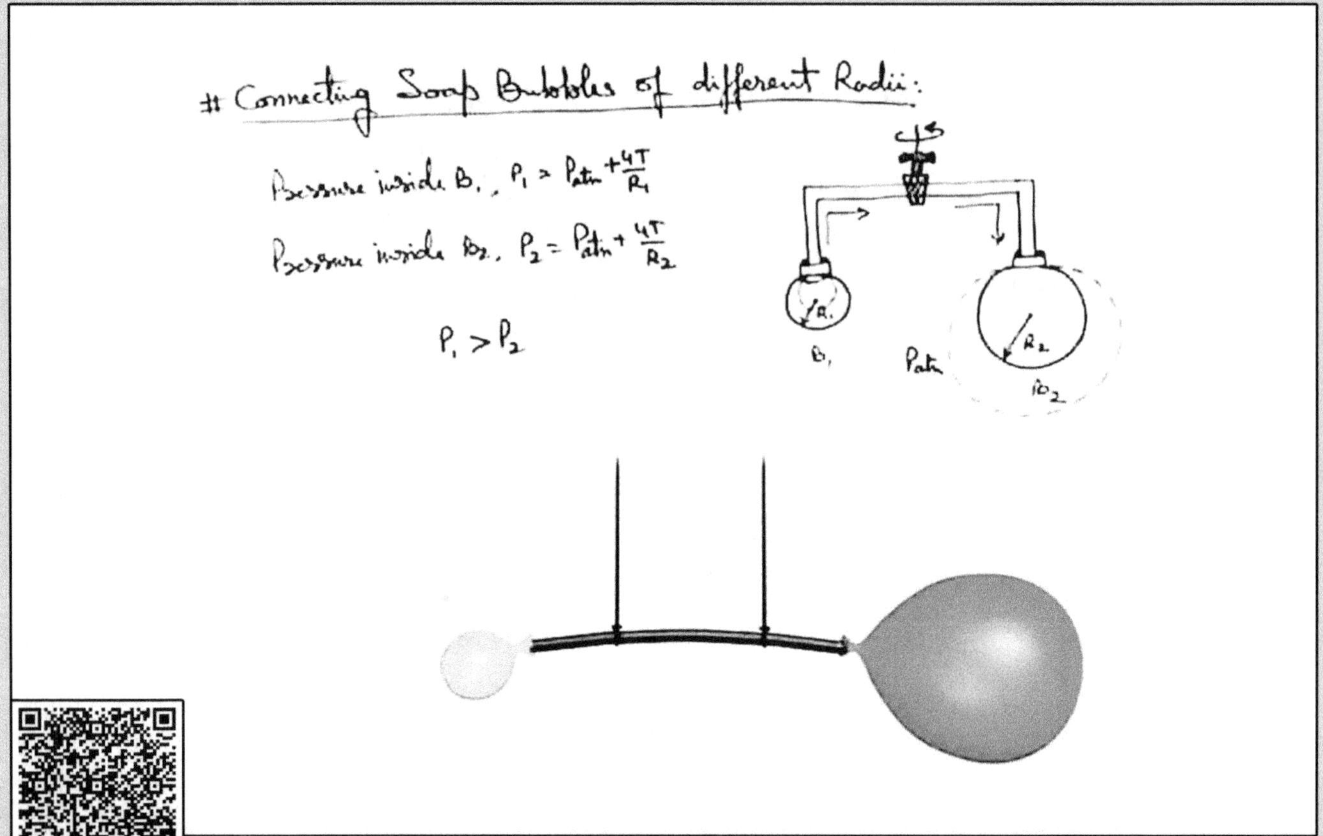


Scan for Video Explanation

Web Reference at www.physicsgalaxy.com

**Module-13 Angle of Contact**

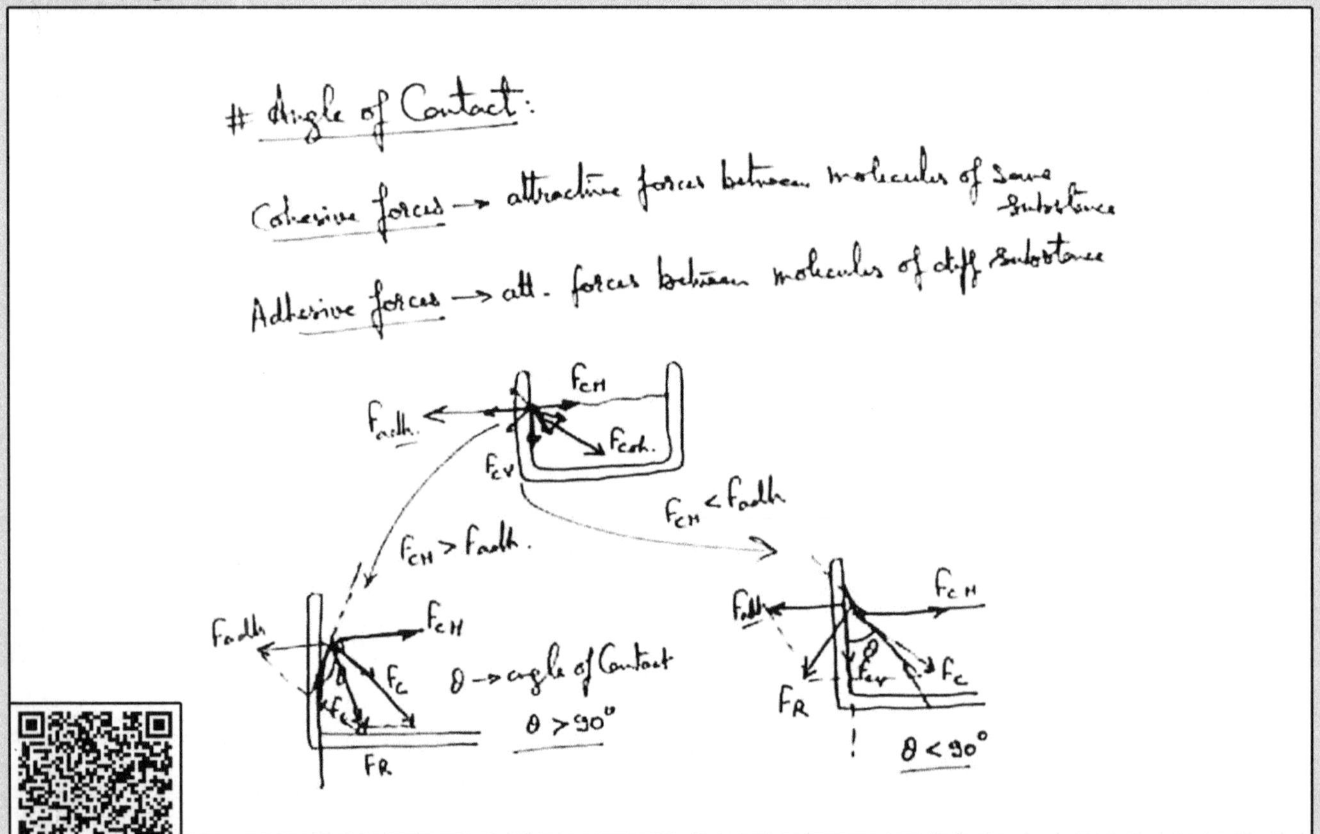


Scan for Video Explanation

Web Reference at www.physicsgalaxy.com

**Module-13 Angle of Contact**

for $\theta < 90^\circ$, liquid will wet the solid surface.

$F_{adh} > F_{coh}$

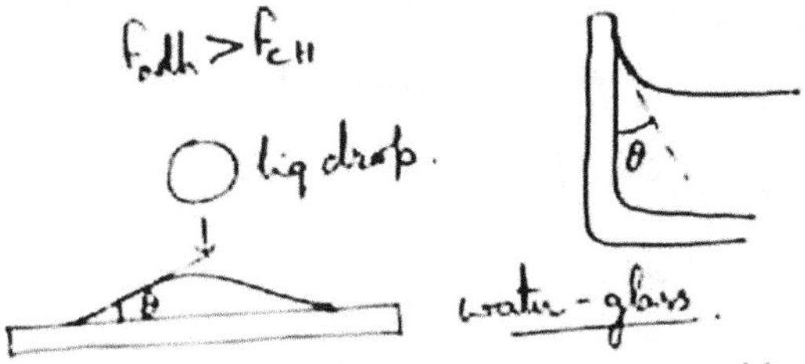


for $\theta > 90^\circ$, liquid do not wet the solid surface

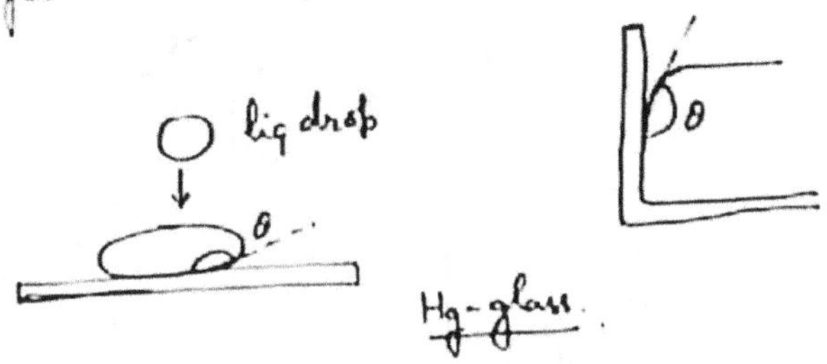




**Module-14 Shape of Liquid Meniscus in a Tube**

# Shape of Liquid Meniscus in a tube:

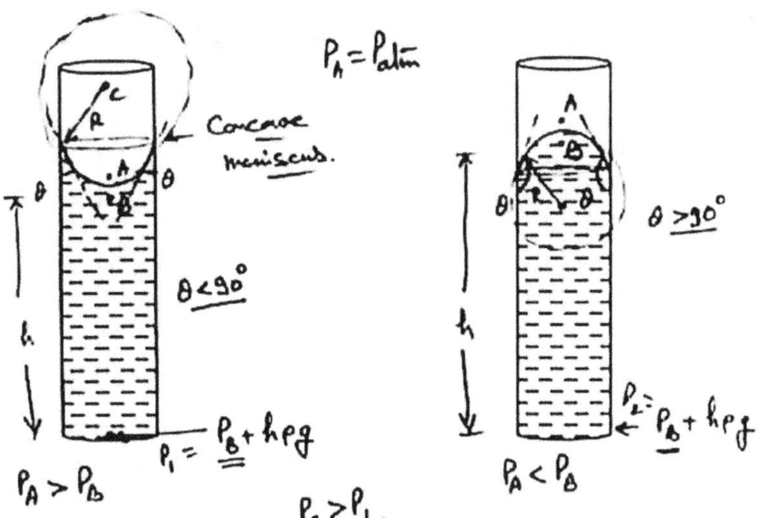


$P_A > P_B$ $\qquad$ $P_2 > P_1$ $\qquad$ $P_A < P_B$

$$P_A - P_B = \frac{2T}{R} \qquad\qquad P_B - P_A = \frac{2T}{R}$$

**Module-15 Capillary Action**

# Capillary Action:

A tube of very narrow bore is called capillary.

When a capillary is dipped normally in a liquid it is found that inside capillary liquid level either ascend or descend relative to surrounding level of liq. This phenomenon is called Capillary action or capillarity.

$P_A = P_{atm}$

$P_B = P_{atm} - \frac{2T}{R}$

$P_C = P_{atm}$

$P_C = P_B + h\rho g = P_{atm}$

$P_{atm} - \frac{2T}{R} + h\rho g = P_{atm} \rightarrow h = \frac{2T}{R\rho g}$

Capillary action height $h = \frac{2T\cos\theta}{r\rho g}$

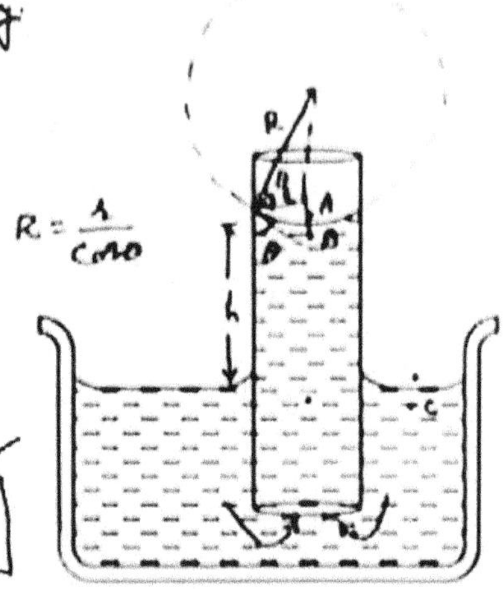


Scan for Video Explanation

Web Reference at www.physicsgalaxy.com

**Module-15 Capillary Action**

Similarly when a liq having $\theta > 90°$ then the liq level in capillary tube will fall.

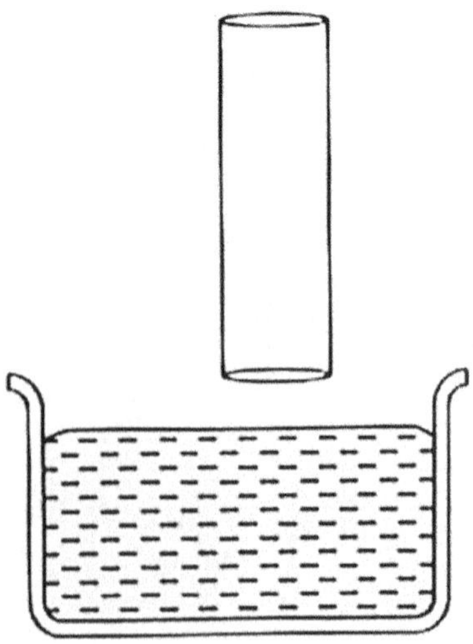

Scan for Video Explanation

Web Reference at www.physicsgalaxy.com

**Module-15 Capillary Action**

Similarly when a liq having $\theta > 90^\circ$ then the liq level in capillary tube will fall.

$P_A = P_{atm}$

$P_B = P_{atm} + \frac{2T}{R}$

$P_C = P_{atm}$

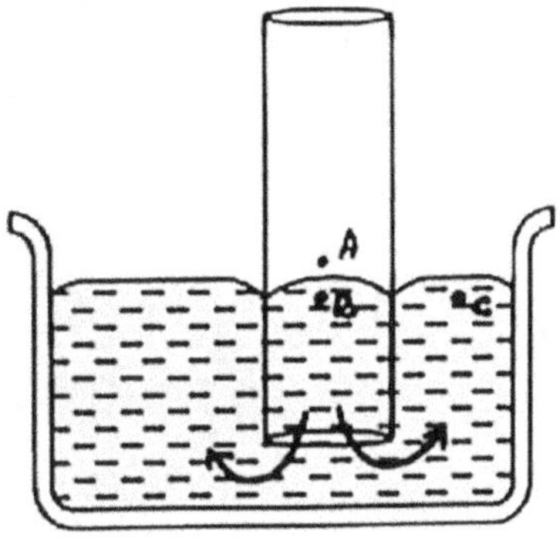


Web Reference at www.physicsgalaxy.com | Scan for Video Explanation

**Module-15 Capillary Action**

Similarly when a liq having $\theta > 90^\circ$ then the liq level in capillary tube will fall.

$P_A = P_{atm}$

$P_B = P_{atm} + \frac{2T}{R}$

$P_C = P_{atm}$

$P_D = P_{atm} + h\rho g = P_{atm} + \frac{2T}{R}$

$$h = \frac{2T}{R\rho g} = \frac{2T\cos\theta}{r\rho g}$$

$R = \frac{r}{\cos\theta}$

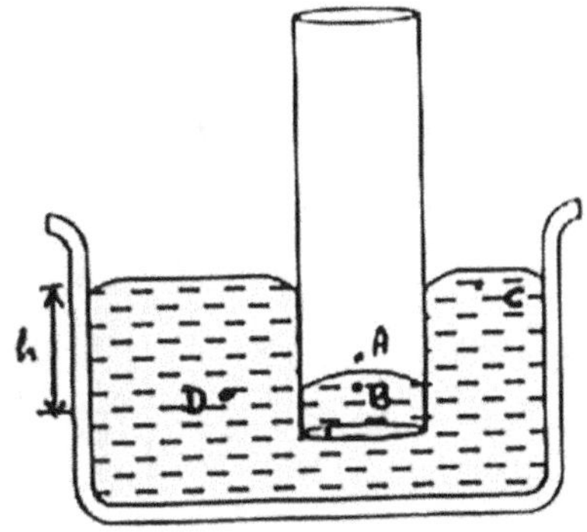


Web Reference at www.physicsgalaxy.com | Scan for Video Explanation

## Module-16 Force Between Two Parallel Plates Due to Liquid Between Them

# Force between parallel plates due to liquid between these:

to separate these plates we need to apply a force

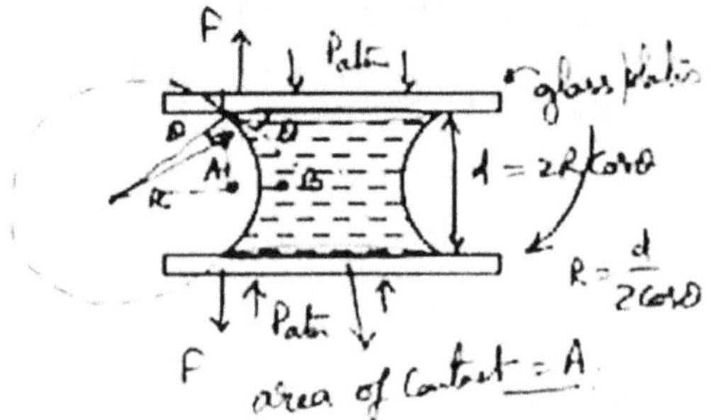


$P_A - P_B = \frac{T}{R}$

$P_B = P_{atm} - T/R$

$F = (P_{atm} - P_B)A$

$$F = \frac{TA}{R} = \frac{2TA\cos\theta}{d}$$

if $\theta = 0° \Rightarrow$ $F = \frac{2TA}{d}$

Scan for Video Explanation

Web Reference at www.physicsgalaxy.com

## Module-17 Capillary Action with a Capillary of Insufficient Height

# Capillary Action with a capillary of insufficient height:

As we know, Capillary height

$h = \frac{2T\cos\theta}{r\rho g}$

$\frac{2T}{R} = h\rho g$

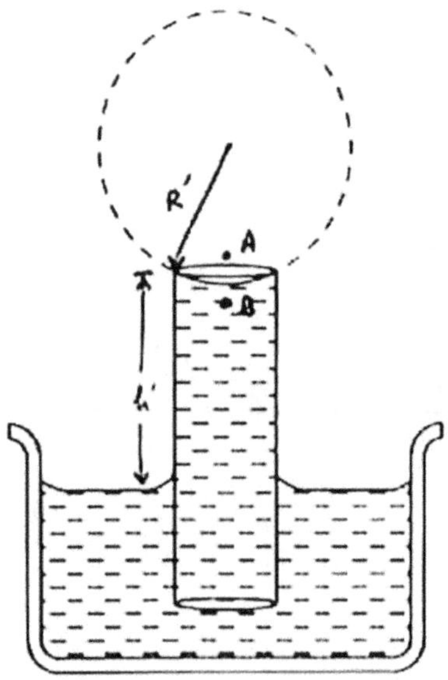


$P_B = P_{atm} - \frac{2T}{R'}$

$P_{atm} - \frac{2T}{R'} + h'\rho g = P_{atm}$

$\frac{2T}{R'} = h'\rho g$

$hR = h'R'$

$h' < h$

$R' > R$

liq will never overflow even if $h' < h$. because $R' > R$.

Scan for Video Explanation

Web Reference at www.physicsgalaxy.com

**Module-18 Solved Example-5**

Ex : A liquid of specific gravity 1.5 is observed to rise 3 cm in a capillary tube of diameter 0.5 mm and the liquid wets the surface of the tube. Calculate the excess pressure inside a spherical bubble of 1 cm diameter blown from the same liquid. Given that angle of contact is 0° and take g = 10 m/s$^2$.

Sol$^n$: As we know the surface tension of a liq is given as.

$$T = \frac{rh\rho g}{2\cos\theta} = \frac{0.25\times10^{-3}\times 3\times10^{-2}\times 1000\times10\times1.5}{2\times 1}$$

$$= 56.25\times10^{-3}\ N/m$$

Excess Pressure inside the bubble is

$$P_{ex} = \frac{4T}{R} = \frac{4\times 56.25\times10^{-3}}{0.5\times10^{-2}} = 45\ Pa.\ \text{Ans.}$$

Web Reference at www.physicsgalaxy.com | Scan for Video Explanation

**Module-19 Solved Example-6**

Ex : Two narrow bores of diameter 3 mm and 6 mm are joined together to make a U-tube open at both ends. If some water is poured in the tube find difference in level in two limbs. Take ST of water is 73 × 10$^{-3}$ N/m, angle of contact 0°, g = 10 m/s$^2$ and water density = 1000 kg/m$^3$.

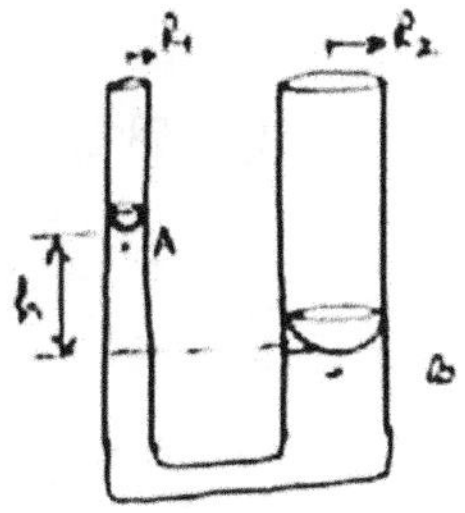


Sol$^n$: Pressure at A $P_A = P_{atm} - \frac{2T}{R_1}$

Pressure at B $P_B = P_{atm} - \frac{2T}{R_2}$

at level diff h, we can use

$$P_B - P_A = h\rho g$$

$$2T\left[\frac{1}{R_1} - \frac{1}{R_2}\right] = h\rho g$$

$$h = \frac{2T}{\rho g}\left[\frac{1}{R_1} - \frac{1}{R_2}\right] = \frac{2\times 73\times10^{-3}}{1000\times10}\left[\frac{1}{1.5} - \frac{1}{3}\right]\times10^{3}$$

$$= 4.86\ mm\ \text{Ans.}$$

Web Reference at www.physicsgalaxy.com | Scan for Video Explanation

**Module-20 Solved Example-7**

**Ex :** In a cylindrical vessel at its bottom a round hole of diameter 1 mm is drilled and water is filled in it. Find the maximum height to which water can be filled in it without leakage. Take ST of water $75 \times 10^{-3}$ N/m, density of water = 1000 kg/m$^3$, g = 10 m/s$^2$, and angle of contact 0°.

Sol$^n$: Water will start leaking out when

$$h\rho g \cdot \pi r^2 \geq T \cdot 2\pi r$$

$$h \geq \frac{2T}{\rho g r} = \frac{2 \times 75 \times 10^{-3}}{1000 \times 10 \times 0.5 \times 10^{-3}}$$

$$= 3 \times 10^{-2}$$

$$= 3 \text{ cm} \quad \text{Ans.}$$

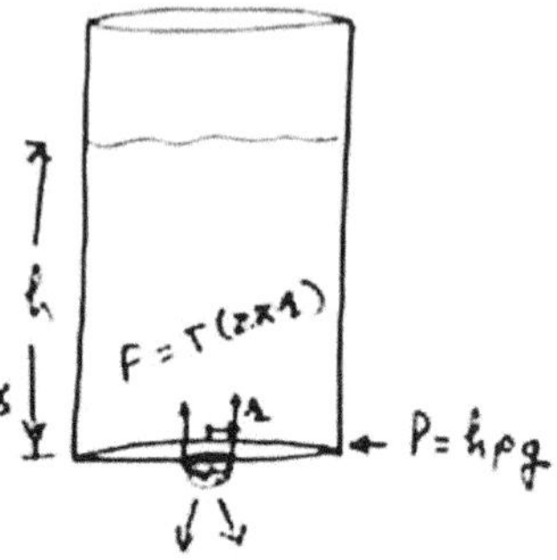


Scan for Video Explanation

Web Reference at www.physicsgalaxy.com

**Module-21 Solved Example-8**

**Ex :** The lower end of a capillary tube of diameter 2 mm is dipped 8 cm below the surface of water in a beaker. What is the pressure required in the tube to blow an air bubble at its end in water. Given that surface tension of water = $73 \times 10^{-3}$ N/m$^2$ and $P_{atm} = 10^5$ Pa, g = 10 m/s$^2$.

Sol$^n$: To blow the air bubble

$$P_A - P_B = \frac{2T}{r}$$

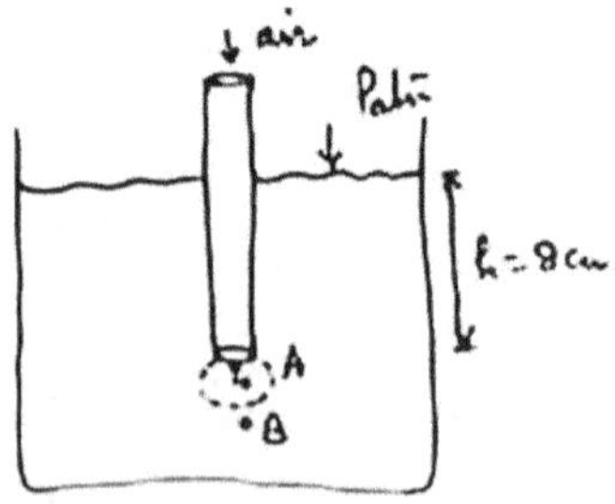


$$P_B = P_{atm} + h\rho g$$

$$P_A = \frac{2T}{r} + P_{atm} + h\rho g$$

$$= \frac{2 \times 73 \times 10^{-3}}{10^{-3}} + 10^5 + 0.08 \times 1000 \times 10$$

$$= 146 + 10^5 + 800 = 946 + 10^5$$

$$= 1.00946 \times 10^5 \text{ Pa.} \quad \text{Ans.}$$

Scan for Video Explanation

Web Reference at www.physicsgalaxy.com

**Module-22 Solved Example-9**

Ex : Two separate air bubbles of radii 2 mm and 4 mm formed of same soap solution of surface tension $70 \times 10^{-3}$ N/m. When these bubble come in contact to form a double bubble, find the radius of curvature of the interface film which is common to both bubbles.

Soln: here we know

$$P_1 = \frac{4T}{r_1} \quad ; \quad P_2 = \frac{4T}{r_2}$$

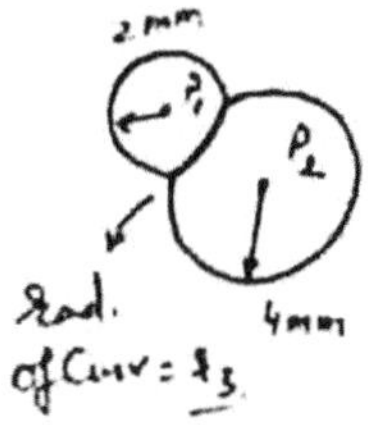


for the common interface of double bubble we can use

$$P_1 - P_2 = \frac{4T}{r_3}$$

$$\frac{1}{r_1} - \frac{1}{r_2} = \frac{1}{r_3} \Rightarrow \quad r_3 = \frac{r_1 r_2}{r_2 - r_1} = \frac{2 \times 4}{2} = 4\text{mm Ans.}$$

# Chapter 15
# Viscosity

## Lecture Notes Modules


**Module-1:** Viscosity
**Module-2:** Some Facts about Viscosity
**Module-3:** Flowing Fluid Relative to a Solid Surface in Contact
**Module-4:** Stoke's Law and Terminal Velocity
**Module-5:** Critical Velocity and Reynold's Number
**Module-6:** Poisueille's Equation
**Module-7:** Solved Example-1
**Module-8:** Solved Example-2
**Module-9:** Solved Example-3
**Module-10:** Solved Example-4
**Module-11:** Solved Example-5

**Module-1: Viscosity**

# Viscosity:

It is the prop of a fluid due to its intermolecular forces by which in case of fluid flow the relative motion betw<sup></sup>n layer of fluid is opposed.

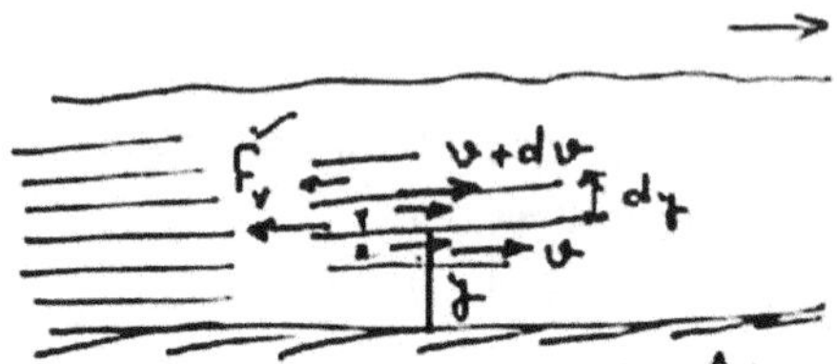


it is found that viscous force on a fluid layer mainly depend on two factors

① $F_v \propto A$ (surface area of layer)

② $F_v \propto \frac{dv}{dy}$ (velocity gradient of layer)

Scan for Video Explanation

Web Reference at www.physicsgalaxy.com

**Module-1: Viscosity**

(1) $F_v \propto A$ [Surface area of layer]

(2) $F_v \propto \frac{dv}{dy}$ [velocity gradient of layer]

viscous force on fluid layer is

$$F_v = \eta A \frac{dv}{dy}$$

$\eta$ → Const of prop. called coefficient of viscosity (depends on intermolecular prop of the fluid).

Scan for Video Explanation

Web Reference at www.physicsgalaxy.com

**Module-2: Some Facts about Viscosity**

# Some Facts about viscosity of a fluid:

- Viscosity depends only on nature of fluid.
- $\eta_{liquids} >> \eta_{gases}$.
- For liquids as density increases, viscosity also increases but in gases viscosity decrease with inc. in density.
- Units used for viscous coefficient $\eta$ is Poiseuille (Pl) in SI unit, in CGS units is poise, the rel. is –

$$1\,Pl = 10\,poise$$

Web Reference at www.physicsgalaxy.com | Scan for Video Explanation

**Module-3: Flowing Fluid Relative to a Solid Surface in Contact**

# Flowing fluid relative to a Solid Surface in Contact:

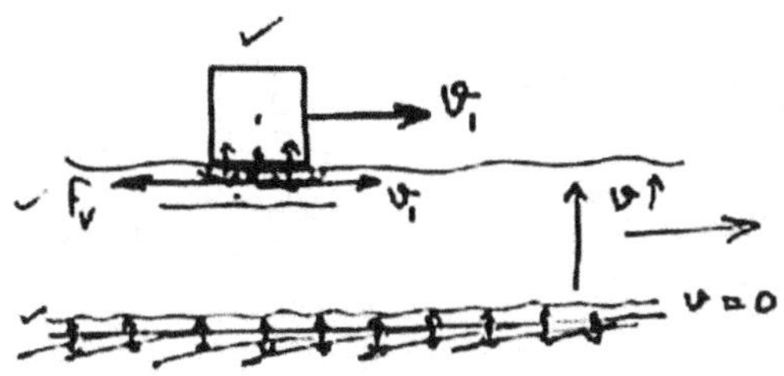

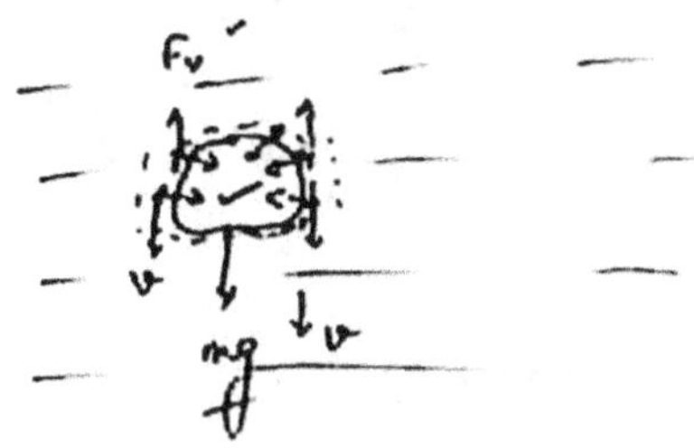

Web Reference at www.physicsgalaxy.com | Scan for Video Explanation

**Module-4: Stoke's Law and Terminal Velocity**

# Stoke's Law and Terminal velocity:

When a body moves through a fluid, the fluid in contact with body is also dragged with it and the relative motion between fluid layer oppose the motion of body.

Acc. to Stokes Law the viscous force on a sphere of radius r moving with speed v through a fluid of coefficient of vis $\eta$ is given as -

$$\boxed{F_v = 6\pi\eta r v}$$

Scan for Video Explanation

Web Reference at www.physicsgalaxy.com

**Module-4: Stoke's Law and Terminal Velocity**

When a sphere is dropped in a fluid. Due to viscous force, it attains a steady uniform speed $v_T$, called terminal speed at which $a = 0$.

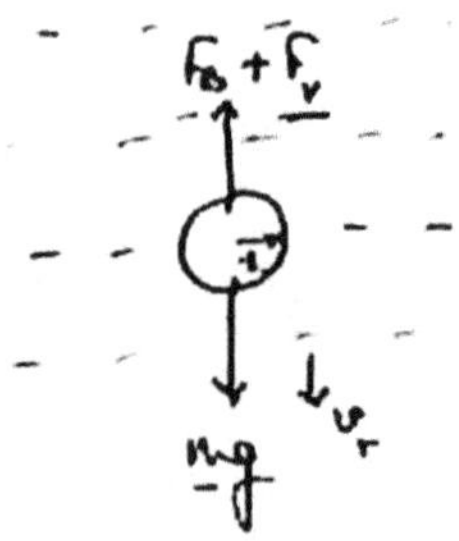


$\Rightarrow$

$$F_B + F_v = mg$$

$$\frac{4}{3}\pi r^3 \rho_l g + 6\pi\eta r v_T = \frac{4}{3}\pi r^3 \rho_s g$$

$$\boxed{v_T = \frac{2}{9} r^2 \left[\frac{\rho_s - \rho_l}{\eta}\right] g.}$$

terminal speed of sphere.

$\rho_s \to$ density of sphere

$\rho_l \leftarrow$ density of liq.

$\eta \to$ vis coeff

Scan for Video Explanation

Web Reference at www.physicsgalaxy.com

**Module-5: Critical Velocity and Reynold's Number**

# Critical Velocity and Reynold's Number:

As we know upto a specific flow vel, fluid flow remains streamline & beyond a critical vel it becomes turbulent.

exp. it was established that this critical vel depends on some factors, these are -

(1) $v_c \propto \eta$

(2) $v_c \propto \frac{1}{\rho}$

(3) $v_c \propto \frac{1}{r}$

$$v_c = N_R\left(\frac{\eta}{r\rho}\right)$$

$N_R$ → Reynold's Number.

Web Reference at www.physicsgalaxy.com

Scan for Video Explanation

**Module-6: Poisueille's Equation**

# Poiseuille's Equation:

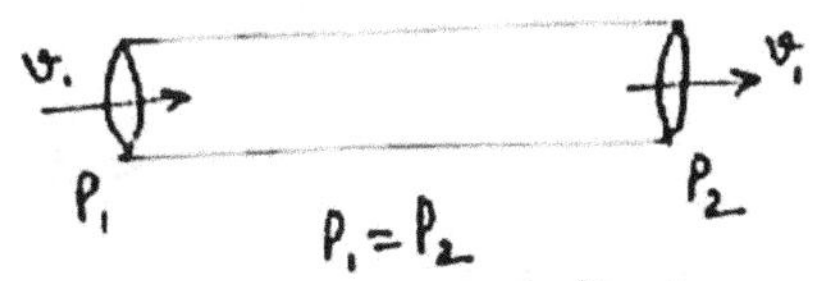


$P_1 = P_2$

for an ideal fluid.

discharge rate

$Q = Av = \pi r^2 v$

$P_A > P_B$

Acc. to poiseuille's eqⁿ the discharge rate of a viscous fluid through a tube can be given as

$$Q = \frac{\pi r^4}{8\eta l}(P_A - P_B) = \frac{\pi r^4}{8\eta l}\Delta P$$

Web Reference at www.physicsgalaxy.com

Scan for Video Explanation

**Module-7: Solved Example-1**

Ex : An air bubble of radius 1 mm is allowed to rise through a long cylindrical column of a viscous liquid of radius 5 cm and travels at a steady rate of 2.1 cm/sec. If density of liquid is 1.5 gm/cm$^3$. Find its viscosity. Take g = 10 m/s$^2$ and neglect air density.

Sol$^n$:

for steady speed of bubble

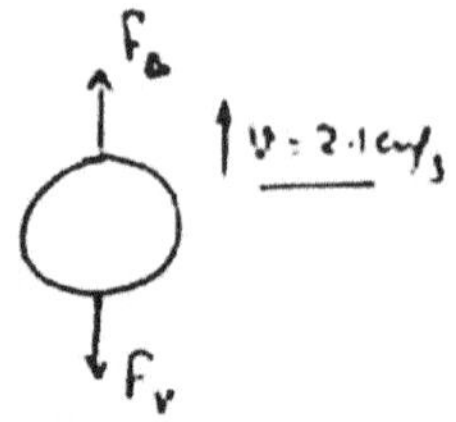


$$F_B = F_v$$

$$\frac{4}{3}\pi r^3 \rho g = 6\pi\eta r v$$

$$\eta = \frac{2}{9}\frac{\rho r^2 g}{v} = \frac{2}{9}\times\frac{1.5\times(10^{-1})^2\times(1000)}{2.1}$$

$$= 1.58 \text{ poise} \quad \text{Ans.}$$

Scan for Video Explanation

Web Reference at www.physicsgalaxy.com

**Module-8: Solved Example-2**

Ex : The velocity water in river is 5 m/s at the surface. If river is 5 m deep, find the shear stress between the horizontal layers of water. The viscosity of water is 10$^{-3}$ poiseuille.

Sol$^n$:

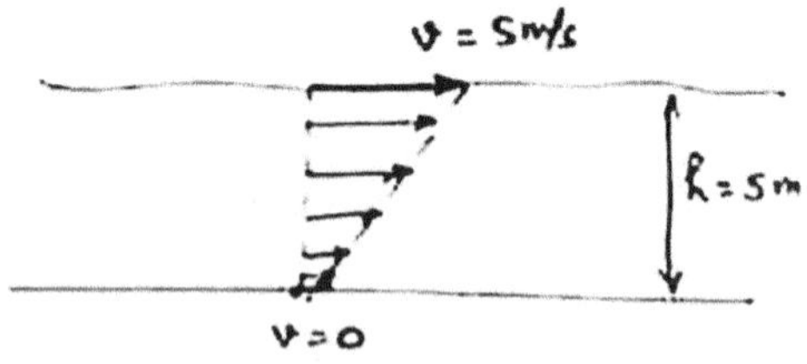


$$\text{Shear stress} = \frac{F_{tg}}{A}$$

vel. gradient

$$\frac{dv}{dy} = \frac{5-0}{5} = 1.$$

$$= \frac{1}{A}\left[\eta A\frac{dv}{dy}\right]$$

$$= \eta\frac{dv}{dy} = 10^{-3}\times 1 = 10^{-3}\ \text{N/m}^2 \quad \text{Ans.}$$

Scan for Video Explanation

Web Reference at www.physicsgalaxy.com

**Module-9: Solved Example-3**

Ex : Figure shows a cylindrical vessel of height H filled with a liquid of density ρ. At its bottom in a small hole a capillary tube of radius r and length L is fitted as shown. Find the flow rate of liquid when flow starts. Viscosity of fluid is η.

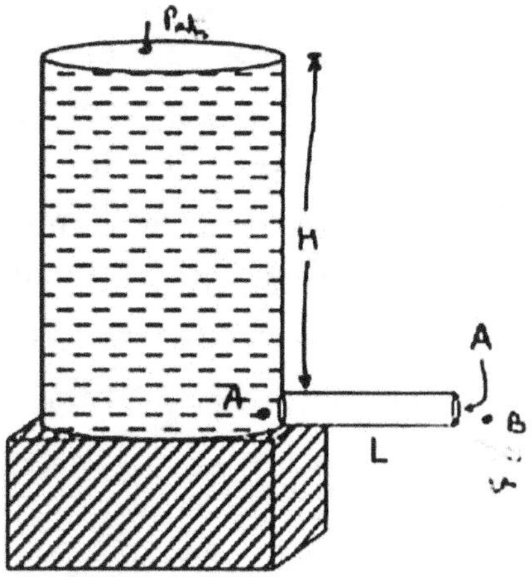


Sol$^n$: Pressure diff across Capillary tube is $P_A - P_B = H\rho g$

$P_A = P_{atm} + H\rho g$

$P_B = P_{atm}$

acc. to Poiseuille's eq$^n$:

flow rate $\frac{dV}{dt} = \frac{\pi r^4}{8\eta l}(P_A - P_B) = \frac{\pi r^4 (H\rho g)}{8\eta L}$ Ans.

$v \pi r^2 = (\ldots) \rightarrow v = \ldots$

Web Reference at www.physicsgalaxy.com | Scan for Video Explanation

**Module-10: Solved Example-4**

Ex : Spherical dust particles of different radii are shaken up in water in a bottle and kept on a flat surface. If height of water in bottle is 10 cm find the radius of largest dust particle remain in suspension after 1 Hour. Take density of water 1000 kg/m$^2$, density of dust particles 1800 kg/m$^3$ and viscosity of water $10^{-2}$ poise and $g = 10$ m/s$^2$.

Sol$^n$: particle vel which reach bottom in 1 Hr is $v = \frac{\Delta s}{\Delta t} = \frac{0.1}{3600} = \frac{10^{-3}}{36}$ m/s

As we know that term speed.

$v = \frac{2}{9}\frac{gr^2}{\eta}(\rho_s - \rho_l)$

$r = \sqrt{\frac{9\eta v}{2g(\rho_s - \rho_l)}} = \sqrt{\frac{9 \times 10^{-3} \times 10^{-3}/36}{2 \times 10 \times 800}} = 3.95\ \mu m$ Ans.

Web Reference at www.physicsgalaxy.com | Scan for Video Explanation

**Module-11: Solved Example-5**

**Ex :** **Water flows through a capillary tube of radius r and length *l* at a rate of 40 ml/sec when pressure difference across tube is due to a height h of water column. Now another tube of same length and radius r/2 is connected in series with this tube with same pressure difference across the combination. Find the new flow rate of water through this combination.**

Sol$^n$

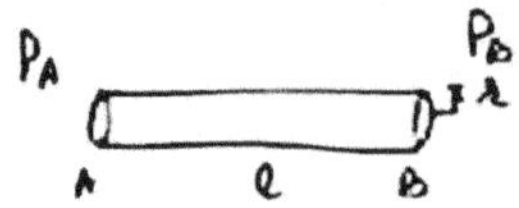


Initial flow rate

$$Q = \frac{\pi r^4}{8\eta l}(h\rho g) = 40\ ml/s$$

$$P_A - P_B = h\rho g$$

Final flow rate

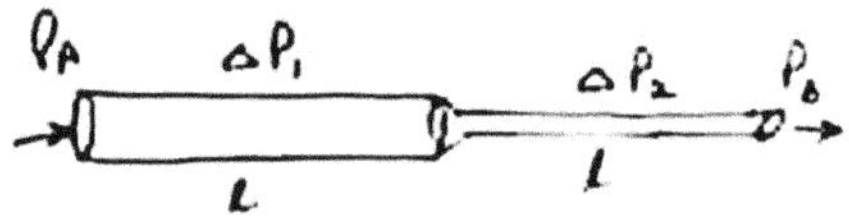


$$Q' = \frac{\pi r^4}{8\eta l}\Delta P_1 = \frac{\pi (r/2)^4}{8\eta l}\Delta P_2$$

$$16\Delta P_1 = \Delta P_2$$

$$\Delta P_1 + \Delta P_2 = h\rho g$$

$$17\Delta P_1 = h\rho g$$

$$\Delta P_1 = h\rho g/17$$

$$Q' = \frac{\pi r^4}{8\eta l}\cdot\frac{h\rho g}{17} = \frac{40}{17}\ ml/s \text{ Ans.}$$

Scan for Video Explanation

Web Reference at www.physicsgalaxy.com

# Chapter 16
# Elasticity

## Lecture Notes Modules


**Module-1:** Introduction to Elasticity
**Module-2:** Rigid Body
**Module-3:** Elastic and Partially Elastic Bodies
**Module-4:** Inelastic or Plastic Bodies
**Module-5:** Stress due to a Force
**Module-6:** Types of Stress
**Module-7:** Strain
**Module-8:** Types of Strain
**Module-9:** Relation Between Stress and Strain
**Module-10:** Hooke's Law and Modulus of Elasticity
**Module-11:** Types of Modulus of Elasticity
**Module-12:** Solved Example-1
**Module-13:** Solved Example-2
**Module-14:** Solved Example-3
**Module-15:** Solved Example-4
**Module-16:** Solved Example-5
**Module-17:** Solved Example-6
**Module-18:** Some Facts about Elasticity
**Module-19:** Elongation of Suspended Rod Under Self Weight
**Module-20:** Metal Wires of Rods as Equivalent Spring
**Module-21:** Elastic Potential Energy Stored due to Elasticity
**Module-22:** Solved Example-7
**Module-23:** Solved Example-8

**Module-1: Introduction to Elasticity**

# Elasticity:

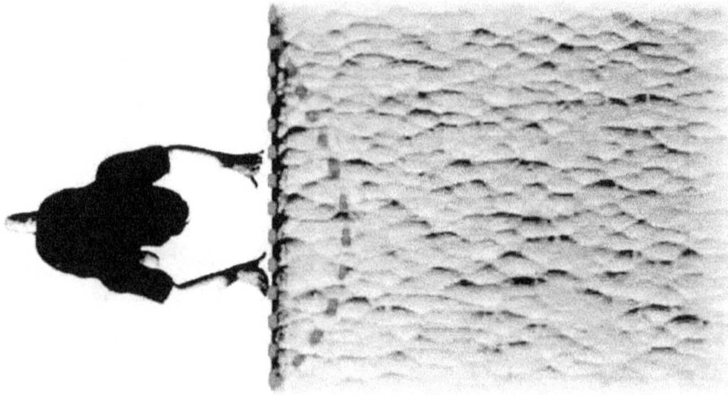

The prop of matter by virtue of which a body tend to regain its original shape & size after deforming forces are removed is called elasticity.

Elasticity is due to intermolecular forces between constituent particles of body.

Scan for Video Explanation

Web Reference at www.physicsgalaxy.com

**Module-2: Rigid Body**

# Rigid body:

A body in which the sep between its constituent particles does not change.

- it never gets deformed.
- in nature no body is perfectly rigid.

Scan for Video Explanation

Web Reference at www.physicsgalaxy.com

## Module-3: Elastic and Partially Elastic Bodies

# Elastic and Partially elastic Bodies:

Elastic body:

it complitly regains its original shape &

WD in def = stored PE in body.

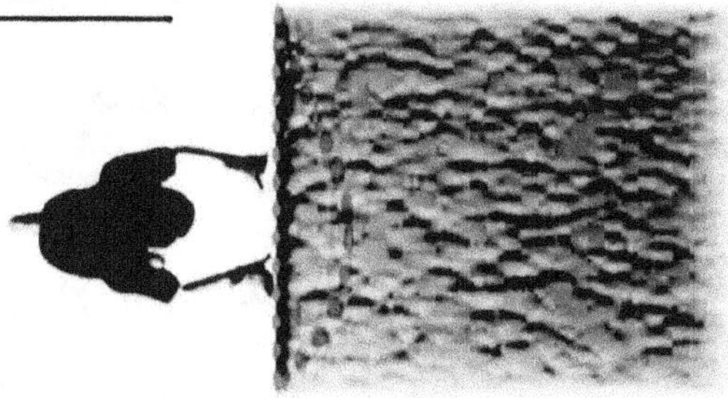

Partially Elastic Body:

it does not complitly regain its original shape &

WD in def → partly dissipated + partly stored as PE.

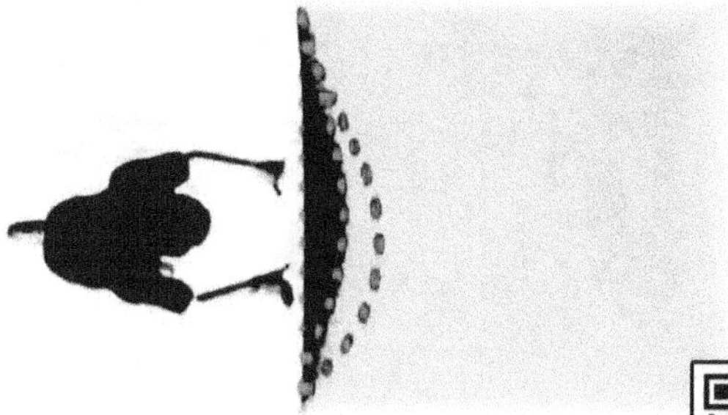

Web Reference at www.physicsgalaxy.com

Scan for Video Explanation

## Module-4: Inelastic or Plastic Bodies

# Inelastic or Plastic Bodies:

Such Bodies do not have any tendency to regain original shape &

WD in def → complitly dissipated during def.

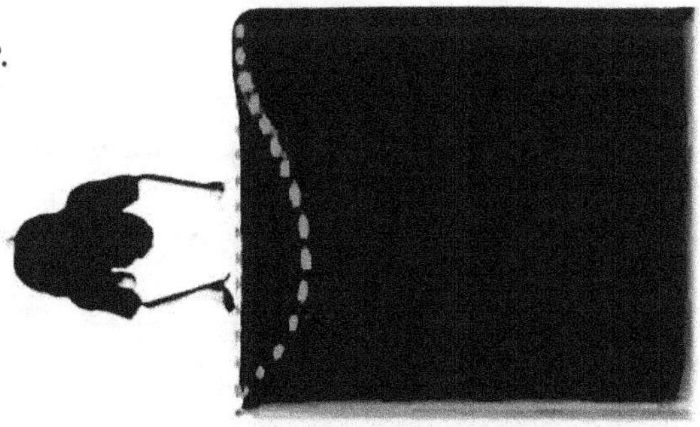

Web Reference at www.physicsgalaxy.com

Scan for Video Explanation

**Module-5: Stress due to a Force**

# Stress due to a Force:

When ext forces are applied on a body, it gets deformed. Due to def, the intermolecular forces inside the body may develop some restoring forces within the volume of body which oppose the external deforming forces.

The internal restoring forces per unit cross-sect area of body is called 'Stress'.

under eqm state,

$$\text{STRESS} = \frac{\text{Deforming force}}{\text{Surface area of cross-section normal to force.}}$$

Scan for Video Explanation

Web Reference at www.physicsgalaxy.com

**Module-6: Types of Stress**

# Types of Stress:

for solid bodies when stress normal to body surface, it is called normal stress & when stress is tangential to a body surface it is called shear stress.

Normal Stress = F/A.

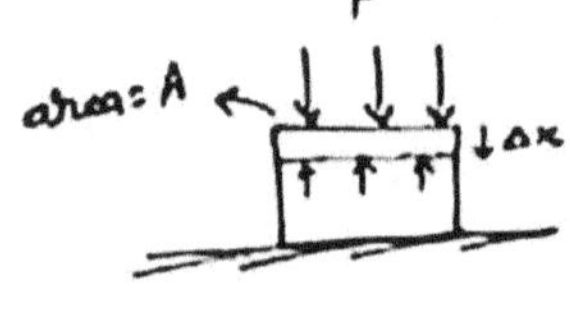


Shear stress = F/A

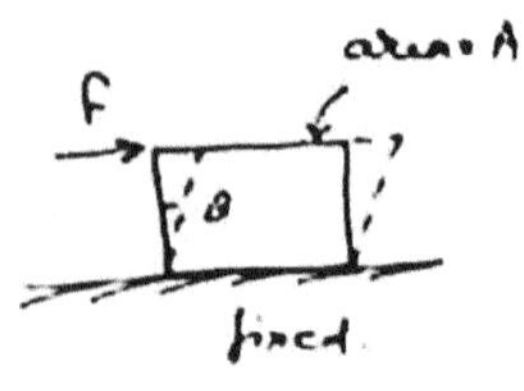


Scan for Video Explanation

Web Reference at www.physicsgalaxy.com

**Module-6: Types of Stress**

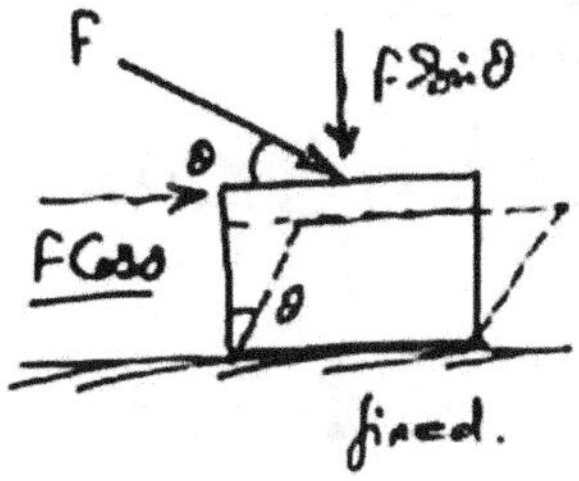


Normal stress = $\frac{F\sin\theta}{A}$

Shear stress = $\frac{F\cos\theta}{A}$

Web Reference at www.physicsgalaxy.com | Scan for Video Explanation

**Module-7: Strain**

# Strain:

When applied forces on a body produce deformation stress is developed inside the body

The relative change in dimension of a body which is subjected to stress is called "Strain".

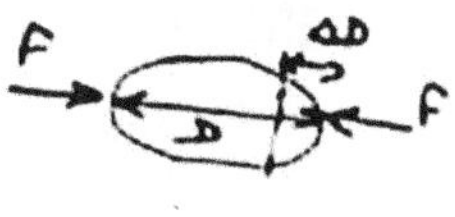


$$\text{Strain} = \frac{\Delta D}{D}$$

↙ Longitudinal Strain

↓ Volume Strain

↘ Shear Strain

Web Reference at www.physicsgalaxy.com | Scan for Video Explanation

**Module-8: Types of Strain**

# Types of Strain:

① Longitudinal Strain:

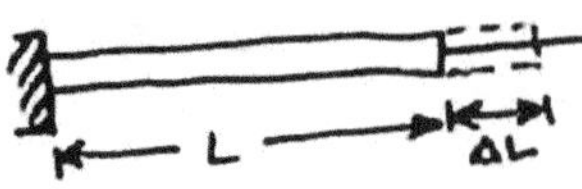


$$\text{longitudinal strain} = \frac{\Delta L}{L}$$ → elongation/Compression

② Volume Strain: generally used in fluids.

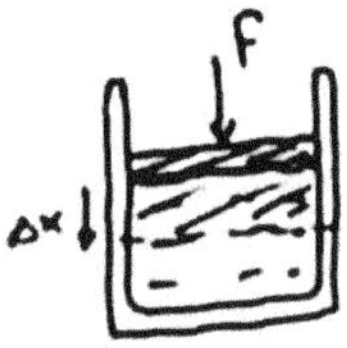


$$\left[\text{volume strain} = \frac{\Delta V}{V}\right]$$

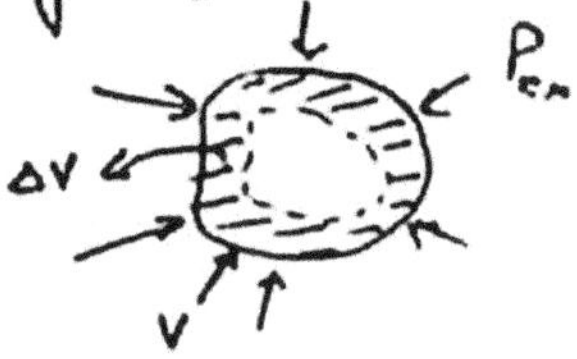


Scan for Video Explanation

Web Reference at www.physicsgalaxy.com

**Module-8: Types of Strain**

③ Shear Strain: tangential def in a body due to shear stress.

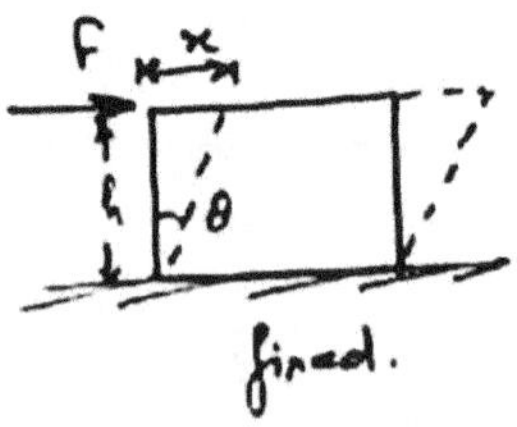


$$\text{Shear strain} = \frac{x}{h} = \theta \quad [\text{small angle}]$$

Scan for Video Explanation

Web Reference at www.physicsgalaxy.com

**Module-9: Relation Between Stress and Strain**

# Relation between Stress and Strain:

It is observed experimentally that for small strain stress is proportional to strain in a body.

This is known as Hooke's law.

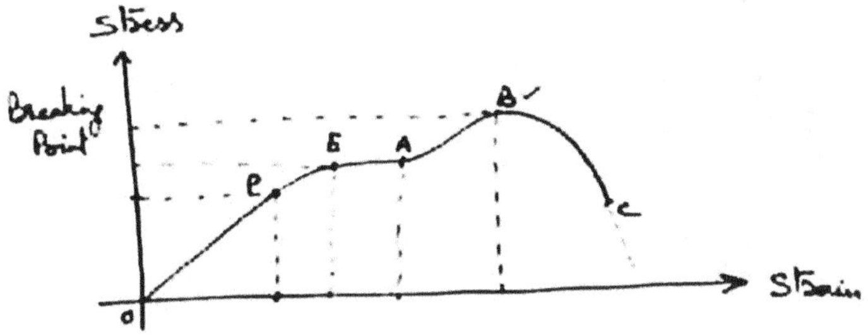


Hooke's law. [stress ∝ strain]

P ← Hooke's limit

Web Reference at www.physicsgalaxy.com | Scan for Video Explanation

**Module-10: Hooke's Law and Modulus of Elasticity**

# Hooke's Law and Modulus of Elasticity:

Within Hooke's limit

$$\text{stress} \propto \text{strain}$$

$$\text{stress} = E(\text{strain})$$

$$\boxed{E = \left(\frac{\text{stress}}{\text{strain}}\right)}$$ { constant

↓

Modulus of elasticity for the material

↓

It is a measure of material elastic properties

Web Reference at www.physicsgalaxy.com | Scan for Video Explanation

**Module-11: Types of Modulus of Elasticity**

# Types of Modulus of Elasticity:

Depending on type of stress applied and resulting strain there are three types of elasticity Moduli.

(1) Young's Modulus: for solids it is given as

$$Y = \frac{\text{longitudinal stress}}{\text{longitudinal strain}} = \frac{F/S}{\Delta L/L}$$

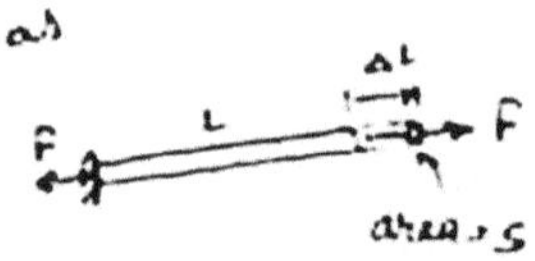


(2) Bulk Modulus: for solids and fluids (liq/gases)

$$B = \frac{\text{Volume Stress}}{\text{Volume Strain}} = \frac{\text{Excess Pressure}}{\text{Vol. strain}} = \frac{\Delta P_{ex}}{|\Delta V/V|} = -\frac{\Delta P_{ex}}{(\Delta V/V)}$$

Scan for Video Explanation

Web Reference at www.physicsgalaxy.com

**Module-11: Types of Modulus of Elasticity**

(3) Shear Modulus: for solids it is defined as

$$\eta = \frac{\text{Shear stress}}{\text{Shear strain}}$$

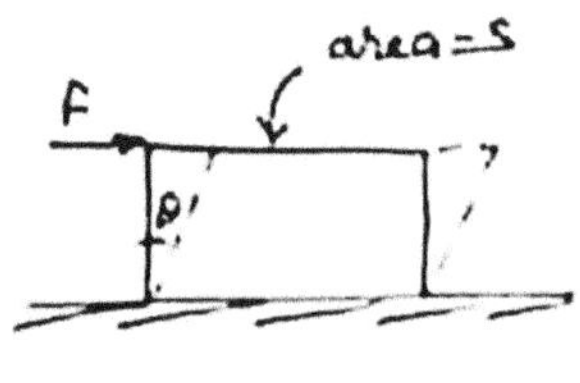


$$\eta = \left(\frac{F/S}{\theta}\right)$$

# Compressibility of a material:

$$C = \frac{1}{B} = -\frac{\Delta V}{V \Delta P_{ex}}$$

$$\text{As } B = -\frac{\Delta P_{ex}}{\Delta V/V} = -\frac{V \Delta P_{ex}}{\Delta V}$$

Scan for Video Explanation

Web Reference at www.physicsgalaxy.com

**Module-12: Solved Example-1**

**Ex :** The diameter of a brass rod is 4 mm and Young's modulus of brass is $9 \times 10^{10}$ N/m$^2$. Find the force required to stretch it by 0.1% of its length.

Sol$^n$: As we know

$$Y = \frac{\text{Stress}}{\text{Strain}} = \frac{F/A}{\Delta L/L}$$

$$\frac{\Delta L}{L} = 0.1 \times \frac{1}{100} = 0.001$$

$$F = \frac{YA\Delta L}{L}$$

$$F = 9\times10^{10} \times \pi(2\times10^{-3})^2 \times 0.001$$

$$F = 360\pi \text{ N} \quad \text{Ans.}$$

Web Reference at www.physicsgalaxy.com | Scan for Video Explanation

**Module-13: Solved Example-2**

**Ex :** Calculate the elongation of the Copper and Aluminium wire shown in figure. Diameters of Cu and Al wires are 1.4 mm and 1 mm respectively. Given that $Y_{Cu} = 11 \times 10^{10}$ N/m$^2$ and $Y_{Al} = 7 \times 10^{10}$ N/m$^2$. ($g = 10 \text{ m/s}^2$)

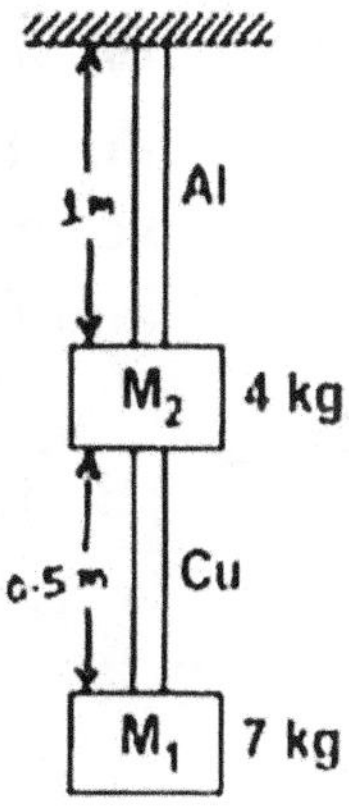


Sol$^n$: Tension in Cu wire $T_1 = M_1 g = 70$ N

$$Y_{Cu} = \frac{\text{Stress}}{\text{Strain}} = \frac{T_1/A_1}{\Delta L_1/L_1} \Rightarrow \Delta L_1 = \frac{T_1 L_1}{Y_1 A}$$

$$= \frac{70 \times 0.5}{11\times10^{10} \times \pi(0.7\times10^{-3})^2}$$

$$= 0.2 \text{ mm} \quad \text{Ans.}$$

Tension in Al wire $T_2 = (M_1 + M_2)g = 110$ N.

elongation in Al wire
$$\Delta L_2 = \frac{T_2 l_2}{Y_2 A_2} = \frac{110 \times 1}{7\times10^{10} \times \pi(0.5\times10^{-3})^2}$$

$$= 2 \text{ mm.} \quad \text{Ans.}$$

Web Reference at www.physicsgalaxy.com | Scan for Video Explanation

**Module-14: Solved Example-3**

**Ex :** **A steel wire is suspended vertically from a rigid support. When it is loaded by a weight in air, it is elongated by 5 mm and when the weight is submerged in water elongation is changed to 4 mm. Find the relative density of the material of weight.**

Sol: As we know

$$Y = \frac{W/S}{\Delta x/L}$$

$$\Delta x = \frac{WL}{YS} \quad \text{---} (1)$$

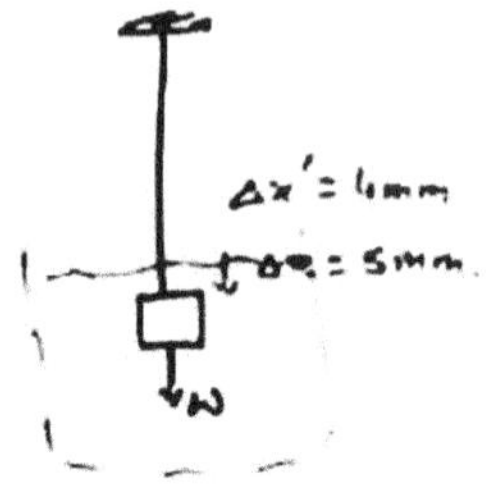


after submerging the weight in liq

$$\Delta x' = \frac{W_{eff} L}{YS} \quad \text{---} (2)$$

$$W_{eff} = W - F_B = W - \frac{W \rho_l}{\rho_s} = W\left(1 - \frac{\rho_l}{\rho_s}\right)$$

from (1) & (2) $\quad \frac{\Delta x}{\Delta x'} = \frac{\rho_s}{\rho_s - \rho_l} = \frac{5}{4}$

$$4\rho_s = 5\rho_s - 5\rho_l \Rightarrow \rho_s = 5\rho_l$$

s.g $\rightarrow$ 5 Ans.

Scan for Video Explanation

Web Reference at www.physicsgalaxy.com

**Module-15: Solved Example-4**

**Ex :** **When a metal block is placed in vacuum chamber from normal atmosphere find the fractional change in its volume. Given that atmospheric pressure is $10^5$ N/m$^2$ and Bulk modulus of metal is $1.25 \times 10^{11}$ N/m$^2$.**

Sol: As Bulk mod. is given as

$$B = \left| \frac{\Delta P_{ex}}{\Delta V/V} \right|$$

$$\frac{\Delta V}{V} = \frac{\Delta P_{ex}}{B} = \frac{10^5}{1.25 \times 10^{11}} = 8 \times 10^{-7} \text{ Ans.}$$

Scan for Video Explanation

Web Reference at www.physicsgalaxy.com

**Module-16: Solved Example-5**

**Ex :** A rubber ball is taken to a depth of 200 m in a lake, its volume decreases by 0.1%. If density of water is 1000 kg/m$^3$ and g = 10 m/s$^2$. Find Bulk modulus of material of ball.

Sol$^n$:

Excess pressure at a depth of 200m is

$$\Delta P_{ex} = h\rho g$$

$$= 200 \times 1000 \times 10 = 2 \times 10^6 \text{ N/m}^2$$

volume dec is $\Delta V = 0.1 \times \frac{V}{100} = 0.001V$

Bulk mod. $B = \left|\frac{\Delta P_{ex}}{\Delta V/V}\right| = \frac{2\times10^6}{0.001} = 2\times10^9 \text{ N/m}^2$ Ans.

Web Reference at www.physicsgalaxy.com

Scan for Video Explanation

**Module-17: Solved Example-6**

**Ex :** A wire of cross-sectional area A, Young's modulus Y and length 2L is stretched between two rigid poles. A weight of mass M is suspended at its middle. Find the angle made by string with horizontal.

Sol$^n$:

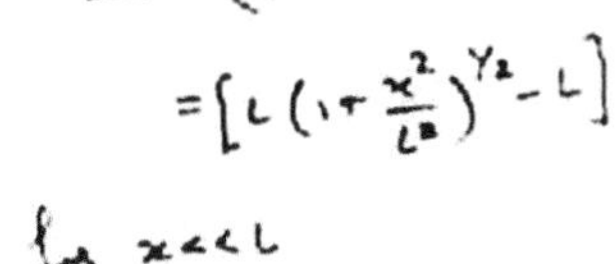


for eq$^n$ of weight

$\sin\theta \simeq \theta = \frac{x}{L}$

$$2T\sin\theta = Mg \quad -(1)$$

using def of Young's Mod -

$$Y = \frac{T/A}{\Delta L/L}$$

$$T = \frac{YAx^2}{2L^2}$$

elongation in wire

$$\Delta L = \left(\sqrt{L^2+x^2} - L\right)$$

$$= \left[L\left(1+\frac{x^2}{L^2}\right)^{1/2} - L\right]$$

for $x << L$

$$\Delta L = L\left(1+\frac{x^2}{2L^2}\right) - L$$

$$\Delta L = \frac{x^2}{2L}$$

from eq$^n$(1) $2\left(\frac{YAx^2}{2L^2}\right)\frac{x}{L} = Mg.$

$$\sin^3\theta = \frac{x^3}{L^3} = \frac{Mg}{YA} \rightarrow \theta = \sin^{-1}\left(\frac{Mg}{YA}\right)^{1/3} \text{ Ans.}$$

Web Reference at www.physicsgalaxy.com

Scan for Video Explanation

**Module-18: Some Facts about Elasticity**

# Some Facts about Elasticity:

(1) For a material, modulus of elasticity is a constant and it is a measure of its stiffness or elasticity.

As $E = \frac{\text{stress}}{\text{strain}}$

for high value of $E$, then for large stress, strain will be less

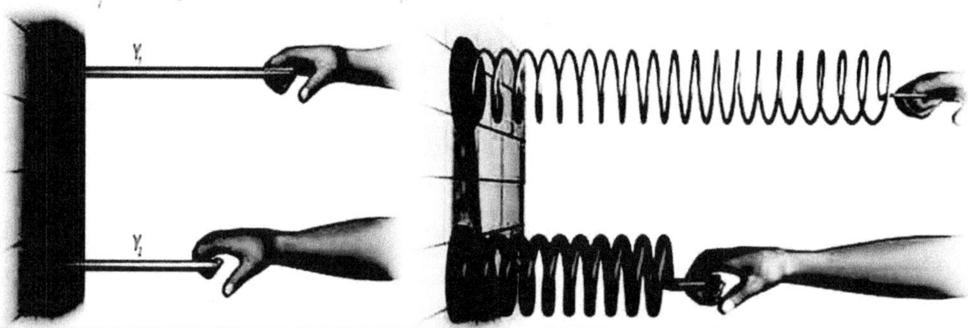

Scan for Video Explanation

Web Reference at www.physicsgalaxy.com

**Module-18: Some Facts about Elasticity**

(2) For solids, liquids & gases, we can use

$$E_{solids} > E_{liq} > E_{gases}$$

(3) As temp increases, intermolecular sep increases

$\Rightarrow$ restoring force decreases

& it lowers the value of $E$ for the material

Scan for Video Explanation

Web Reference at www.physicsgalaxy.com

**Module-19: Elongation of Suspended Rod Under Self Weight**

# Elongation of a Suspended Rods under self Weight:

tension in rod at a dist x from bottom is

$$T = \left(\frac{M}{L}x\right)g$$

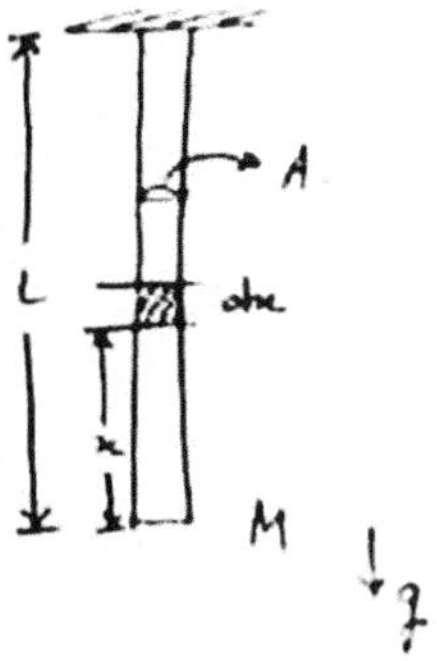


If dy is the elongation in elemental segment dx then we have

Strain = dy/dx
Stress = T/A

$$\longrightarrow Y = \frac{\text{Stress}}{\text{Strain}} = \frac{T/A}{dy/dx}$$

$$dy = \frac{Mg}{YAL}x\,dx$$

Total elongation $y = \int dy = \frac{Mg}{YAL}\int_0^L x\,dx = \frac{Mg}{YAL}\cdot\frac{L^2}{2} = \boxed{\frac{MgL}{2YA}}$

Web Reference at www.physicsgalaxy.com | Scan for Video Explanation

**Module-20: Metal Wires of Rods as Equivalent Spring**

# Metal wires or Rods as equivalent spring:

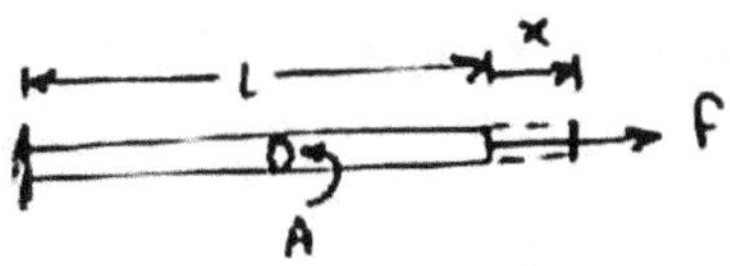


Strain = x/L
Stress = F/A

$$Y = \frac{F/A}{x/L}$$

$$F = \left(\frac{YA}{L}\right)x = kx.$$

eq. force constant for rod $k = \frac{YA}{L}$

Web Reference at www.physicsgalaxy.com | Scan for Video Explanation

**Module-20: Metal Wires of Rods as Equivalent Spring**

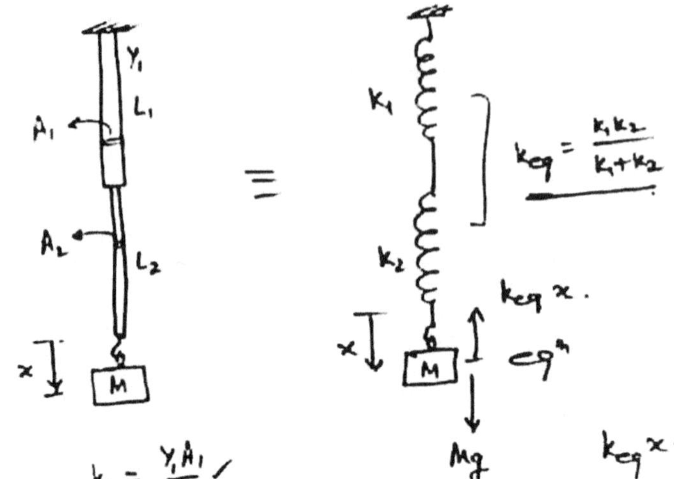


$$k_1 = \frac{Y_1 A_1}{L_1}$$

$$k_2 = \frac{Y_2 A_2}{L_2}$$

$$k_{eq}x = Mg$$

$$x = \frac{Mg}{K_{eq}}$$

Scan for Video Explanation

Web Reference at www.physicsgalaxy.com

**Module-21: Elastic Potential Energy Stored due to Elasticity**

# Elastic Potential Energy stored due to elasticity:

When a solid is elongated, some work is done in this process & a part of this work is stored as PE in the material due to which restoring forces exist

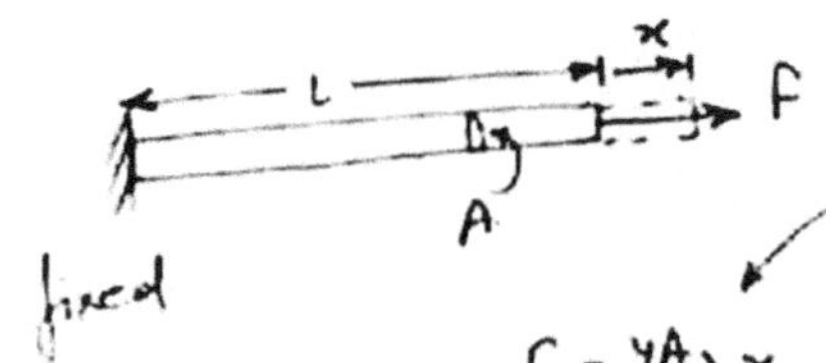


$$Y = \frac{F/A}{x/L}$$

$$x = \frac{FL}{YA}$$

$$F = \left(\frac{YA}{L}\right)x$$

equivalent force constant

Rod as if it is behaving like a spring is $\boxed{k = \frac{YA}{l}}$

PE stored in rod is $U = \frac{1}{2}kx^2 = \frac{1}{2}\left(\frac{YA}{l}\right)\times\left(\frac{Fl}{YA}\right)^2 = \frac{1}{2}\frac{F^2 l}{YA}$

Scan for Video Explanation

Web Reference at www.physicsgalaxy.com

**Module-21: Elastic Potential Energy Stored due to Elasticity**

$$\text{Density of elastic PE} = \frac{PE}{\text{Volume}}$$

$$u_e = \frac{\frac{1}{2}\frac{F^2 l}{YA}}{Al} = \frac{1}{2}\frac{F^2}{A^2}\cdot\frac{1}{Y} = \boxed{\frac{1}{2}\frac{(\text{Stress})^2}{Y}}\ J/m^3$$

$$\boxed{u_e = \frac{1}{2}(\text{stress})\times(\text{strain})}\ J/m^3$$

Web Reference at www.physicsgalaxy.com

Scan for Video Explanation

**Module-22: Solved Example-7**

Ex : Calculate the elastic potential energy stored in a brass rod of length 20 cm and area 1 cm$^2$ when it is elongated by a load of 5 kg mass. Given that Young's modulus of brass is $10^{11}$ N/m$^2$. $g = 10\ m/s^2$

Sol$^n$

$$\text{Energy stored in rod} = u_e \times \text{volume}$$

$$= \frac{1}{2}\frac{(\text{Stress})^2}{Y}\times\text{volume}$$

$$= \frac{1}{2}\frac{\left[50/10^{-4}\right]^2}{10^{11}}\times 0.2\times 10^{-4}\ J$$

$$= 2.5\times 10^{-5}\ J \quad \text{Ans}$$

Web Reference at www.physicsgalaxy.com

Scan for Video Explanation

**Module-23: Solved Example-8**

**Ex :** **A 4m long cylindrical wire with cross section 1 cm$^2$ is suspended vertically from a rigid support and carries a bob of mass 100 kg at the other end. If the bob gets snapped, calculate the change in temperature of wire. Given that density of steel is 7800 kg/m$^3$, Young's modulus of steel is 2 × 10$^{11}$ N/m$^2$ and specific heat of steel is 420 J/kg-K and g = 10 m/s$^2$.**

Sol$^n$: Total energy stored in wire

$$E = \frac{1}{2}\frac{(\text{stress})^2}{Y} \times \text{volume}$$

$$= \frac{1}{2} \times \frac{(1000/10^{-4})^2}{2 \times 10^{11}} \times 4 \times 10^{-4} = 0.1\ J$$

If temp of rod inc by $\Delta T$ $\Rightarrow$ $0.1 = ms\Delta T$

$$\Delta T = \frac{0.1}{L.A.\rho s} = \frac{0.1}{4 \times 10^{-4} \times 7800 \times 420} = 7.63 \times 10^{-5}\ K \quad \text{Ans}$$

Scan for Video Explanation

Web Reference at www.physicsgalaxy.com

# Chapter 17
# Gravitation Theory

## Lecture Notes Modules


**Module-1:** Introduction to Gravitation
**Module-2:** Limitations of Newton's Law of Gravitation
**Module-3:** Vector Representation of Gravitational Force
**Module-4:** Principle of Superposition of Forces
**Module-5:** Solved Example-1
**Module-6:** Solved Example-2
**Module-7:** Solved Example-3
**Module-8:** Gravitational Field
**Module-9:** Gravitational Field Strength due to a Point Mass and Extended Bodies
**Module-10:** Gravitational Field Strength due to a Uniform Rod
**Module-11:** Gravitational Field Strength due to a Uniform Ring
**Module-12:** Gravitational Field Strength due to a Circular Arc
**Module-13:** Gravitational Field Strength due to a Hollow Sphere
**Module-14:** Gravitational Field Strength due to a Solid Sphere
**Module-15:** Gravitational Field Strength due to a Long Cylinder
**Module-16:** Solved Example-4
**Module-17:** Solved Example-5
**Module-18:** Solved Example-6
**Module-19:** Solved Example-7
**Module-20:** Gravitational Lines of Forces
**Module-21:** Gravitational Field on Earth Surface
**Module-22:** Value of g at a Height h Above Earth Surface
**Module-23:** Value of g at a Height h Below Earth Surface
**Module-24:** Effect of Earth's Rotation on Value of g
**Module-25:** Effect of Shape of Earth on Value of g
**Module-26:** Solved Example-8
**Module-27:** Solved Example-9
**Module-28:** Solved Example-10
**Module-29:** Solved Example-11
**Module-30:** Gravitational Interaction Energy
**Module-31:** Gravitational Interaction Energy for a System of Particles
**Module-32:** Solved Example-12
**Module-33:** Solved Example-13
**Module-34:** Gravitational Potential
**Module-35:** Relation in g and V in Gravitational Field
**Module-36:** Gravitational Potential due to a Point Mass
**Module-37:** Gravitational Potential due to a Uniform Ring
**Module-38:** Work Done in Gravitational Field
**Module-39:** Gravitational Potential due to a Hollow Sphere
**Module-40:** Gravitational Potential due to a Solid Sphere
**Module-41:** Gravitational Potential Energy of a Body on Earth
**Module-42:** Solved Example-14
**Module-43:** Solved Example-15
**Module-44:** Solved Example-16
**Module-45:** Solved Example-17
**Module-46:** Solved Example-18
**Module-47:** Gravitational Self Energy of a Body
**Module-48:** Gravitational Self Energy of a Uniform Hollow Sphere
**Module-49:** Gravitational Self Energy of a Uniform Solid Sphere

## Module-1: Introduction to Gravitation

# Gravitation:

Study, effects and analysis related to grav. field & forces in surrounding of masses & Earth.

# Newton's law of Gravitation:

Acc. to this law -

(i) $F \propto m_1 m_2$

(ii) $F \propto \frac{1}{r^2}$

$$\Rightarrow \boxed{F = G\frac{m_1 m_2}{r^2}}$$

G ← universal grav. Constant

$G = 6.67 \times 10^{-11}\ N \cdot m^2/kg^2$

Scan for Video Explanation

Web Reference at www.physicsgalaxy.com

## Module-2: Limitations of Newton's Law of Gravitation

# Limitations of Newton's law of Gravitation:

This law is only valid for pt. masses. On extended bodies it can not be directly applied.

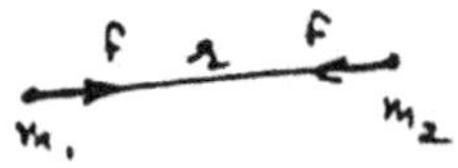


$$F = \frac{G m_1 m_2}{r^2}$$

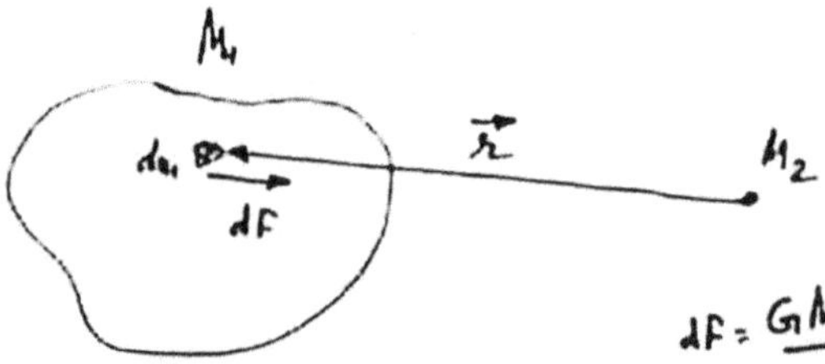


$$dF = \frac{G M_2 dm}{r^2}$$

$$\boxed{\vec{dF} = -\frac{G M_2 dm}{r^2}\hat{r} = -\frac{G M_2 dm}{r^3}\cdot\vec{r}}$$

$$\vec{F} = \int \vec{dF} = -\int \frac{G M_2 dm}{r^3}\cdot\vec{r}$$

Scan for Video Explanation

Web Reference at www.physicsgalaxy.com

**Module-2: Limitations of Newton's Law of Gravitation**

NOTE: For extended bodies NLG can be applied by taking small elements in extended body and integrate the exp. of force on element over the whole body.

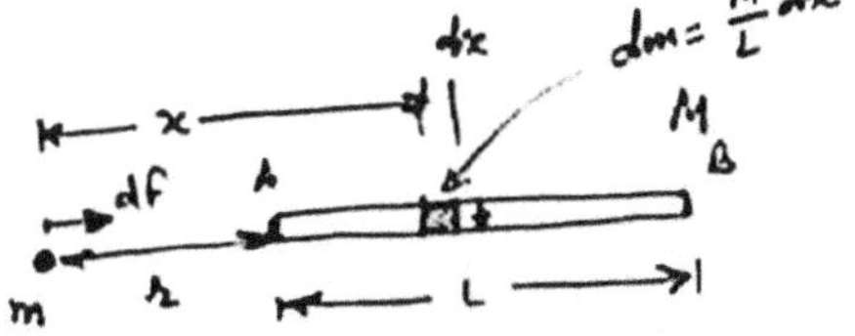


$$F \neq \frac{GmM}{(r+L/2)^2}$$

If dF be the grav force on m by dm $\Rightarrow$ $dF = \frac{Gm\,dm}{x^2}$

Total force on m due to rod $F = \int dF = \int_r^{r+L} \frac{GmM\,dx}{Lx^2}$

$$F = \frac{GmM}{L}\left[-\frac{1}{x}\right]_r^{r+L}$$



**Module-2: Limitations of Newton's Law of Gravitation**

$$F = \frac{GmM}{L}\left[-\frac{1}{x}\right]_r^{r+L}$$

$$F = \frac{GmM}{L}\left[\frac{1}{r} - \frac{1}{r+L}\right]$$

**Module-3: Vector Representation of Gravitational Force**

# Vector Representation of gravitational force:

$$F = \frac{Gm_1m_2}{r^2}$$

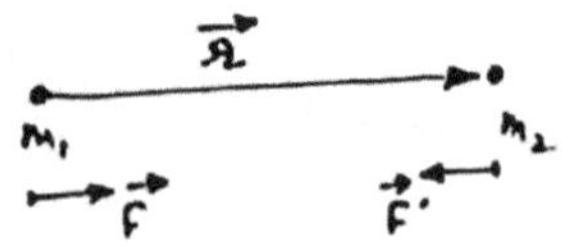


$$\vec{F}' = -\frac{Gm_1m_2}{r^2}\hat{r}$$

$$\boxed{\vec{F}' = -\frac{Gm_1m_2}{r^3}\cdot\vec{r}}$$

$$\boxed{\vec{F} = \frac{Gm_1m_2}{r^3}\vec{r}}$$

$$\vec{F} = -\vec{F}'$$

$$|\vec{F}| = |\vec{F}'| = F$$

$$\hat{r} = \frac{\vec{r}}{r}$$

Scan for Video Explanation

Web Reference at www.physicsgalaxy.com

**Module-4: Principle of Superposition of Forces**

# Principle of Superposition of Forces:

When there are more than two masses are there in a system then grav force on any mass can be obtained by vector sum of all individual forces acting on the mass due to all other masses of system.

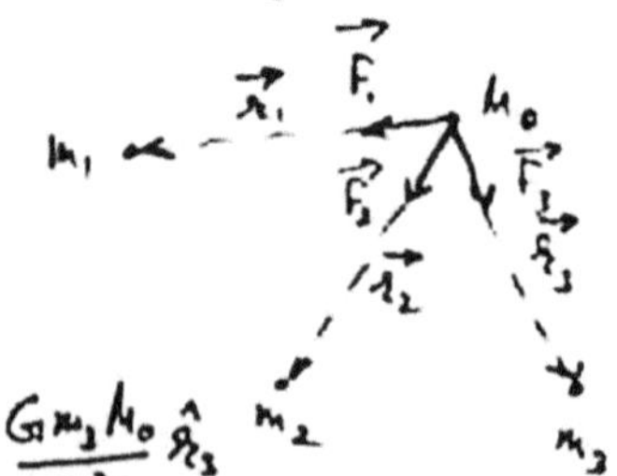


Net grav force on $M_0$ as shown is

$$\vec{F}_{net} = \vec{F}_1 + \vec{F}_2 + \vec{F}_3$$

$$\vec{F}_{net} = \frac{Gm_1M_0}{r_1^2}\hat{r}_1 + \frac{Gm_2M_0}{r_2^2}\hat{r}_2 + \frac{Gm_3M_0}{r_3^2}\hat{r}_3$$

Scan for Video Explanation

Web Reference at www.physicsgalaxy.com

**Module-5: Solved Example-1**

**Ex :** Two balls of mass m each are hung side by side by two long threads of equal length $l$. If the distance between upper ends is r, show that the distance r' between the centres of balls is given by

$$gr'^2 (r - r') = 2lGm$$

Sol$^n$: For eq$^n$ of balls, we use

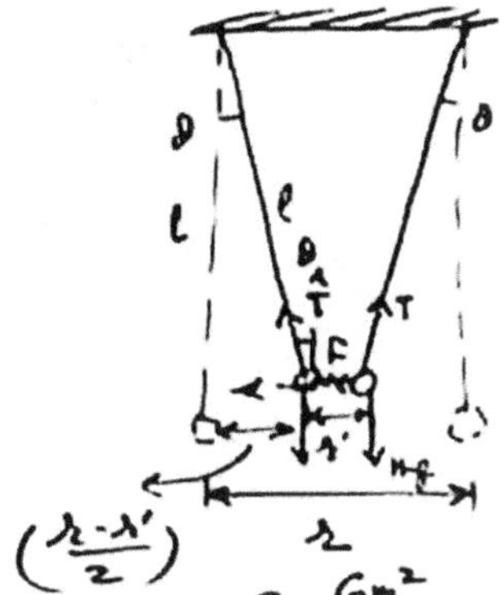

$$T\cos\theta = mg \quad \text{---(1)}$$

$$T\sin\theta = \frac{Gm^2}{r'^2} \quad \text{---(2)}$$

$$\frac{(2)}{(1)} \Rightarrow \tan\theta = \frac{Gm}{gr'^2} = \frac{r-r'}{2l}$$

$$F = \frac{Gm^2}{r'^2}$$

$$gr'^2(r-r') = 2Gml.$$

hence proved.

Web Reference at www.physicsgalaxy.com

Scan for Video Explanation

**Module-6: Solved Example-2**

**Ex :** Two particles of equal mass go round a circle of radius R under the action of their mutual gravitational force. Find the speed of each particle.

Sol$^n$:

For circular motion of masses

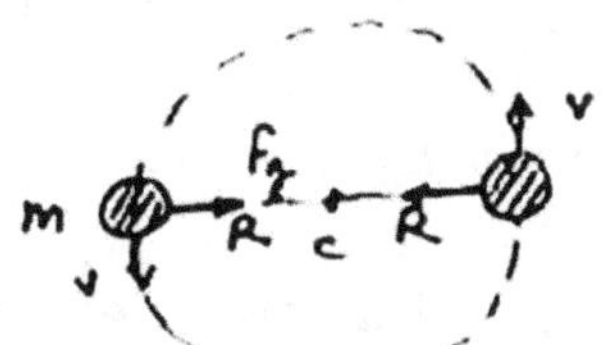

$$\frac{Gm^2}{(2R)^2} = \frac{mv^2}{R}$$

$$v = \sqrt{\frac{Gm}{4R}} \quad \text{Ans.}$$

Web Reference at www.physicsgalaxy.com

Scan for Video Explanation

**Module-7: Solved Example-3**

**Ex :** Find the gravitational attraction on mass m due to the quarter ring of same mass m and radius R as shown in figure.

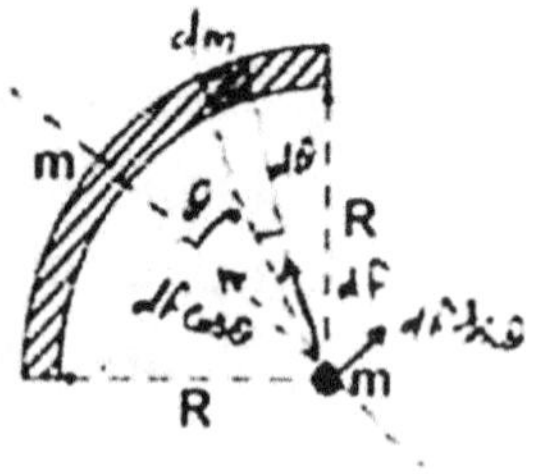


**Sol$^n$:** force on m due to dm on quarter ring is

$$dF = \frac{Gm\,dm}{R^2} = \frac{2Gm^2}{\pi R^2}\,d\theta \qquad dm = \frac{m}{\pi/2}\,d\theta = \frac{2m}{\pi}\,d\theta$$

net force on m will be

$$F = \int dF\cos\theta$$

$$= \frac{2Gm^2}{\pi R^2}\int_{-\pi/4}^{+\pi/4}\cos\theta\,d\theta$$

$$= \frac{2Gm^2}{\pi R^2}\left[\sin\theta\right]_{-\pi/4}^{+\pi/4}$$

$$= \frac{2Gm^2}{\pi R^2}\left[\frac{1}{\sqrt{2}} + \frac{1}{\sqrt{2}}\right]$$

$$F = \frac{2\sqrt{2}\,Gm^2}{\pi R^2} \quad \text{Ans.}$$

Scan for Video Explanation

Web Reference at www.physicsgalaxy.com

**Module-8: Gravitational Field**

# Gravitational Field:

It is the region in surrounding of a mass where other masses experience a grav. force.

# Strength of Grav. Field:

At a pt in grav. field, it is meas as force experienced by a unit mass. It is a vector qty & denoted by $\vec{g}$.

grav. field strength at pt P is

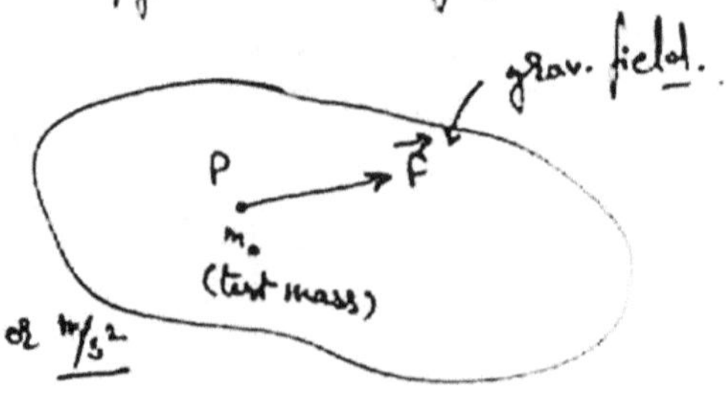


$$\vec{g}_P = \frac{\vec{F}}{m_0} \quad N/kg \text{ or } m/s^2$$

acceleration due to grav in the grav. field.

Scan for Video Explanation

Web Reference at www.physicsgalaxy.com

**Module-9: Gravitational Field Strength due to a Point Mass and Extended Bodies**

# Gravitational Field Strength due to a point mass:

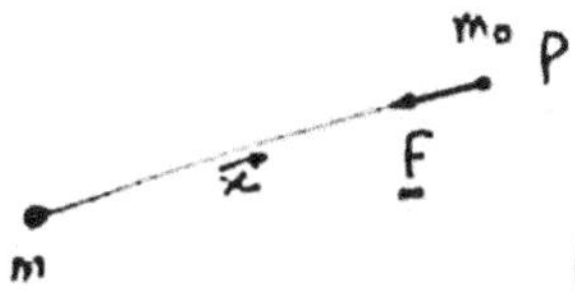


$$F = \frac{Gmm_0}{x^2}$$

grav. field strength at P due to pt mass m is

$$g_P = \frac{F}{m_0} = \frac{Gm}{x^2}$$

vectorially

$$\vec{g}_P = -\frac{Gm}{x^2}\hat{x} = -\frac{Gm}{x^3}\cdot\vec{x}$$

Web Reference at www.physicsgalaxy.com

Scan for Video Explanation

**Module-9: Gravitational Field Strength due to a Point Mass and Extended Bodies**

# Gravitational Field Strength due to extended bodies:

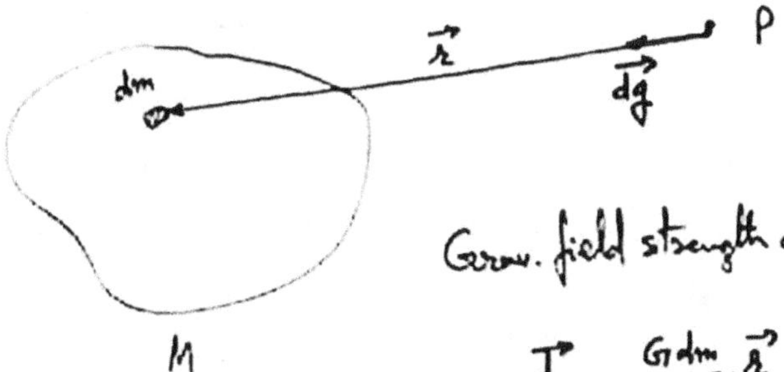


Grav. field strength at P due to dm is

$$\vec{dg} = \frac{G\,dm}{r^3}\cdot\vec{r}$$

grav. field strength at P due to whole body is

$$\vec{g}_P = \int \vec{dg} = \int_M \frac{G\,dm}{r^3}\cdot\vec{r}$$

Web Reference at www.physicsgalaxy.com

Scan for Video Explanation

**Module-10: Gravitational Field Strength due to a Uniform Rod**

# Gravitational Field Strength due to a uniform Rod:

Case-I: At an axial point of Rod.

grav. field strength at P due to dm is

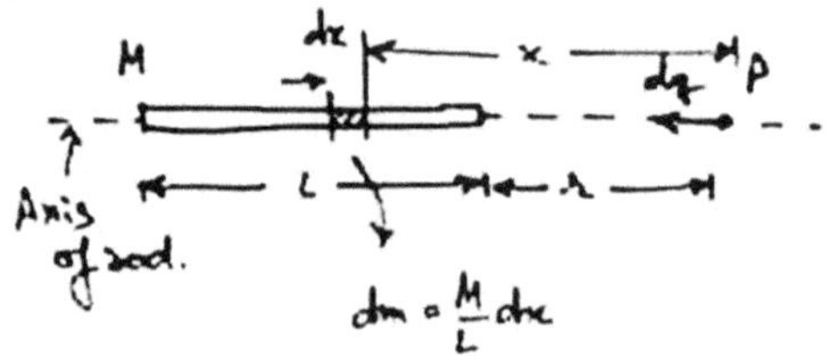


$$dg = \frac{G\,dm}{x^2} = \frac{GM}{Lx^2}dx$$

$$g_P = \int dg = \int_r^{r+L} \frac{GM}{Lx^2}dx$$

$$g_P = \frac{GM}{L}\left[-\frac{1}{x}\right]_r^{r+L}$$

$$\boxed{g_P = \frac{GM}{L}\left[\frac{1}{r} - \frac{1}{r+L}\right]}$$

Scan for Video Explanation

Web Reference at www.physicsgalaxy.com

**Module-10: Gravitational Field Strength due to a Uniform Rod**

Case-II: At an equatorial pt of rod.

If dg is grav field strength at P due to dm -

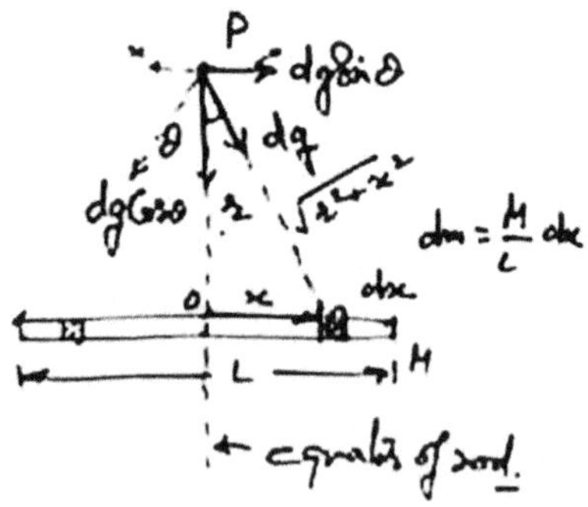


$$dg = \frac{G\,dm}{(r^2+x^2)}$$

net grav field at pt P is

$$g_P = \int dg\cos\theta$$

$$g_P = \int_{-L/2}^{+L/2} \frac{GM\,dx}{L(r^2+x^2)}\cdot\frac{r}{\sqrt{r^2+x^2}}$$

$$g_P = \frac{GMr}{L}\int_{-L/2}^{+L/2}\frac{dx}{(r^2+x^2)^{3/2}}$$

Scan for Video Explanation

Web Reference at www.physicsgalaxy.com

**Module-10: Gravitational Field Strength due to a Uniform Rod**

$$g_P = \frac{GMr}{L}\int_{-L/2}^{+L/2}\frac{dx}{(x^2+r^2)^{3/2}}$$

let $x = r\tan\theta$

$dx = r\sec^2\theta\, d\theta$

$$= \frac{GMr}{L}\int\frac{r\sec^2\theta\, d\theta}{r^3\sec^3\theta}$$

$(r^2\tan^2\theta + r^2)^{3/2}$

$= (r^2\sec^2\theta)^{3/2}$

$= r^3\sec^3\theta$

$$= \frac{GM}{Lr}\int\cos\theta\, d\theta$$

$$= \frac{GM}{Lr}[\sin\theta]$$

$$= \frac{GM}{Lr}\left[\frac{x}{\sqrt{x^2+r^2}}\right]_{-L/2}^{+L/2}$$

$$= \frac{GM}{Lr}\left[\frac{L/2}{\sqrt{L^2/4+r^2}}\times 2\right]$$

$$\boxed{g_P = \frac{2GM}{r\sqrt{L^2+4r^2}}}$$

Web Reference at www.physicsgalaxy.com | Scan for Video Explanation

**Module-11: Gravitational Field Strength due to a Uniform Ring**

# Gravitational Field Strength due to a uniform Ring:

Case-I: At a point at Centre of Ring.

$$\boxed{g_c = 0}$$

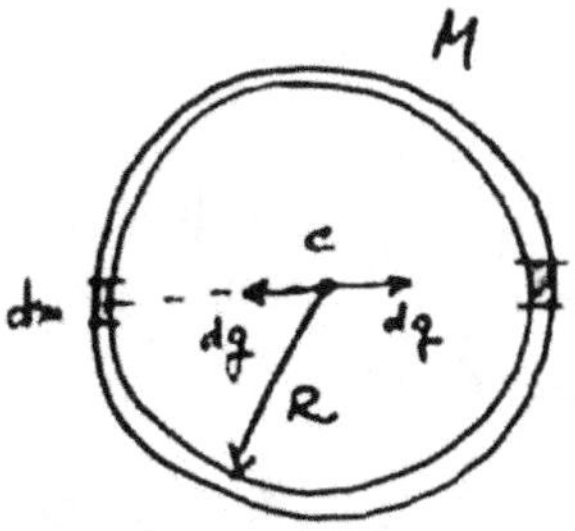


Case-II: At an axial pt of ring

grav. field strength at P due to dm is

$dm = \frac{M}{2\pi R}\cdot dl$

$$dg = \frac{G\,dm}{(R^2+z^2)}$$

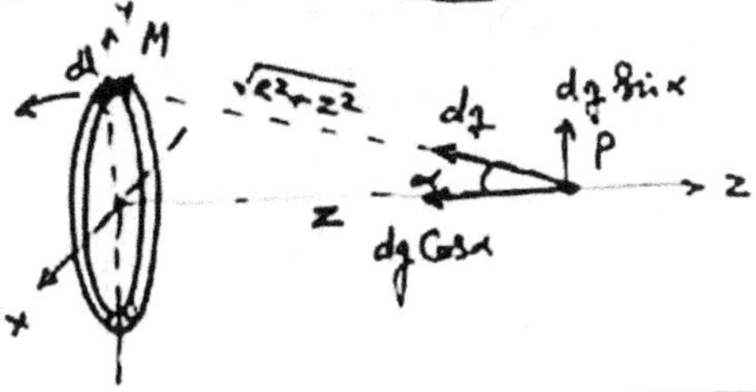


net grav field strength at P is -

$$g_P = \int dg\cos\alpha = \int_0^{2\pi R}\frac{GM\,dl}{2\pi R(R^2+z^2)}\cdot\frac{z}{\sqrt{R^2+z^2}} = \boxed{\frac{GMz}{(R^2+z^2)^{3/2}}}$$

Web Reference at www.physicsgalaxy.com | Scan for Video Explanation

**Module-12: Gravitational Field Strength due to a Circular Arc**

# Gravitational Field due to a Circular Arc:

$dm = \frac{M}{R\phi} R d\theta = \frac{M}{\phi} d\theta$

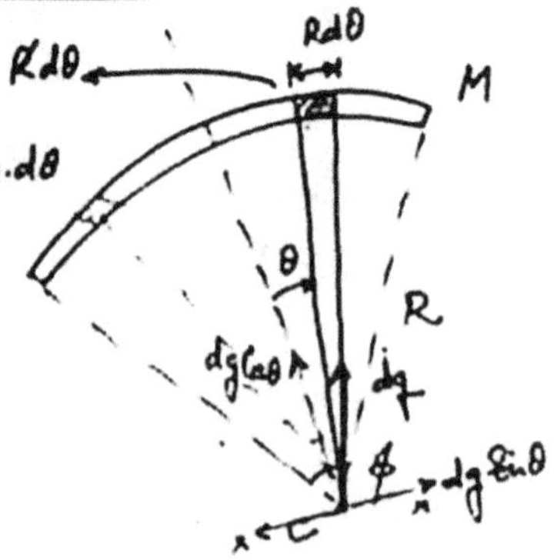


If dg be the grav field strength due to mass dm

$$dg = \frac{G dm}{R^2}$$

net grav field at pt C is

$$g_c = \int dg\cos\theta = \int_{-\phi/2}^{+\phi/2} \frac{GM d\theta}{\phi R^2} . \cos\theta$$

$$g_c = \frac{GM}{\phi R^2} \int_{-\phi/2}^{+\phi/2} \cos\theta d\theta = \frac{GM}{\phi R^2} \left[\sin\theta\right]_{-\phi/2}^{+\phi/2}$$

$$\boxed{g_c = \frac{2GM\sin(\phi/2)}{\phi R^2}}$$

Scan for Video Explanation

Web Reference at www.physicsgalaxy.com

**Module-13: Gravitational Field Strength due to a Hollow Sphere**

# Gravitational Field Strength due to a Uniform Hollow Sphere:

In surr at all pts $\vec{g}$ exist in radially inward dir.

for a uniform shell we can assume that all of its mass is concentrated at its centre (for outer pt.)

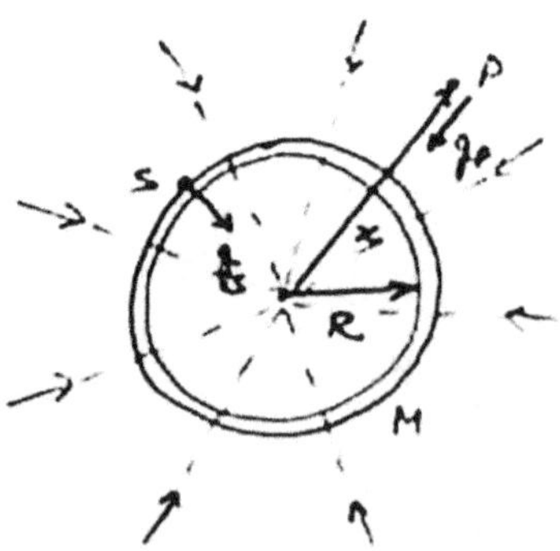


for outer pts $x > R$ $\boxed{g_P = \frac{GM}{x^2}}$

for surface pts $x = R$ $\boxed{g_s = \frac{GM}{R^2}}$

Scan for Video Explanation

Web Reference at www.physicsgalaxy.com

**Module-13: Gravitational Field Strength due to a Hollow Sphere**

grav field at pt P can be given

$$g_p = \int (dg_1 - dg_2)$$

$$= \int \frac{G\,dm_1}{x_1^2} - \frac{G\,dm_2}{x_2^2} = 0$$

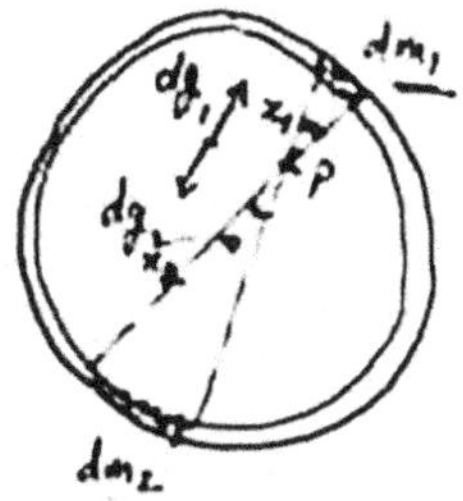

$dm_1 = k x_1^2$

$dm_2 = k x_2^2$

at all interior pts of a uniform shell

for $x < R$. $\boxed{g_{in} = 0}$

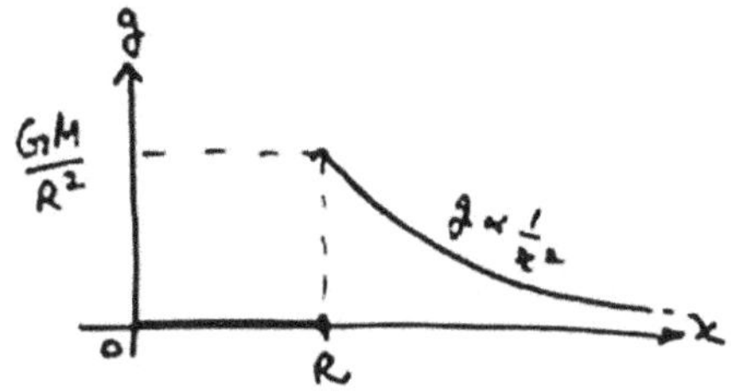

Web Reference at www.physicsgalaxy.com | Scan for Video Explanation

**Module-14: Gravitational Field Strength due to a Solid Sphere**

# Gravitational Field Strength due to a uniform Solid Sphere:

for outer pts $x > R$

$$\boxed{g_p = \frac{GM}{x^2}}$$

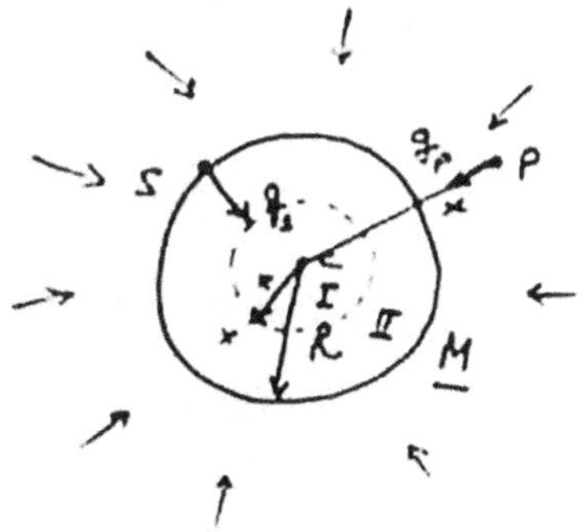

for surface pts $x = R$

$$\boxed{g_s = \frac{GM}{R^2}}$$

for interior pts $x < R$

$$g_{in} = \frac{Gm}{x^2} = \frac{G}{x^2}\left[\frac{Mx^3}{R^3}\right]$$

$$\boxed{g_{in} = \frac{GMx}{R^3}}$$

mass enclosed in the inner sphere of rad x.

$$m = \frac{M}{\frac{4}{3}\pi R^3} \times \frac{4}{3}\pi x^3 = \frac{Mx^3}{R^3}$$

Web Reference at www.physicsgalaxy.com | Scan for Video Explanation

**Module-14: Gravitational Field Strength due to a Solid Sphere**

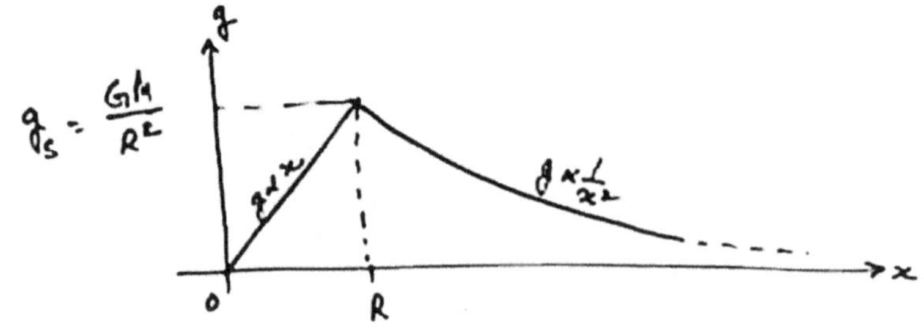


$$g_{in} = \frac{GMx}{R^3} \qquad g_{outer} = \frac{GM}{x^2}$$

Scan for Video Explanation

Web Reference at www.physicsgalaxy.com

**Module-15: Gravitational Field Strength due to a Long Cylinder**

# Gravitational Field Strength due to a long thread:

grav. field strength at P due to dm is

$$dg = \frac{G\,dm}{(r^2+y^2)}$$

net grav. field at pt. P will be

$$g_P = \int dg \cos\theta$$

$$= \int_{-\infty}^{+\infty} \frac{G\lambda dy}{(r^2+y^2)} \cdot \frac{r}{\sqrt{r^2+y^2}} = \int_{-\infty}^{+\infty} \frac{G\lambda r\,dy}{(r^2+y^2)^{3/2}}$$

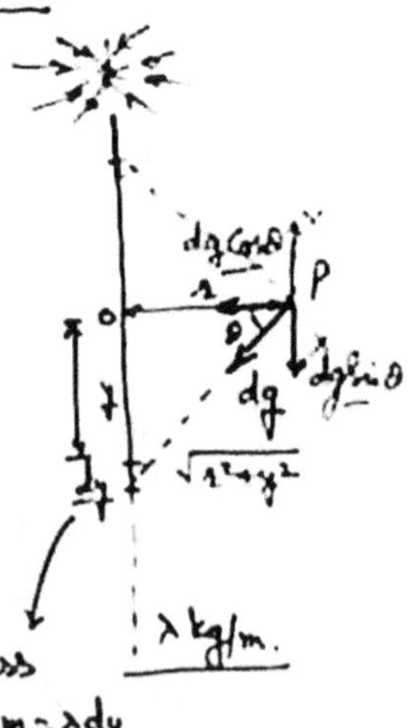


mass $dm = \lambda dy$

let $y = r\tan\theta$

$dy = r\sec^2\theta\, d\theta$

$$g_P = \int \frac{G\lambda r \cdot r\sec^2\theta\, d\theta}{r^3\sec^3\theta}$$

at $y = -\infty \rightarrow \theta = -\pi/2$

$y = \infty \rightarrow \theta = +\pi/2$

$$g_P = \frac{G\lambda}{r}\int_{-\pi/2}^{+\pi/2} \cos\theta\, d\theta = \frac{G\lambda}{r}\left[\sin\theta\right]_{-\pi/2}^{+\pi/2} = \frac{2G\lambda}{r}$$

$$\boxed{g_P = \frac{2G\lambda}{r}}$$

Scan for Video Explanation

Web Reference at www.physicsgalaxy.com

**Module-16: Solved Example-4**

**Ex :** Find the distance of a point from the earth's centre where the resultant gravitational field due to earth and moon is zero. The mass of earth is $6 \times 10^{24}$ kg and that of moon is $7.4 \times 10^{22}$ kg and the separation between earth and moon is $4 \times 10^5$ km.

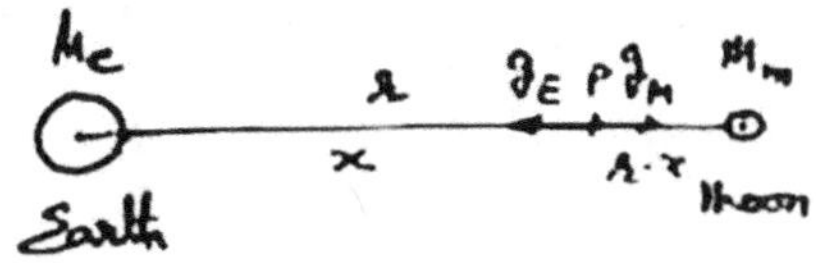


**Sol$^n$:** As at P $g_{net} = 0$

$$\Rightarrow \quad g_E = g_M$$

$$\frac{GM_e}{x^2} = \frac{GM_m}{(r-x)^2}$$

$$\frac{6 \times 10^{24}}{x^2} = \frac{7.4 \times 10^{22}}{(4 \times 10^5 - x)^2}$$

$$\frac{x}{4 \times 10^5 - x} = \sqrt{\frac{600}{7.4}} = 9$$

Solving for x we get

$x = 3.6 \times 10^5$ km Ans.

Web Reference at www.physicsgalaxy.com | Scan for Video Explanation

**Module-17: Solved Example-5**

**Ex :** A spherical hollow cavity is made in a lead sphere of radius R such that its surface touches the outside surface of lead sphere and passes through its centre. Find the gravitational field strength at the point P as shown. Given that mass of sphere before cavity is made was M.

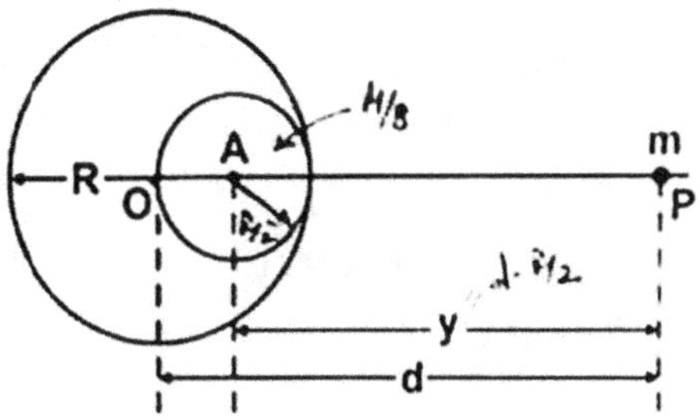


**Sol$^n$:**

$$g_P = g_{sphere} - g_{cavity\ mass}$$

$$g_P = \frac{GM}{d^2} - \frac{G(M/8)}{(d - R/2)^2}$$

$$g_P = \frac{GM}{d^2}\left[1 - \frac{1}{8(1 - R/2d)^2}\right] \text{ Ans.}$$

Web Reference at www.physicsgalaxy.com | Scan for Video Explanation

**Module-18: Solved Example-6**

**Ex :** Inside a uniform sphere of density ρ, there is a spherical cavity whose centre is at a distance *l* from the centre of sphere. Find the strength of gravitational field inside the cavity.

**Sol$^n$:**

gravity at P due to complete sphere is $\vec{g}_1 = -\frac{GM\vec{x}}{R_1^3}$

gravity at P only due to m is

$$\vec{g}_2 = -\frac{Gm\vec{y}}{R_2^3}$$

at P $\vec{g}_P = \vec{g}_1 - \vec{g}_2$

$$\vec{g}_P = G\left(\frac{4\pi}{3}\rho\right)(\vec{y} - \vec{x})$$

$$\boxed{\vec{g}_P = -\frac{4\pi\rho G}{3}\vec{l}}$$ Ans.

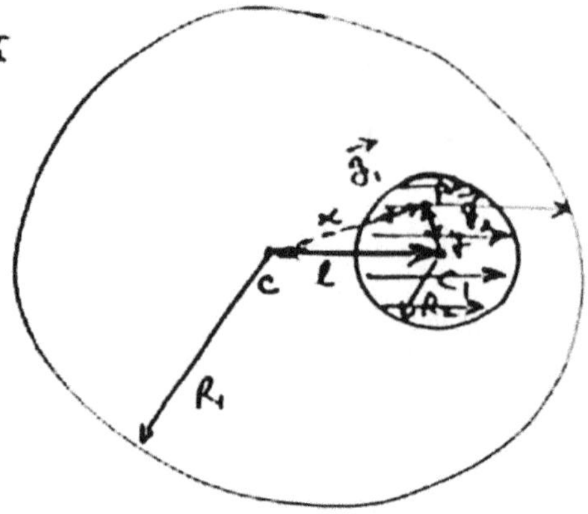


$M = \rho \times \frac{4}{3}\pi R_1^3$

$m = \rho \times \frac{4}{3}\pi R_2^3$

Scan for Video Explanation

Web Reference at www.physicsgalaxy.com

**Module-19: Solved Example-7**

**Ex :** A small point mass m is to be thrown with such a speed at a distance x from the axis of a long cylinder of radius R and density ρ, so that m starts revolving around the cylinder in circular orbit. Find the projection speed of mass.

**Sol$^n$:** for m to move in circular orbit,

$$F_g = \frac{mv^2}{x}.$$

$$m\left[\frac{2G\rho\pi R^2}{x}\right] = \frac{mv^2}{x}$$

$$v = \sqrt{2G\rho\pi R^2}$$

$$\underline{v = R\sqrt{2G\rho\pi}}$$ Ans.

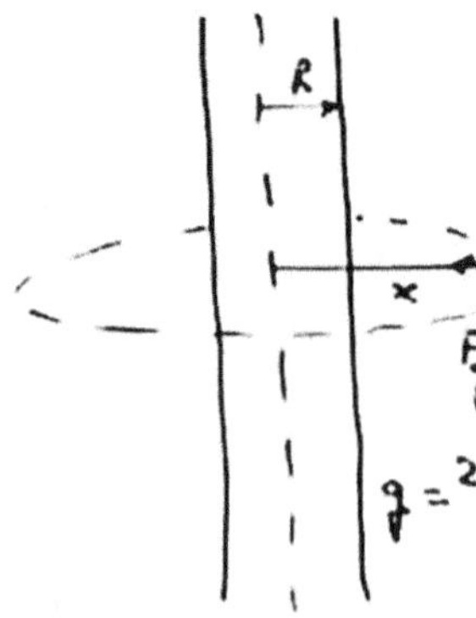


$F_g = m.g$

$g = \frac{2G\rho\pi R^2}{x}$

Scan for Video Explanation

Web Reference at www.physicsgalaxy.com

**Module-20: Gravitational Lines of Forces**

# Gravitational Lines of Forces:

– These are hypothetical lines in grav. field which gives info about direction and magnitude of grav field strength at a point in the region.

– In a region these lines are drawn such that tangent to any point on a line gives the dir of $\vec{g}$ at that point & density of lines in the neighbourhood of a point are prop. to the magnitude of $\vec{g}$.

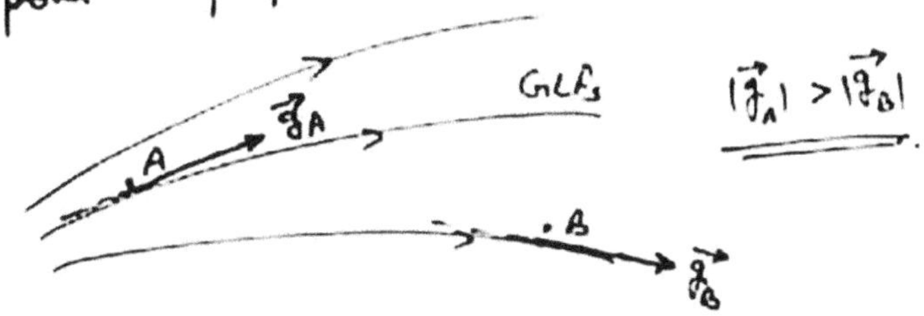


Web Reference at www.physicsgalaxy.com | Scan for Video Explanation

**Module-20: Gravitational Lines of Forces**

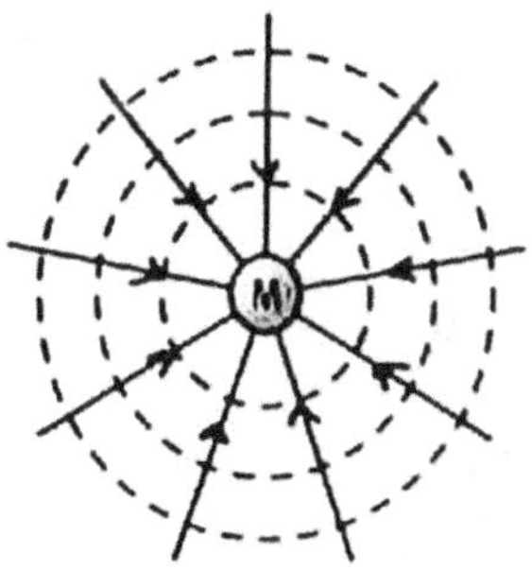


dir of $\vec{g}$ → tangent to GLF

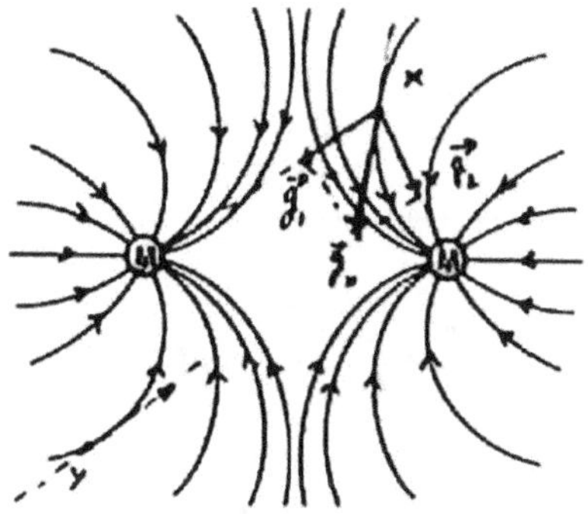


Web Reference at www.physicsgalaxy.com | Scan for Video Explanation

**Module-21: Gravitational Field on Earth Surface**

# Gravitational Field on Earth Surface:

Earth → very large solid sphere.

$$g_s = \frac{GM_e}{R_e^2} = 9.8\ m/s^2$$

↓ acc$^n$ due to gravity

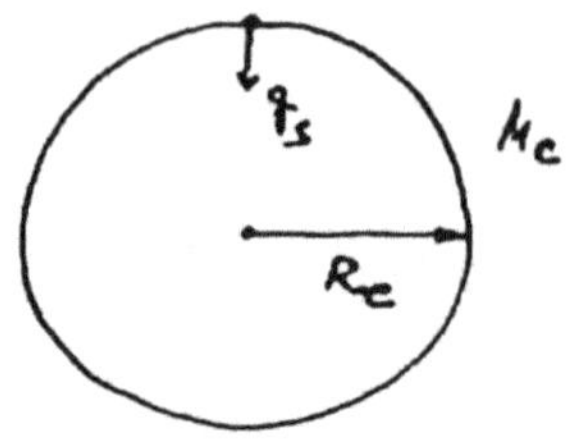


Scan for Video Explanation

Web Reference at www.physicsgalaxy.com

**Module-22: Value of g at a Height h Above Earth Surface**

# Value of g at a height h above the Earth Surface:

$$g_P = \frac{GM_e}{(R_e+h)^2}$$

$$g_h = \frac{GM_e}{R_e^2\left(1+\frac{h}{R_e}\right)^2}$$

$$\boxed{g_h = \frac{g_s}{\left(1+h/R_e\right)^2}}$$

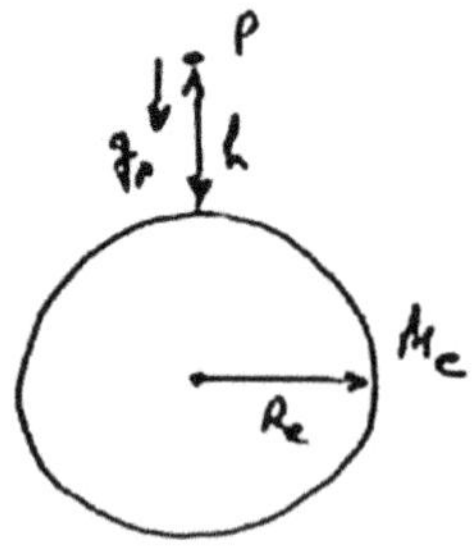


$R_e = 6400$ km.

for $h << R_e$

$$g_h = g_s\left(1+\frac{h}{R_e}\right)^{-2} = g_s\left(1-\frac{2h}{R_e}\right)$$

$$(1+x)^n = 1+nx \quad \text{for } x<<1$$

Scan for Video Explanation

Web Reference at www.physicsgalaxy.com

**Module-23: Value of g at a Height h Below Earth Surface**

# Value of g at a depth h inside the Earth Surface:

inside a solid sphere at a dist x

$$g = \frac{GMx}{R^3}$$

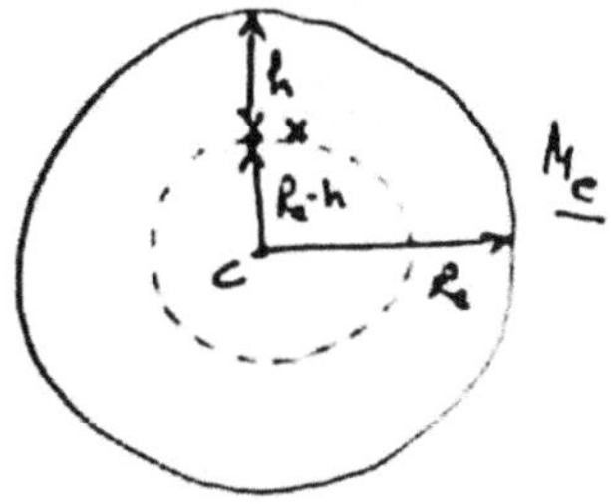


$$g_{in\,x} = \frac{GM_e (R_e - h)}{R_e^3}$$

$$\boxed{g_{in\,x} = g_s \left(1 - \frac{h}{R_e}\right)}$$

Web Reference at www.physicsgalaxy.com

Scan for Video Explanation

**Module-24: Effect of Earth's Rotation on Value of g**

# Effect of Earth's Rotation on value of g:

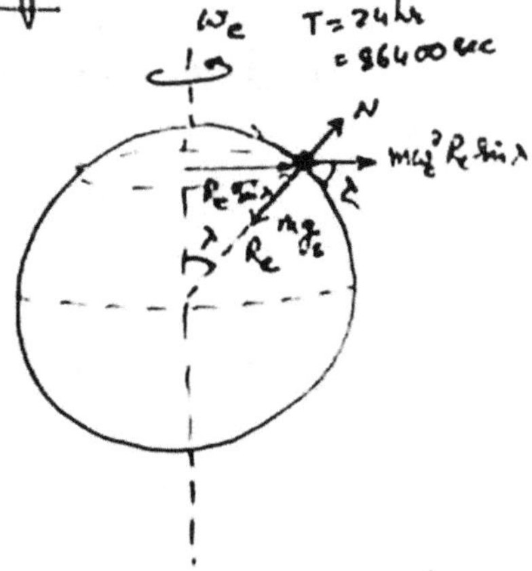


for equilibrium of this body

$$N + m\omega_e^2 R_e \sin^2\lambda = mg_s$$

$$N = mg_s - m\omega_e^2 R_e \sin^2\lambda$$

$$mg_{eff} = W_{eff} = m(g_s - \omega_e^2 R_e \sin^2\lambda)$$

$$\omega_e = \frac{2\pi}{86400} \text{ rad/sec}$$

$$\boxed{g_{eff} = g_s - \omega_e^2 R_e \sin^2\lambda}$$

at poles $\lambda = 0 \Rightarrow g_{Poles} = g_s = 9.83\ m/s^2$

at equator $\lambda = \pi/2 \Rightarrow g_{eq} = g_s - \omega_e^2 R_e = 9.78\ m/s^2$

$g_{eq} < g_{Poles}$

Web Reference at www.physicsgalaxy.com

Scan for Video Explanation

**Module-25: Effect of Shape of Earth on Value of g**

# Effect of Shape of Earth on Value of g:

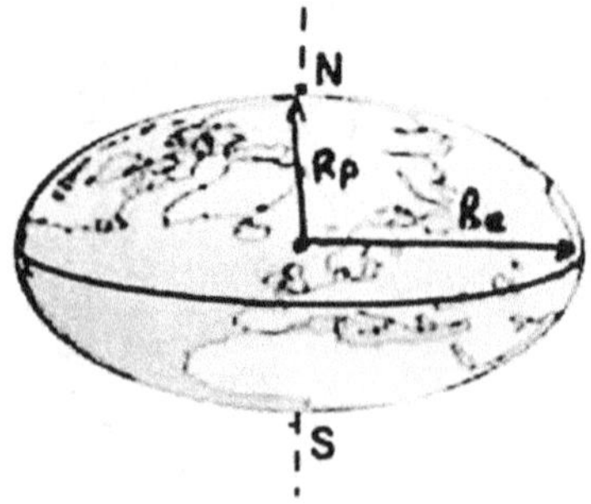


at variation location on Earth Surface

$R_E - R_P \sim 21$ to $34$ km.

at diff pos. on earth $g_P \geq g_E$

at various locations we find $g_P - g_E \sim 0.02$ to $0.04$ m/s$^2$

NOTE: Due to shape of Earth variation in $g$ is very small and for most of analysis this effect can be ignored.

Scan for Video Explanation

Web Reference at www.physicsgalaxy.com

**Module-26: Solved Example-8**

**Ex :** If the radius of earth were to shrink by one percent, its mass remaining the same. What would be new acceleration due to gravity on earth surface.

Sol$^n$:

As we know $g_s = \dfrac{GM_e}{R_e^2}$

new gravity $g_s' = \dfrac{GM_e}{(0.99R_e)^2} = \dfrac{g_s}{(0.99)^2}$

$g_s' = g_s(1-0.01)^{-2} = 1.02\, g_s$

new value of gravity will be increased by 2%. Ans.

Scan for Video Explanation

Web Reference at www.physicsgalaxy.com

**Module-27: Solved Example-9**

**Ex :** At what angular speed should earth rotate so that apparent value of g at equator becomes zero ? What will be the length of the day in this situation [Take $g_s = 9.8\ m/s^2$ and $r_e = 6400$ km]

Sol$^n$: due to rot$^n$ of earth the effective gravity at equator

$$g_{eq} = g_s - \omega^2 R_e = 0$$

$$\omega = \sqrt{\frac{g_s}{R_e}} = \sqrt{\frac{9.8}{6400\times10^3}}$$

$$= 1.24\times10^{-3}\ rad/s.\ Ans.$$

length of the day will be $T = \frac{2\pi}{\omega} = \frac{2\times3.14}{1.24\times10^{-3}} = 5065$ sec

$= 84.4$ min Ans.

Web Reference at www.physicsgalaxy.com

Scan for Video Explanation

**Module-28: Solved Example-10**

**Ex :** Find the height over the earth surface at which weight of body becomes half of its value of surface.

Sol$^n$: at Earth Surface $W = mg_s$

at a height h above $W' = \frac{mg_s}{(1+h/R_e)^2}$

given that $W' = W/2$

$$\frac{mg_s}{(1+h/R_e)^2} = \frac{1}{2} mg_s$$

$$1 + \frac{h}{R_e} = \sqrt{2} \Rightarrow h = R_e(\sqrt{2}-1)\ Ans.$$

Web Reference at www.physicsgalaxy.com

Scan for Video Explanation

**Module-29: Solved Example-11**

**Ex :** **If diameter of a planet is four times that of earth. Find the period of a pendulum on the planet, if it is a seconds pendulum on earth. Take mean density of planet equal to that of earth.**

Solⁿ: for a seconds pendulum on Earth $T = 2\pi\sqrt{\frac{l}{g_s}} = 2\text{ sec}$

on the planet $T' = 2\pi\sqrt{\frac{l}{g_p}} = 2\pi\sqrt{\frac{l}{4g_s}} = \frac{2}{2} = 1\text{ sec}$ Ans.

on planet surface $g_p = \frac{GM_p}{R_p^2} = \frac{G}{(4R_e)^2} \times \rho \cdot \frac{4}{3}\pi(4R_e)^3$

$= G\rho\left(\frac{4\pi}{3}\right) \cdot 4R_e = 4 \cdot \frac{GM_e}{R_e^2}$

$\rho = \frac{M_e}{\frac{4}{3}\pi R_e^3}$

$g_p = 4g_s$

Scan for Video Explanation

Web Reference at www.physicsgalaxy.com

**Module-30: Gravitational Interaction Energy**

# Gravitational Interaction Energy:

- This is the PE in a SOP due to their grav interaction.
- It is the W/D in assembling the SOP in given configuration from ∞ sep against grav. forces of system.
- As grav forces are always attractive, W/D against these forces in assembling a SOP is always −ve hence grav interaction energy of a system is always −ve as we take zero reference at ∞ separation.

Scan for Video Explanation

Web Reference at www.physicsgalaxy.com

**Module-30: Gravitational Interaction Energy**

# Grav. Potential Energy of Two point masses:

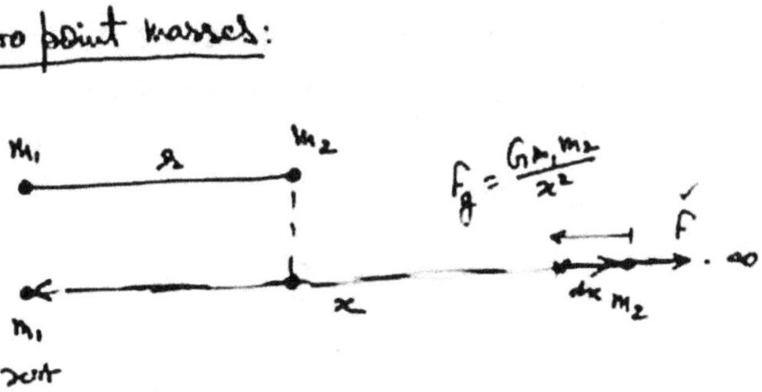


W/D in bringing $m_2$ by a dist $dx$

$$dW = \vec{F}.\vec{dx} = -\frac{Gm_1m_2}{x^2}dx$$

W/D in bringing it from $\infty$ to $r$ will be

$$W = \int dW = \int_{\infty}^{r} \frac{Gm_1m_2}{x^2}dx = Gm_1m_2\int_{\infty}^{r}\frac{1}{x^2}dx$$

$$= Gm_1m_2\left[-\frac{1}{x}\right]_{\infty}^{r}$$

$$\boxed{W = -\frac{Gm_1m_2}{r} = U_g}$$

Web Reference at www.physicsgalaxy.com

Scan for Video Explanation

**Module-31: Gravitational Interaction Energy for a System of Particles**

# Grav Interaction Energy for a System of Particles:

For a given N mass system - $m_1, m_2 - - - m_N$ separated from each other by distances $r_{12}, r_{13} - - - r_{ij}$ The interaction energy of such a system can be given as -

$$\boxed{U_g = -\frac{1}{2}\sum_{i=1}^{N}\sum_{j=1}^{N}\frac{Gm_im_j}{r_{ij}}}$$

$$\boxed{U_g = -\frac{Gm_1m_2}{r_{12}} - \frac{Gm_1m_3}{r_{13}} - \frac{Gm_2m_3}{r_{23}}}$$

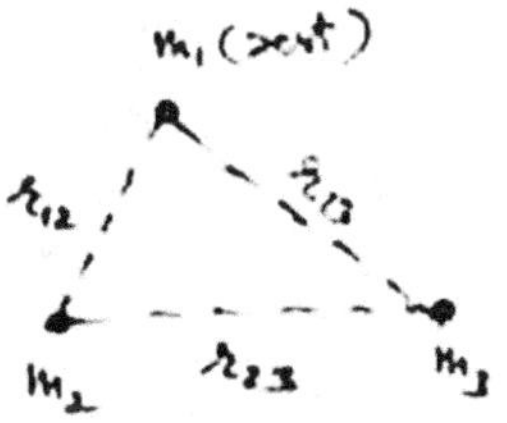


Web Reference at www.physicsgalaxy.com

Scan for Video Explanation

**Module-32: Solved Example-12**

**Ex :** **Three particles each of mass m are placed at the corners of an equilateral triangle of side d. Calculate the work done in increasing the sides of triangle from d to 2d.**

Sol$^n$: W/D in inc. sides of triangle from d to 2d is

$$W = U_f - U_i$$

$$W = \left(-\frac{3}{2}\frac{Gm^2}{d}\right) - \left(-\frac{3Gm^2}{d}\right)$$

$$W = \frac{3}{2}\frac{Gm^2}{d} \quad \text{Ans.}$$

$$U_i = -\frac{3Gm^2}{d}$$

$$U_f = -\frac{3Gm^2}{2d}$$

Scan for Video Explanation

Web Reference at www.physicsgalaxy.com

**Module-33: Solved Example-13**

**Ex :** **Two particles of masses $m_1$ and $m_2$ are initially at rest at infinite separation. When released, find their relative velocity of approach when they are at a separation d.**

Sol$^n$: Initial energy of system $E_i = 0$

As $F_{ext} = 0$

$\Rightarrow m_1 v_1 = m_2 v_2$

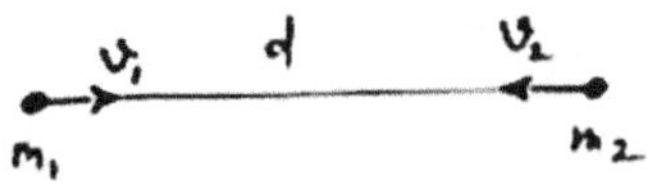


at sep d, the total energy of system $E_f = -\frac{Gm_1m_2}{d} + \frac{1}{2}m_1v_1^2 + \frac{1}{2}m_2v_2^2$

$$-\frac{Gm_1m_2}{d} + \frac{1}{2}m_1v_1^2 + \frac{1}{2}m_2v_2^2 = 0$$

$$\frac{1}{2}m_1v_1^2 + \frac{1}{2}m_2\left(\frac{m_1v_1}{m_2}\right)^2 = \frac{Gm_1m_2}{d}$$

on solving

$$v_1 = \sqrt{\frac{2G}{d(m_1+m_2)}} \cdot m_2$$

$$v_2 = \sqrt{\frac{2G}{d(m_1+m_2)}} \cdot m_1$$

$$v_{ap} = v_1 + v_2$$

$$v_{ap} = \sqrt{\frac{2G(m_1+m_2)}{d}} \quad \text{Ans.}$$

Scan for Video Explanation

Web Reference at www.physicsgalaxy.com

**Module-34: Gravitational Potential**

# Gravitational Potential:

It is defined as grav. interaction energy of unit mass placed at a pt in grav. field.

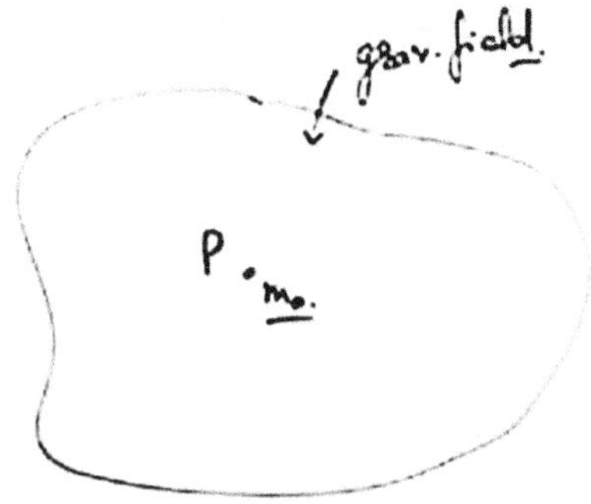


If interaction energy of mass $m_0$ at pt P is $U_g$ then grav. potential at pt P is -

$$V_P = \frac{U_g}{m_0} \text{ J/kg.}$$

At a pt in grav. field grav. potential is defined as the W/D in bringing a unit mass from ∞ to that pt. (at rest)

Web Reference at www.physicsgalaxy.com Scan for Video Explanation

**Module-34: Gravitational Potential**

If at a pt in grav. field, we know the grav potential say V then on placing a mass m at this pt, the grav interaction energy of this mass with the field can be given as

$$U_g = mV$$

NOTE: In a grav field region, grav. potⁿ is a characteristic at each pt in the region.

Web Reference at www.physicsgalaxy.com Scan for Video Explanation

**Module-35: Relation in g and V in Gravitational Field**

# Relation in $\vec{g}$ and V in a gravitational Field:

$$V = -\int_{\infty}^{P} \vec{F_g} \cdot \vec{dx}$$

$$\boxed{V = -\int_{\infty}^{P} \vec{g} \cdot dx}$$

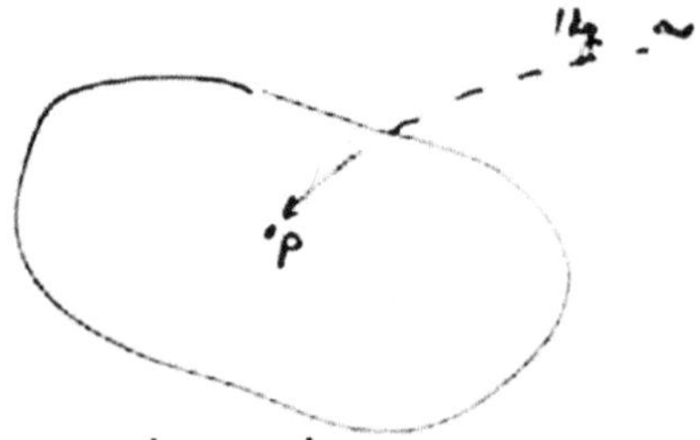

for a unit mass.

$$\vec{F_g} = \vec{g}$$

For a region $\vec{g}$ & V can be related as

$$\vec{g} = -\nabla V = -\text{ grad of } V \quad [\text{for Cons. fields}]$$

Scan for Video Explanation

Web Reference at www.physicsgalaxy.com

**Module-35: Relation in g and V in Gravitational Field**

$$\vec{g} = -\nabla V$$

Where $\nabla \equiv \frac{\partial}{\partial x}\hat{i} + \frac{\partial}{\partial y}\hat{j} + \frac{\partial}{\partial z}\hat{k}$

for unidirection variation in field, we can use

$$\boxed{\vec{g} = -\frac{dV}{dx}\hat{i}}$$

$\vec{g}$ is the -ve of slope of curve of V vs x.

for 3D variation in field, we can use

$$\boxed{\vec{g} = -\left(\frac{\partial V}{\partial x}\hat{i} + \frac{\partial V}{\partial y}\hat{j} + \frac{\partial V}{\partial z}\hat{k}\right)}$$

Scan for Video Explanation

Web Reference at www.physicsgalaxy.com

**Module-36: Gravitational Potential due to a Point Mass**

# Gravitational Potential due to a point mass:

Grav. interaction energy of test mass $m_0$ in the field of m.

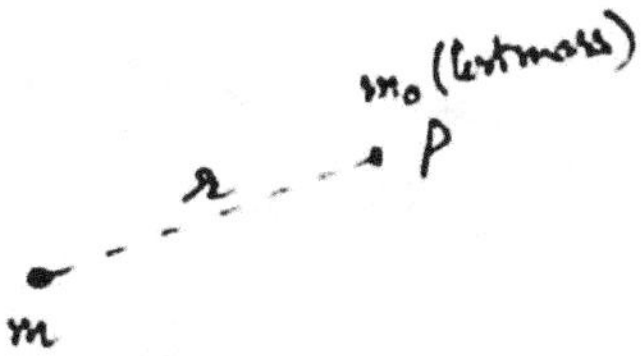


$$U_g = -\frac{Gmm_0}{r}$$

The grav. potential at pt P due to the field of m is

$$\boxed{V = \frac{U_g}{m_0} = -\frac{Gm}{r}}$$ ✓

Alternatively

$$V_P = -\int_{\infty}^{P} \vec{g}\cdot d\vec{x} = \int_{\infty}^{P} \frac{Gm}{x^2}dx = Gm\left[-\frac{1}{x}\right]_{\infty}^{r}$$

$$\boxed{V_P = -\frac{Gm}{r}}$$ ✓

Web Reference at www.physicsgalaxy.com | Scan for Video Explanation

**Module-37: Gravitational Potential due to a Uniform Ring**

# Gravitational Potential due to a Uniform Ring:

Case-I: V at Centre of Ring.

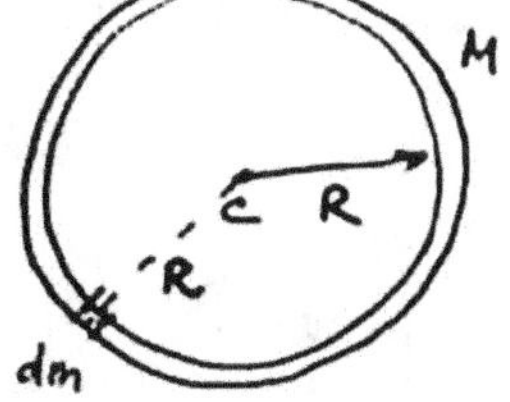


grav pot$^n$ at C due to dm is

$$dV = -\frac{Gdm}{R}$$

grav pot$^n$ due to Complete ring at C –

$$V_c = \int dV = -\int \frac{Gdm}{R} = -\frac{G}{R}\int dm$$

$$\boxed{V_c = -\frac{GM}{R}}$$ ✓✓

Web Reference at www.physicsgalaxy.com | Scan for Video Explanation

## Module-37: Gravitational Potential due to a Uniform Ring

Case-II: V at an axial pt of ring

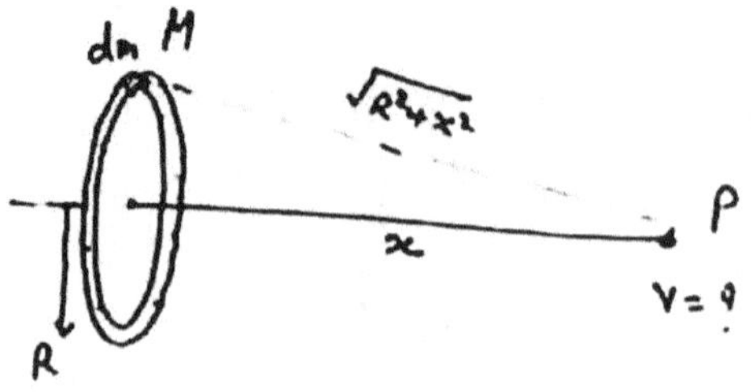


Grav pot^n due to dm at pt P is

$$dV = -\frac{G\,dm}{\sqrt{R^2+x^2}}$$

Net pot^n at P due to whole ring is

$$V_P = \int dV = -\int \frac{G\,dm}{\sqrt{R^2+x^2}}$$

$$\boxed{V_P = -\frac{GM}{\sqrt{R^2+x^2}}}$$



## Module-38: Work Done in Gravitational Field

# Work Done in displacement of a body in Gravitational Field:

W/D in disp of body from A to B is –

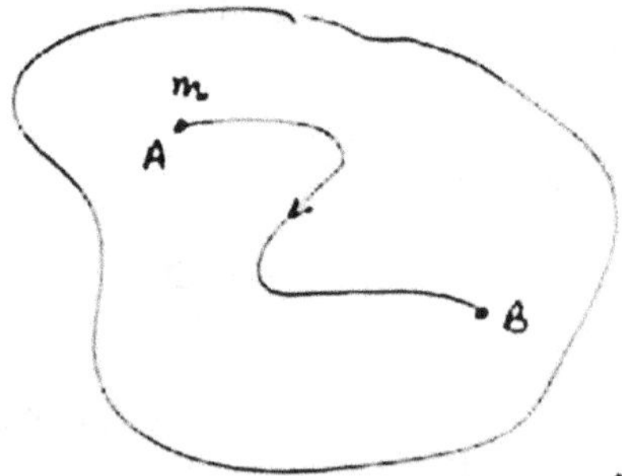


$$\boxed{W = U_B - U_A}$$

$$W = m(V_B - V_A)$$

work = mass × grav. pot^n diff.

$$\boxed{V_B - V_A = \frac{W}{m}}$$ W/D per unit mass.

$$\boxed{V_B - V_A = -\int_A^B \vec{g}\cdot d\vec{r}}$$

If $V_A$ and $V_B$ are grav. potential at pts A and B

$$U_A = mV_A$$

$$U_B = mV_B$$

**Module-39: Gravitational Potential due to a Hollow Sphere**

# Gravitational Potential due to a Hollow Sphere:

at outer pts $x > R$

$$V_{out} = -\frac{GM}{x}$$

at Surface pt $x = R$

$$V_S = -\frac{GM}{R}$$

at any interior $x < R$

$$V_{in} = -\frac{GM}{R}.$$

$$V_B - V_A = -\int_A^B \vec{g}\cdot d\vec{x} = 0$$

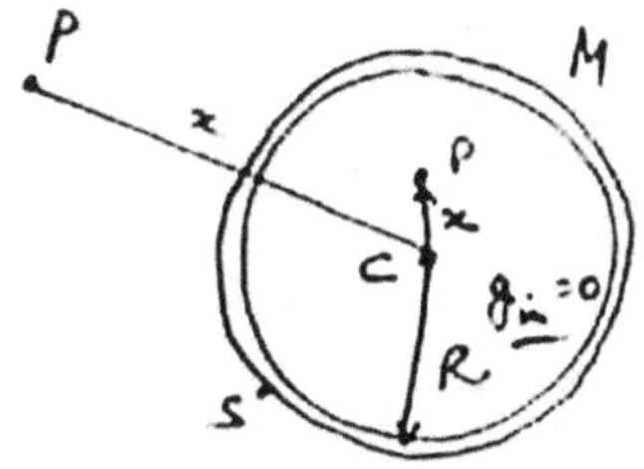


$$V_{in} = V_S$$

NOTE: In a region if at every pt $g = 0$ then this region can be taken as an equipotential region.

Web Reference at www.physicsgalaxy.com | Scan for Video Explanation

**Module-39: Gravitational Potential due to a Hollow Sphere**

# Variation of gravitational potential due to a hollow sphere:

for $x > R$ $\quad V_{out} = -\frac{GM}{x}$

for $x = R$ $\quad V_S = -\frac{GM}{R}$

for $x < R$ $\quad V_{in} = -\frac{GM}{R}$

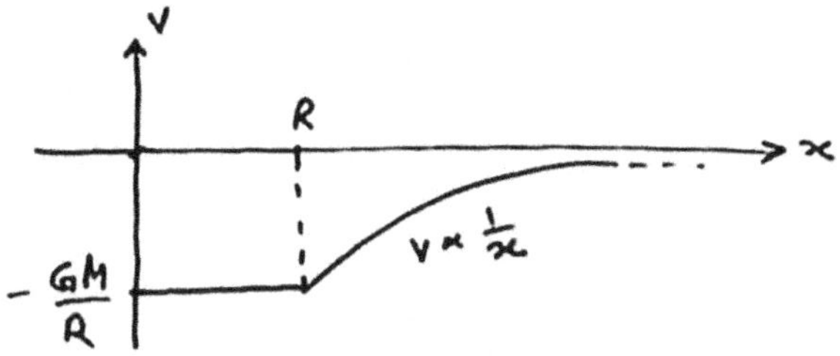


Web Reference at www.physicsgalaxy.com | Scan for Video Explanation

**Module-40: Gravitational Potential due to a Solid Sphere**

# Gravitational Potential due to a Solid Sphere:

for outer pts $x > R$

$$V_{outer} = -\frac{GM}{x}$$

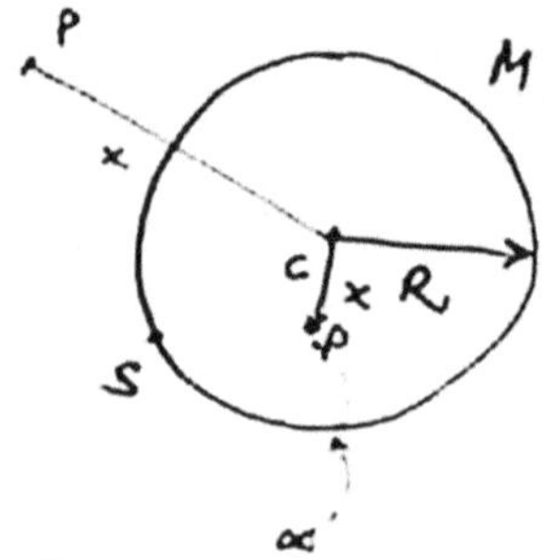


for surface pts $x = R$

$$V_s = -\frac{GM}{R}$$

at inside pts $x < R$

$$V_P = -\int_{\infty}^{x} \vec{g}\cdot\vec{dx}$$

$$= -\left[\int_{\infty}^{R} \vec{g}_{out}\cdot\vec{dx} + \int_{R}^{x} \vec{g}_{in}\cdot\vec{dx}\right]$$

$$V_{in} = -\frac{GM}{2R^3}[3R^2 - x^2]$$

$$= \int_{\infty}^{R} \frac{GM}{x^2}\,dx + \int_{R}^{x} \frac{GMx}{R^3}\,dx$$

at Centre $x = 0$

$$V_C = -\frac{3}{2}\frac{GM}{R} = \frac{3}{2}V_s$$

$$= -\frac{GM}{R} + \frac{GM}{2R^3}[x^2 - R^2] = -\frac{GM}{2R^3}[3R^2 - x^2]$$

Scan for Video Explanation Web Reference at www.physicsgalaxy.com

**Module-40: Gravitational Potential due to a Solid Sphere**

# Variation of grav. potential due to a Solid Sphere:

$$V_{out} = -\frac{GM}{x}$$

$$V_s = -\frac{GM}{R}$$

$$V_{in} = -\frac{GM}{2R^3}[3R^2 - x^2]$$

$$V_c = -\frac{3}{2}\frac{GM}{R} = \frac{3}{2}V_s$$

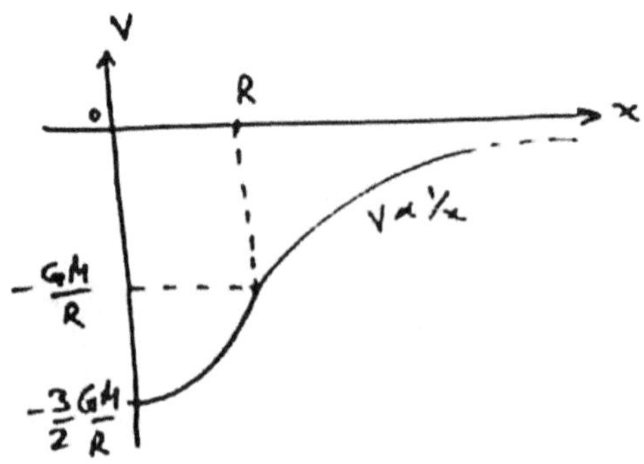


Scan for Video Explanation Web Reference at www.physicsgalaxy.com

## Module-41: Gravitational Potential Energy of a Body on Earth

# Gravitational Potential Energy of a Body in Earth's Grav. Field:

**Case-I:** On Earth Surface.

$$U_g = mV = m\left[-\frac{GM_e}{R_e}\right] = -\frac{GM_e m}{R_e}$$

**Case-II:** At a height h above the Earth surface

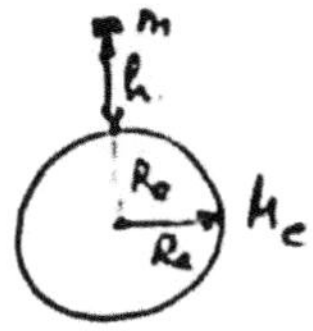


$$U_g = mV = m\left[-\frac{GM_e}{R_e+h}\right] = -\frac{GM_e m}{R_e+h}$$

**Case-III:** At a depth h from Earth surface inside the earth.

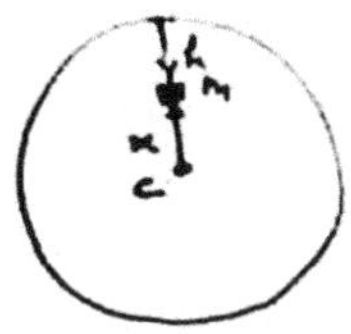


$$U_g = mV = m\left[-\frac{GM_e}{2R_e^3}\left[3R_e^2 - x^2\right]\right]$$

$$U_g = -\frac{GM_e m}{2R_e^3}(3R_e^2 - x^2) \leftarrow x = R_e - h$$



## Module-42: Solved Example-14

**Ex :** Find workdone in shifting a body of mass m from a height h above the earth surface to a height 2h above the surface.

**Sol$^n$:**

$$W = U_{2h} - U_h$$

When body is at a height h $\quad U_h = -\frac{GM_e m}{R_e+h}$

When body is at a height 2h $\quad U_{2h} = -\frac{GM_e m}{R_e+2h}$

$$W = GM_e m\left[\frac{1}{R_e+h} - \frac{1}{R_e+2h}\right] \quad \text{Ans.}$$

**Module-43: Solved Example-15**

Ex : Find the gravitational interaction energy of a system consisting of a disc of mass M, radius R and a point mass m situated at a distance x from disc centre on its axis as shown.

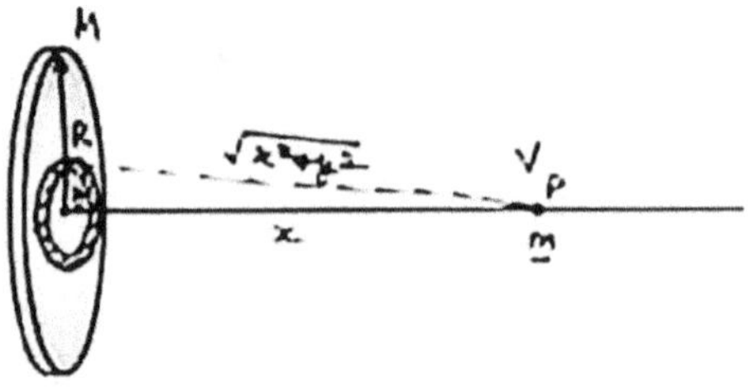


Sol$^n$ $U_g = mV_P$

mass of elemental ring in the disc is $dm = \frac{M}{\pi R^2} \cdot 2\pi y\,dy = \frac{2M}{R^2} y\,dy$

grav. pot$^n$ at P due to ring is $dV = -\frac{G\,dm}{\sqrt{x^2+y^2}} = -\frac{2GMy\,dy}{R^2\sqrt{x^2+y^2}}$

Total pot$^n$ at pt P due to disc is $V = \int dV = -\frac{2GM}{R^2}\int_0^R \frac{y\,dy}{\sqrt{x^2+y^2}} = -\frac{2GM}{R^2}\left[\sqrt{x^2+y^2}\right]_0^R$

$V_P = -\frac{2GM}{R}\left[\sqrt{x^2+R^2} - x\right]$

PE of mass m placed at pt P is $U_g = mV_P = -\frac{2GMm}{R}\left[\sqrt{x^2+R^2} - x\right]$ Ans.

Scan for Video Explanation

Web Reference at www.physicsgalaxy.com

**Module-44: Solved Example-16**

Ex : Figure shows a fixed solid sphere of mass M and radius R with a narrow smooth tunnel along its diameter. Another particle of mass m is placed at point A at a distance 2R from its centre along the line of tunnel as shown. If the particle is released from rest find its speed when it reaches the centre of sphere.

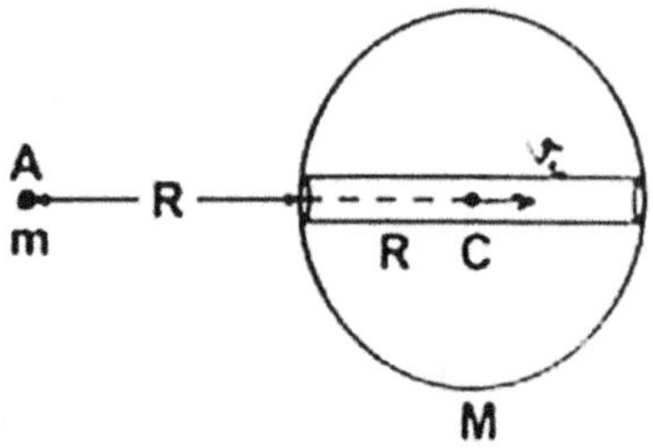


Sol$^n$ Pot$^n$ at A is $V_A = -\frac{GM}{2R}$

Pot$^n$ at C is $V_c = -\frac{3}{2}\frac{GM}{R}$

using Cons of energy for motion of m from A to C, we have.

$mV_A = mV_c + \frac{1}{2}mv_c^2$

$v_c = \sqrt{2(V_A - V_c)} = \sqrt{2\left(-\frac{GM}{R} + \frac{3}{2}\frac{GM}{R}\right)} = \sqrt{\frac{2GM}{R}}$ Ans.

Scan for Video Explanation

Web Reference at www.physicsgalaxy.com

**Module-45: Solved Example-17**

**Ex :** On the pole of earth, a body is imparted a velocity $v_0$ in vertically upward direction. If mass of earth is $M_0$ and radius $R_0$, find the maximum height to which body will ascend. Neglect air drag.

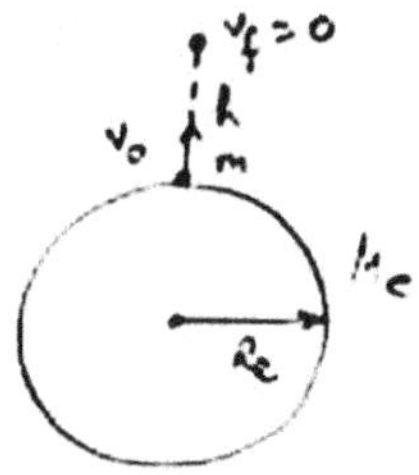


Sol$^n$: by energy Cons.

$$-\frac{GM_e m}{R_e} + \frac{1}{2} m v_0^2 = -\frac{GM_e m}{R_e + h}$$

$$\frac{2GM_e}{R_e + h} = \frac{2GM_e - v_0^2 R_e}{R_e}.$$

$$R_e + h = \frac{2GM_e R_e}{2GM_e - v_0^2 R_e}$$

$$h = \left(\frac{v_0^2 R_e^2}{v_0^2 R_e - 2GM_e}\right) \text{ Ans.}$$

Web Reference at www.physicsgalaxy.com | Scan for Video Explanation

**Module-46: Solved Example-18**

**Ex :** Find the speed with which a small body of mass m is projected from the centre of a ring of mass M and radius R along its axis so that it will never come back.

Sol$^n$

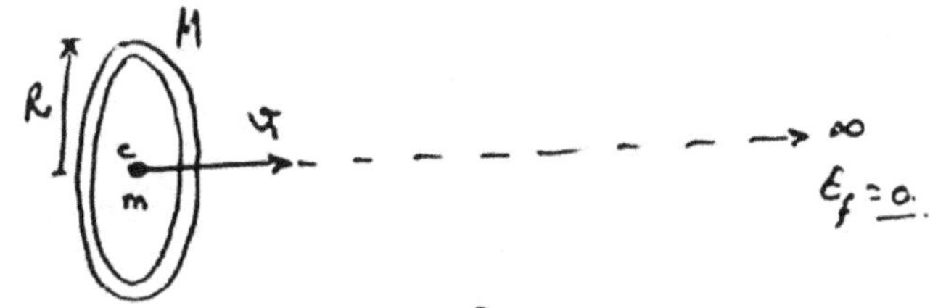


Initial energy of m. $E_i = \frac{1}{2} m v^2 + m V_c$

$$= \frac{1}{2} m v^2 - \frac{GMm}{R} \geq 0$$

$$v \geq \sqrt{\frac{2GM}{R}} \text{ Ans.}$$

Web Reference at www.physicsgalaxy.com | Scan for Video Explanation

**Module-47: Gravitational Self Energy of a Body**

# Gravitational Self Energy of a body:

- PE of a body due to grav. interaction forces inside the body among its particles is called its grav. self energy.

- It is the W/D in assembling the body by bringing all its constituent particles from $\infty$ to the given shape & size against grav. force between these particles.

Scan for Video Explanation

Web Reference at www.physicsgalaxy.com

**Module-48: Gravitational Self Energy of a Uniform Hollow Sphere**

# Gravitational Self Energy of a Hollow Sphere:

When mass of shell is m and W/D in bringing dm from $\infty$ to R is

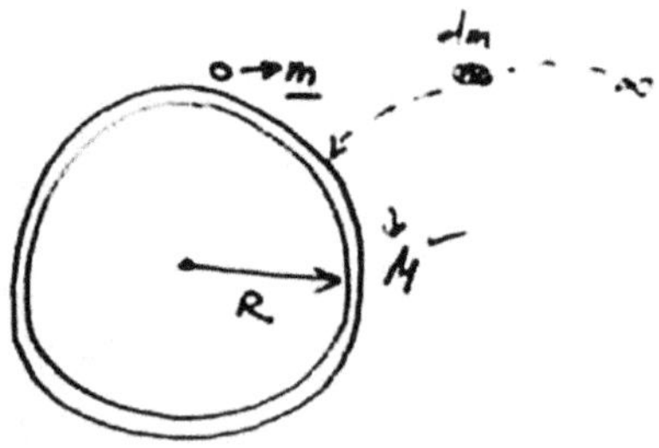


$$dW = dm(V_S - V_\infty) \quad (V_\infty \to 0)$$

$$dW = dm\left(-\frac{Gm}{R}\right) = -\frac{Gm\,dm}{R}$$

$$W = \int dW = \int_0^M -\frac{Gm\,dm}{R}$$

$$W = -\frac{G}{R}\left[\frac{m^2}{2}\right]_0^M = -\frac{GM^2}{2R}$$

$$\boxed{U_{self} = -\frac{GM^2}{2R}}$$

Scan for Video Explanation

Web Reference at www.physicsgalaxy.com

**Module-49: Gravitational Self Energy of a Uniform Solid Sphere**

## # Gravitational Self Energy of a Uniform Solid Sphere:

W/D in bringing $dm$ from $\infty$ to the surface of radius $x$ is

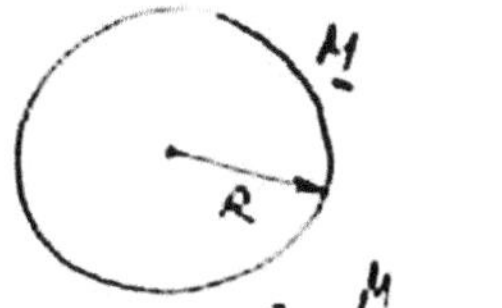


$$\rho = \frac{M}{\frac{4}{3}\pi R^3}$$

$$dW = dm\,(V_x - V_\infty) \quad (V_\infty \to 0)$$

$$dW = dm \cdot \left[-\frac{Gm}{x}\right]$$

$$dW = -\frac{G}{x} \cdot \rho \cdot \frac{4}{3}\pi x^3 \times \rho \cdot 4\pi x^2 dx$$

$$dW = -\frac{16\pi^2}{3} G\rho^2 x^4 \cdot dx$$

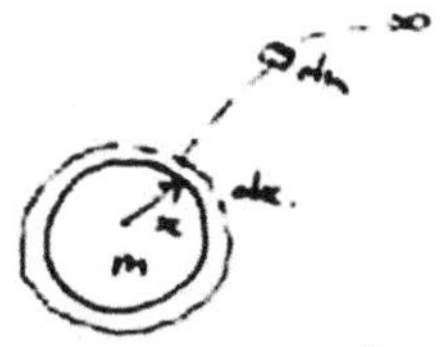


$$dm = \rho \cdot 4\pi x^2 dx$$

Total work in assembling sphere of rad. $R$:

$$W = \int dW = -\frac{16\pi^2 G\rho^2}{3}\int_0^R x^4 dx = -\frac{16}{15}\pi^2 G\rho^2 R^5 = -\frac{16}{15}\pi^2 G R^5\left[\frac{M}{\frac{4}{3}\pi R^3}\right]^2$$

$$\boxed{U_{self} = W = -\frac{3}{5}\frac{GM^2}{R}}$$

# Chapter 18
# Satellite Motion

## Lecture Notes Modules


**Module-1:** Introduction to Satellite Motion
**Module-2:** Energy of a Satellite in Circular Orbit
**Module-3:** Solved Example-1
**Module-4:** Solved Example-2
**Module-5:** Solved Example-3
**Module-6:** Bounded Motion and Escape Velocity
**Module-7:** Escaping a Satellite
**Module-8:** Solved Example-4
**Module-9:** Solved Example-5
**Module-10:** Motion of a Satellite in Elliptical Path
**Module-11:** Conservation of Angular Momentum in Satellite Motion
**Module-12:** Kepler's Laws of Planetary Motion-First Law
**Module-13:** Kepler's Laws of Planetary Motion-Second Law
**Module-14:** Kepler's Laws of Planetary Motion-Third Law
**Module-15:** Projection of Spaceships and Satellites from Earth
**Module-16:** Solved Example-6
**Module-17:** Solved Example-7
**Module-18:** Solved Example-8
**Module-19:** Solved Example-9
**Module-20:** Communication Satellites
**Module-21:** Geostationary Satellite
**Module-22:** Broadcasting Region of a Satellite
**Module-23:** Solved Example-10
**Module-24:** Solved Example-11
**Module-25:** Solved Example-12

**Module-1: Introduction to Satellite Motion**

# Satellite Motion:

for a circular orbit

Satellite motion.

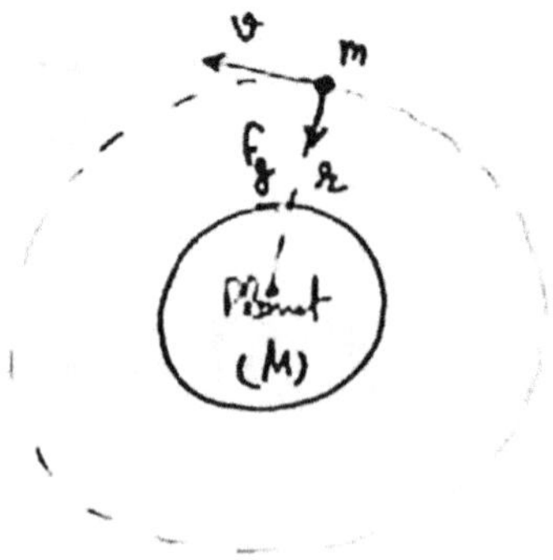


$$\frac{GMm}{r^2} = \frac{mv^2}{r}$$

$$\boxed{v = \sqrt{\frac{GM}{r}}}$$ orbital speed in circular orbit of rad. r.

NOTE: orbital speed of a satellite does not depend on mass of satellite

Scan for Video Explanation

Web Reference at www.physicsgalaxy.com

**Module-1: Introduction to Satellite Motion**

# Orbital Period of a Satellite in circular orbit:

orbital speed $v = \sqrt{\frac{GM}{r}}$

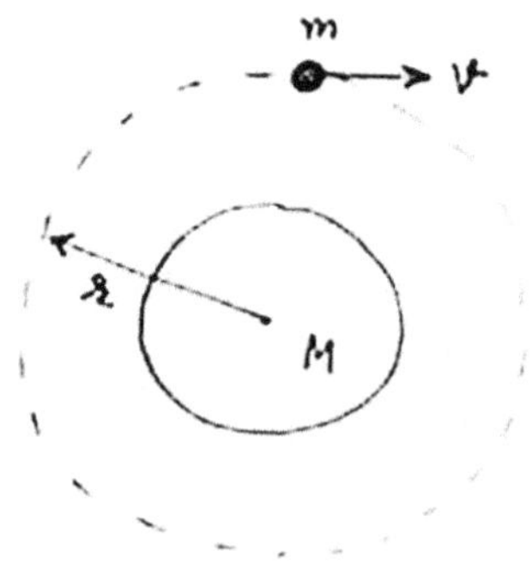


time period of rev. of satellite

$$T = \frac{2\pi r}{v}$$

$$T = \frac{2\pi r}{\sqrt{GM/r}}$$

$$\boxed{T = \frac{2\pi}{\sqrt{GM}} r^{3/2}} \Rightarrow T \propto r^{3/2}$$

Scan for Video Explanation

Web Reference at www.physicsgalaxy.com

**Module-2: Energy of a Satellite in Circular Orbit**

# Energy of a Satellite in circular orbit:

$$v = \sqrt{\frac{GM_p}{r}}$$

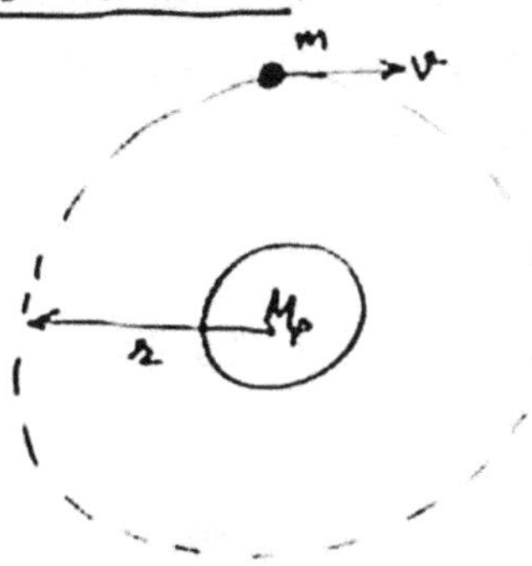


K.E of Satellite is

$$K = \frac{1}{2}mv^2 = \frac{1}{2}\left(m\sqrt{\frac{GM_p}{r}}\right)^2$$

$$\boxed{K = \frac{1}{2}\frac{GM_p m}{r}}$$

PE of Satellite $\boxed{P = -\frac{GM_p m}{r}}$

Total energy $E = K + P = \frac{1}{2}\frac{GM_p m}{r} - \frac{GM_p m}{r} = \boxed{-\frac{1}{2}\frac{GM_p m}{r}}$

$$\boxed{|E| = |K| = \frac{1}{2}|P|}$$

Web Reference at www.physicsgalaxy.com | Scan for Video Explanation

**Module-3: Solved Example-1**

Ex : A satellite is to revolve in an orbit of radius 8000 km around earth. What should be its speed and orbital period. Take $R_0$ = 6400 km and $g_s$ = 10 m/s$^2$.

Solⁿ: orbital speed of a Sat $v = \sqrt{\frac{GM_e}{r}} = \sqrt{\frac{g_s R_e^2}{r}}$

$$g_s = \frac{GM_e}{R_e^2}$$

$$v = \sqrt{\frac{10 \times (6.4 \times 10^6)^2}{8 \times 10^6}}$$

$$v = 7.1 \times 10^3 \text{ m/s} = 7.1 \text{ km/s}$$

Period of Sat. $T = \frac{2\pi}{\sqrt{GM_e}} \cdot (r)^{3/2}$

$$T = \frac{2\pi}{\sqrt{g_s R_e^2}} \cdot r^{3/2} = \frac{2 \times 3.14}{\sqrt{10 \times (6.4 \times 10^6)^2}} \times (8 \times 10^6)^{3/2}$$

$$= 7021 \text{ sec} \simeq 117 \text{ min} \quad \underline{\text{Ans.}}$$

Web Reference at www.physicsgalaxy.com | Scan for Video Explanation

**Module-4: Solved Example-2**

**Ex : Two satellites A and B of same mass are orbiting earth at attitudes R and 3R respectively, where R is the radius of earth. Taking their orbits to be circular, calculate ratio of their kinetic energies.**

Sol:

KE of a Sat $K = \frac{GM_e m_s}{2r}$ $r \to$ rad of orbit

for Sat A & B.

$$\frac{K_A}{K_B} = \frac{R_B}{R_A} = \frac{4R}{2R} = \frac{2}{1}$$ Ans.

Scan for Video Explanation

Web Reference at www.physicsgalaxy.com

**Module-5: Solved Example-3**

**Ex : A satellite of mass $2 \times 10^3$ kg is to be shifted from an orbit of radius $2R_e$ to another orbit of radius $3R_e$ under ideal conditions. Calculate the minimum energy required for this. $R_e$ is radius of earth.**

Sol: Initial energy of Sat $E_i = -\frac{GM_e m}{2(2R_e)} = -\frac{GM_e m}{4R_e}$

Final energy of Sat $E_f = -\frac{GM_e m}{2(3R_e)} = -\frac{GM_e m}{6R_e}$

energy reqd to change orbit of Sat -

$$\Delta E = E_f - E_i = \frac{GM_e m}{R_e}\left[\frac{1}{4} - \frac{1}{6}\right] = \frac{GM_e m}{12R_e}$$

$g_s = \frac{GM_e}{R_e^2}$

$$= \frac{1}{12} m g_s R_e$$

$$= \frac{1}{12} \times 2 \times 10^3 \times 10 \times (6.4 \times 10^6)$$

$$= 1.066 \times 10^{10} \text{ J}$$ Ans.

Scan for Video Explanation

Web Reference at www.physicsgalaxy.com

**Module-6: Bounded Motion and Escape Velocity**

# Bounded Motion and Escape Velocity:

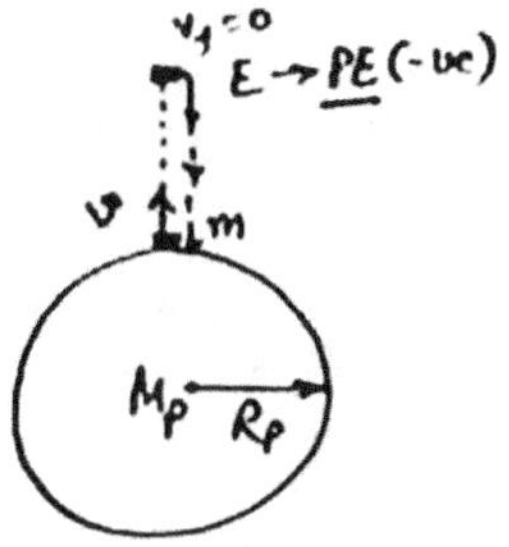


PE of body on surface of planet

$$U = -\frac{GM_pm}{R_p}$$

$-ve$ sign $\Rightarrow$ Boundness.

if some KE is imparted to body, then

$$E = \left(\frac{1}{2}mv^2 - \frac{GM_pm}{R_p}\right) < 0$$

If a body is projected such that on surface of planet its TE = 0 $\Rightarrow$ finally the body will be able to overcome the grav forces of planet & it will escape.

Web Reference at www.physicsgalaxy.com | Scan for Video Explanation

**Module-6: Bounded Motion and Escape Velocity**

if body is projected at speed $v$ such that

$$\frac{1}{2}mv^2 - \frac{GM_pm}{R_p} = 0$$

$$v = \sqrt{\frac{2GM_p}{R_p}} \leftarrow \text{escape velocity on surface of planet.}$$

for earth ($M_e$ & $R_e$)

escape vel. $v_e = \sqrt{\frac{2GM_e}{R_e}}$ $\qquad g_s = \frac{GM_e}{R_e^2} = 9.8\ m/s^2$

$$v_s = \sqrt{2g_sR_e} = 11.2\ km/s$$

NOTE: As planet is taken at rest in this calculation, the escape vel. does not depend on dir of projection.

Web Reference at www.physicsgalaxy.com | Scan for Video Explanation

**Module-7: Escaping a Satellite**

# Escaping a Satellite:

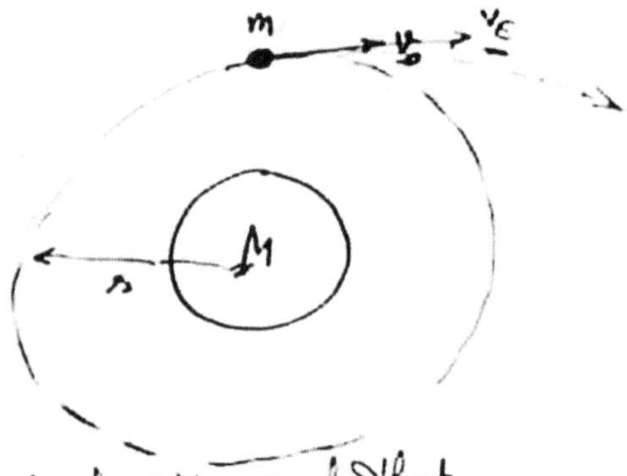


orbital speed $v_0 = \sqrt{\frac{GM}{r}}$

TE of Satellite $E = -\frac{1}{2}\frac{GMm}{r}$

if speed of Sat. is suddenly increased to $v_E$ then to escape it should be such that

$$\frac{1}{2}mv_E^2 - \frac{GMm}{r} = 0 \Rightarrow v_E = \sqrt{\frac{2GM}{r}} = \sqrt{2}v_0$$

increase in orbital speed for a Sat to escape $= \sqrt{2}v_0 - v_0$
$= (\sqrt{2}-1)v_0$
$= 0.414\, v_0$
$= 41.4\%$ of $v_0$.

NOTE: For a sat to escape from gravity of planet its orbital speed should be inc by 41.4% (Imp)

Scan for Video Explanation

Web Reference at www.physicsgalaxy.com

**Module-8: Solved Example-4**

**Ex :** An artificial satellite is moving in a circular orbit around earth with a speed equal to half the escape velocity from earth surface. Find the height of satellite above the earth surface.

Sol: orbital speed $v = \frac{1}{2}\sqrt{2g_s R_e} = \sqrt{\frac{g_s R_e}{2}}$

if Sat is at a height above earth surface.

$$v_0 = \sqrt{\frac{GM_e}{(R_e+h)}} = \sqrt{\frac{g_s R_e}{2}}$$

$$\sqrt{\frac{GM_e}{(R_e+h)}} = \sqrt{\frac{GM_e}{2R_e}}$$

$$R_e + h = 2R_e$$

$$h = R_e \text{ Ans.}$$

Scan for Video Explanation

Web Reference at www.physicsgalaxy.com

**Module-9: Solved Example-5**

Ex : A mass of $6 \times 10^{24}$ kg is to be compressed in a sphere in such a way that the escape velocity from its surface will be $3 \times 10^8$ m/s. What should be the radius of such a sphere ?

Sol: escape vel $v_{esc} = \sqrt{\frac{2GM}{R}}$

$$R = \frac{2GM}{v_{esc}^2}$$

$$R = \frac{2 \times 6.67 \times 10^{-11} \times 6 \times 10^{24}}{(3 \times 10^8)^2}$$

$$= 8.89 \times 10^{-3} \text{ m}$$

$$= 8.89 \text{ mm} \quad \text{Ans.}$$

Web Reference at www.physicsgalaxy.com | Scan for Video Explanation

**Module-10: Motion of a Satellite in Elliptical Path**

# Motion of a Satellite in elliptical path:

orbital speed $v_0 = \sqrt{\frac{GM_e}{x}}$

escape vel $v_e = \sqrt{\frac{2GM_e}{x}}$

if $v_s > \sqrt{\frac{GM_e}{x}}$ but $v_s < \sqrt{\frac{2GM_e}{x}}$

if $v_s < \sqrt{\frac{GM_e}{x}}$

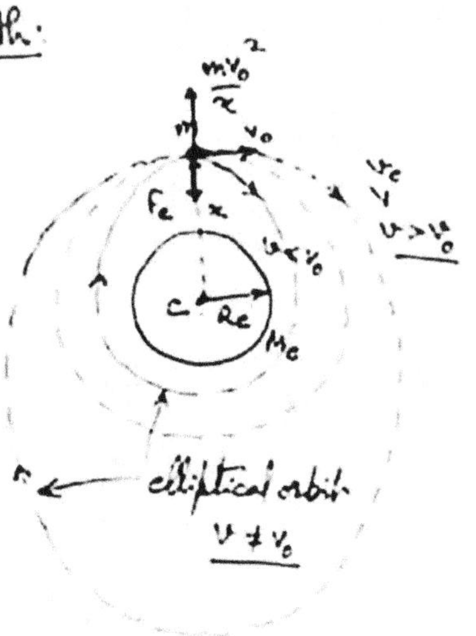


Web Reference at www.physicsgalaxy.com | Scan for Video Explanation

## Module-10: Motion of a Satellite in Elliptical Path

If $v < \sqrt{\frac{GM_e}{x}}$ and having such a value that elliptical trajectory will intersect the earth surface.

In this situation path of body/sat will be an arc of ellipse as shown.

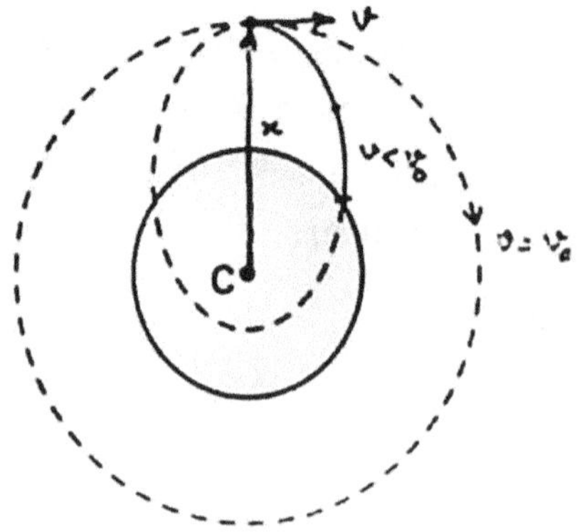


Scan for Video Explanation

Web Reference at www.physicsgalaxy.com

## Module-11: Conservation of Angular Momentum in Satellite Motion

# Conservation of Angular momentum in Satellite Motion:

In motion of a Sat around a planet, only force on it is gravity directed toward the centre of planet.

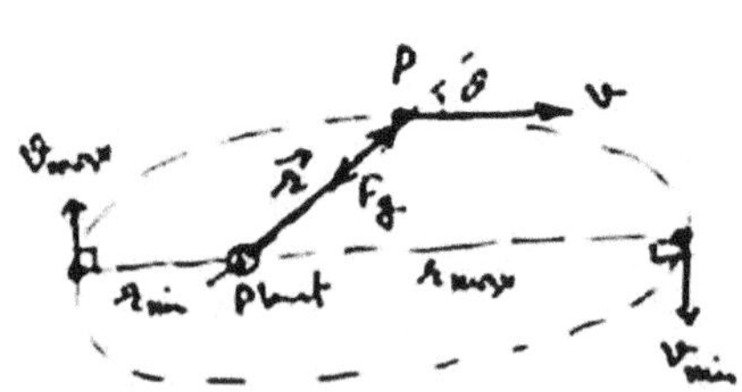


As $F_g$ is passing through centre of planet hence no torque act on Sat w.r.to planet.

⇒ Ang. momentum of Sat w.r.to planet will remain conserved during its motion

$$L = mvr\sin\theta = \text{Const}$$

$$\left[ L = mv_{max}r_{min} = mv_{min}r_{max} = mvr\sin\theta \right]$$

by energy Cons we can write

$$\left[ \frac{1}{2}mv_{max}^2 - \frac{GMm}{r_{min}} = \frac{1}{2}mv_{min}^2 - \frac{GMm}{r_{max}} = \frac{1}{2}mv^2 - \frac{GMm}{r} \right]$$

Scan for Video Explanation

Web Reference at www.physicsgalaxy.com

**Module-12: Kepler's Laws of Planetary Motion-First Law**

# Kepler's Laws of Planetary Motion:

These laws are formulated to understand basics of Planetary motion around Sun and these are also applicable for Satellite motion around a planet.

(1) Kepler's First law: (Law of Orbits)

"All the planets move around Sun in elliptical orbits with Sun located at one of the foci of ellipse."

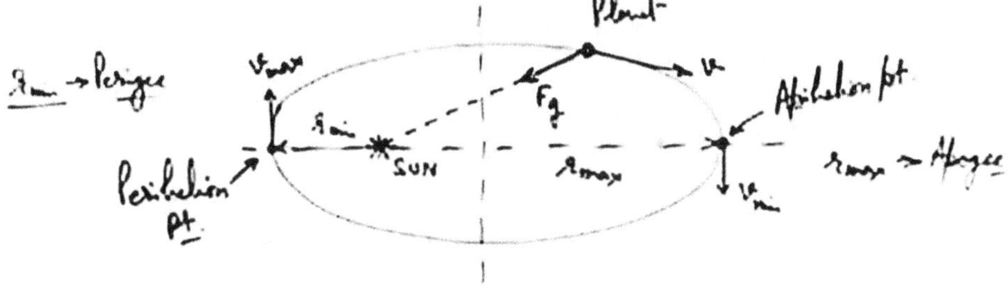


Web Reference at www.physicsgalaxy.com | Scan for Video Explanation

**Module-13: Kepler's Laws of Planetary Motion-Second Law**

(2) Kepler's Second Law: (law of Area)

"The line joining the Sun and planet sweeps out equal areas in equal time intervals or the rate of sweeping area by the position vector of planet w.r.t. to Sun remains Constant."

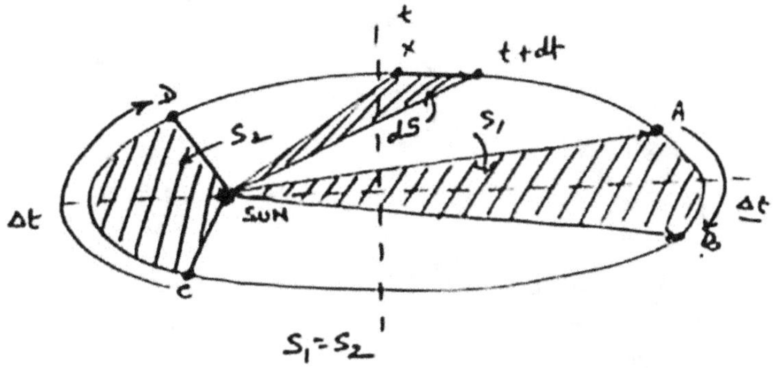


rate of sweeping area by p.v. of planet is $\frac{dS}{dt}$ = Constant

Web Reference at www.physicsgalaxy.com | Scan for Video Explanation

## Module-13: Kepler's Laws of Planetary Motion-Second Law

Area of orbit = $\pi ab$.

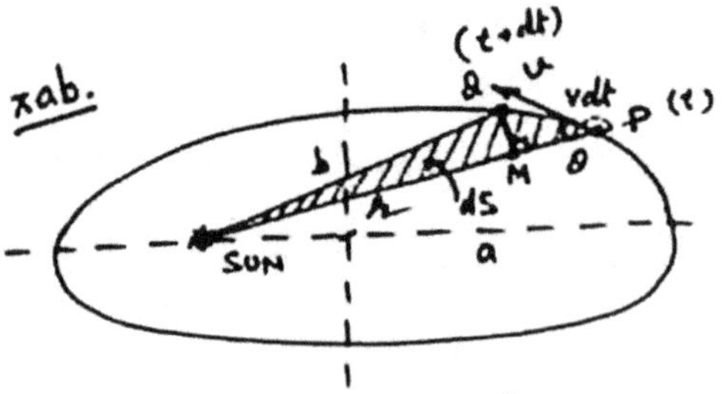


$QM = v\,dt\sin\theta$

Area swept by p.v of planet in time dt

$$dS = \frac{1}{2} \times r \times v\,dt\sin\theta = \frac{1}{2} r v\,dt\sin\theta$$

rate of sweeping area $\frac{dS}{dt} = \frac{1}{2} r v \sin\theta = \frac{L}{2m} = \text{Const.}$

Angular momentum of planet at pt. P is $L = mvr\sin\theta$

if T is the orbital period of planet $\Rightarrow$ $\boxed{\frac{dS}{dt} = \frac{\pi ab}{T} = \frac{L}{2m}}$

$$L = \frac{2\pi abm}{T} \quad (\text{Const})$$

Scan for Video Explanation

Web Reference at www.physicsgalaxy.com

## Module-14: Kepler's Laws of Planetary Motion-Third Law

③ Kepler's Third Law: (Law of Period)

" The square of period of rev. of a planet in its orbit is dir proportional to the cube of semi-major axis of the elliptical orbit around sun."

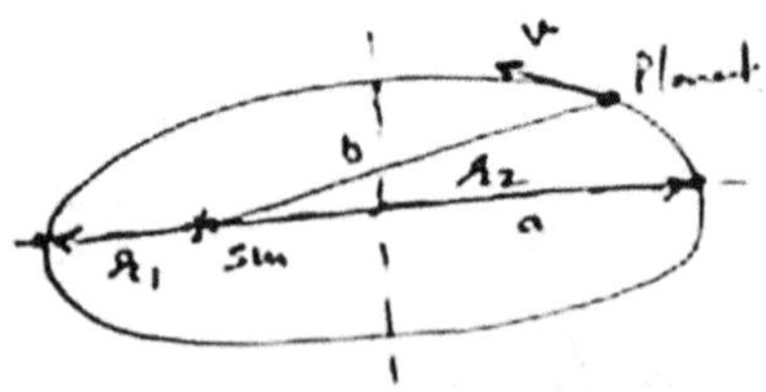


$T^2 \propto a^3$

$$\boxed{T^2 = \frac{4\pi^2}{GM_s} \cdot a^3}$$

$$a = \left(\frac{r_1 + r_2}{2}\right)$$

for circular orbits $a = b = R$ (orbital rad)

$$\boxed{T^2 = \frac{4\pi^2}{GM_s} \cdot R^3}$$

Total energy of planet in elliptical orbit $\boxed{E = -\frac{GM_s m}{2a}}$

Scan for Video Explanation

Web Reference at www.physicsgalaxy.com

**Module-15: Projection of Spaceships and Satellites from Earth**

# Projection of Spaceships and Satellites from Earth:

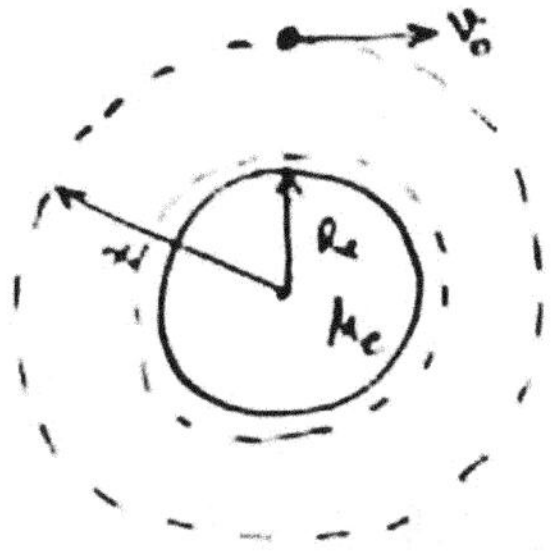

$$v_0 = \sqrt{\frac{GM_e}{x}}$$

if $x \approx R_e$ [very close to Earth Surface]

$$v_0 = \sqrt{\frac{GM_e}{R_e}} = \sqrt{g_e R_e} \approx 7.93 \text{ km/s}$$

first cosmic velocity

The vel reqd to escape from grav attraction of Earth is

$$v_{escape} = \sqrt{2 g_e R_e} \approx 11.2 \text{ km/s}$$

$$= \sqrt{2}\, v_0$$



**Module-15: Projection of Spaceships and Satellites from Earth**

# Trajectories of Projected bodies:

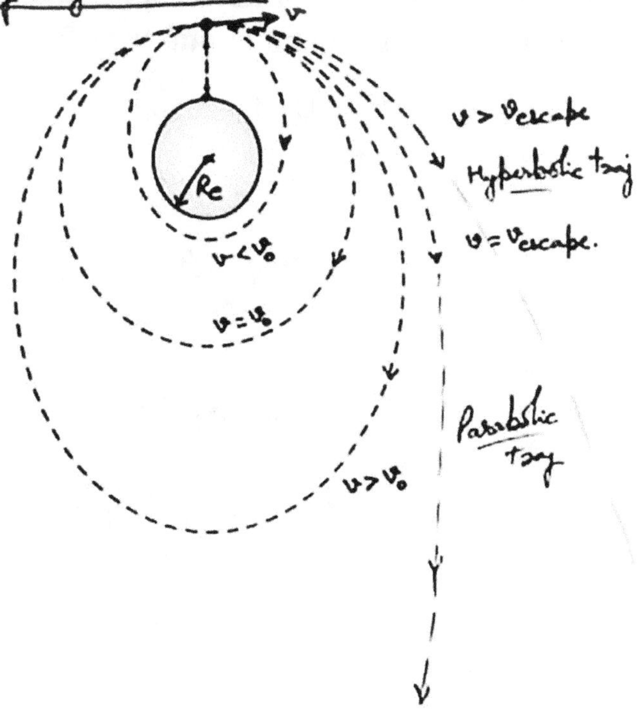

**Module-16: Solved Example-6**

Ex : The radius of a planet is $R_1$ and a satellite revolves around it in circular orbit of radius $R_2$. The period of revolution is T. Find the acceleration due to gravity on the surface of planet.

Sol:

$$g_s = \frac{GM_p}{R_1^2} \qquad M_p \rightarrow \text{mass of planet}$$

from Kepler's III law -

$$T^2 = \frac{4\pi^2}{GM_p} \cdot R_2^3 \longrightarrow GM_p = \frac{4\pi^2 R_2^3}{T^2}$$

$$g_s = \frac{4\pi^2 R_2^3}{R_1^2 T^2} \quad \text{Ans.}$$

Scan for Video Explanation

Web Reference at www.physicsgalaxy.com

**Module-17: Solved Example-7**

Ex : If earth be one half of its present distance from sum, how many days will be in one year ?

Sol: As we know for planetary motion

$$T^2 \propto R^3$$

$$\left(\frac{T_1}{T_2}\right)^2 = \left(\frac{R_1}{R_2}\right)^3$$

if $R_2 = R_1/2$

$$T_2 = T_1 \times \left(\frac{R_2}{R_1}\right)^{3/2}$$

$$\Rightarrow \frac{T_1}{2^{3/2}} = \frac{365}{2^{3/2}} = 129 \text{ days} \quad \text{Ans.}$$

Scan for Video Explanation

Web Reference at www.physicsgalaxy.com

**Module-18: Solved Example-8**

**Ex :** A satellite revolves around a planet in an elliptical orbit. Its perigee and apogee are $0.5 \times 10^7$ m and $1.5 \times 10^7$ m respectively. If the minimum speed of satellite in orbit is $5 \times 10^3$ m/s find its maximum speed in orbit.

Sol$^n$:

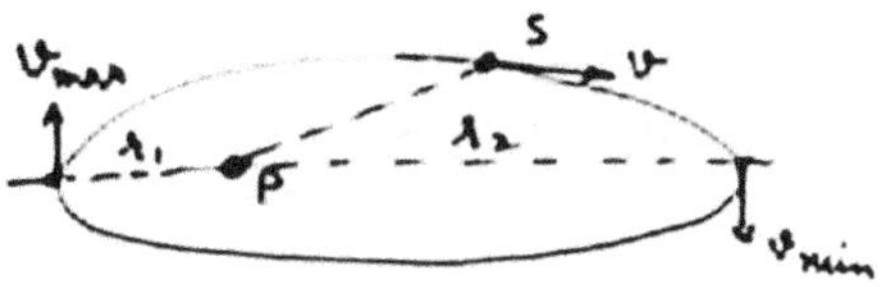


by cons of ang momentum, we can use

$$m v_{max} r_1 = m v_{min} r_2$$

$$v_{max} = v_{min} \times \frac{r_2}{r_1}$$

$$= 5 \times 10^3 \times \frac{1.5 \times 10^7}{0.5 \times 10^7} = 1.5 \times 10^4 \text{ m/s} \quad \text{Ans.}$$

Web Reference at www.physicsgalaxy.com | Scan for Video Explanation

**Module-19: Solved Example-9**

**Ex :** A satellite is revolving around a planet of mass M in an elliptical orbit of semimajor axis a. Show that the orbital speed of the satellite when it is at a distance r from the focus will be given by

$$v = \sqrt{GM\left[\frac{2}{r} - \frac{1}{a}\right]}$$

Sol$^n$: Total energy of Sat in elliptical motion is

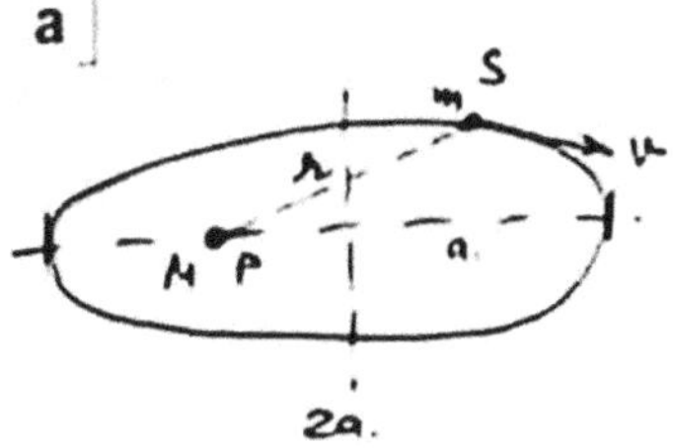


$$E = -\frac{GMm}{2a} = \text{Constant}$$

at a distance r from planet, the energy of Sat -

$$-\frac{GMm}{2a} = -\frac{GMm}{r} + \frac{1}{2} m v^2$$

$$v^2 = GM\left[\frac{2}{r} - \frac{1}{a}\right] \Rightarrow v = \sqrt{GM\left(\frac{2}{r} - \frac{1}{a}\right)}$$

hence proved.

Web Reference at www.physicsgalaxy.com | Scan for Video Explanation

## Module-20: Communication Satellites

# Communication Satellites:

Such Sat are used for communication of information from one pt to everywhere over the globe.

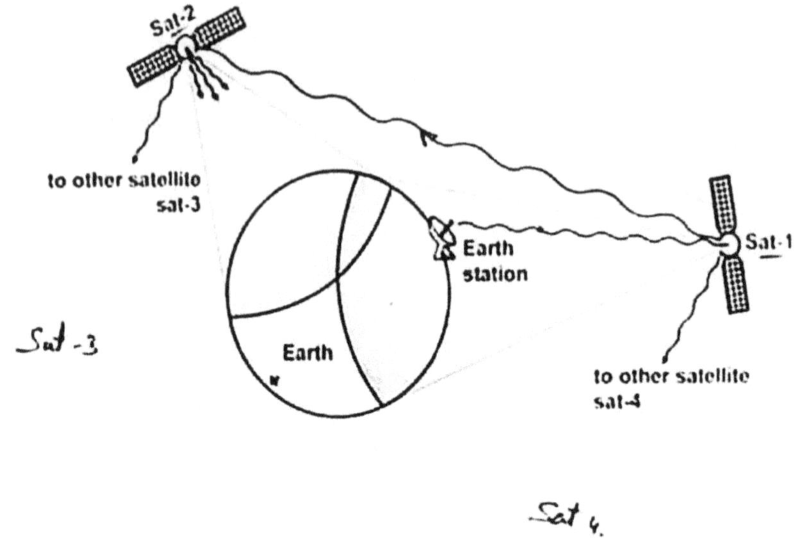


Scan for Video Explanation

Web Reference at www.physicsgalaxy.com

## Module-21: Geostationary Satellite

# Geostationary Satellite:

Sat which appear stationary as seen from Earth surface is called a Geostationary Satellite

orbital speed

$$v = \sqrt{\frac{GM_e}{R_e+h}}$$

for a G.S. Sat $T = 24$ hrs.

Sat. period. $$T = \frac{2\pi}{\sqrt{GM_e}} \cdot (R_e+h)^{3/2}$$

$$\left(\frac{GM_e}{R_e^2} = g_s\right) \quad T^2 = \frac{4\pi^2}{GM_e}(R_e+h)^3$$

$$h = \left(\frac{g_s R_e^2 T^2}{4\pi^2}\right)^{1/3} - R_e$$

$$h \simeq 35950 \simeq 36 \times 10^3 \text{ km}.$$

Scan for Video Explanation

Web Reference at www.physicsgalaxy.com

**Module-21: Geostationary Satellite**

NOTE: Parking orbit must be in equatorial plane of Earth.

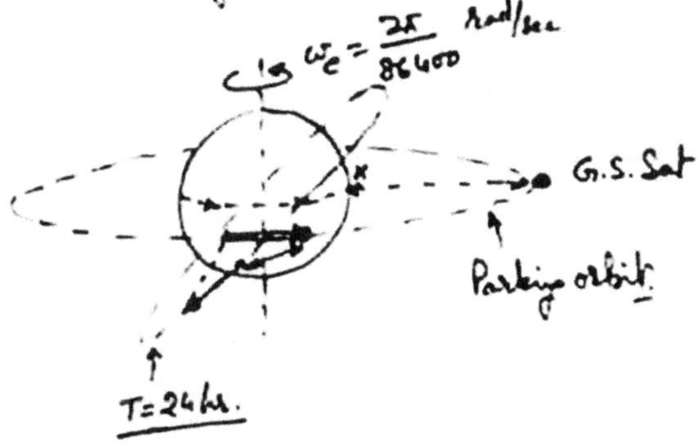


Web Reference at www.physicsgalaxy.com | Scan for Video Explanation

**Module-22: Broadcasting Region of a Satellite**

# Broadcasting Region of a Satellite:

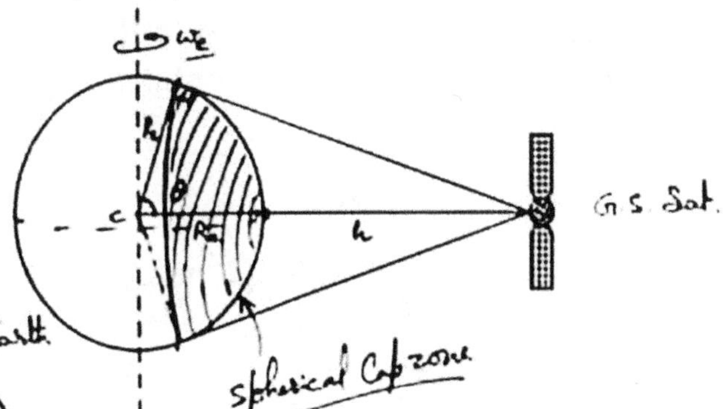


$$\theta = \cos^{-1}\left(\frac{R_e}{R_e+h}\right)$$

Solid angle subtended by cap zone at Earth Centre

$$\Omega = 2\pi(1-\cos\theta) = 2\pi\left(1-\frac{R_e}{R_e+h}\right) = \frac{2\pi h}{R_e+h}$$

Reception area on Earth by Sat is $S = \Omega R_e^2$

$$S = \frac{2\pi h R_e^2}{(R_e+h)}$$

Web Reference at www.physicsgalaxy.com | Scan for Video Explanation

**Module-23: Solved Example-10**

**Ex :** If a satellite is revolving around the earth in a circular orbit in a plane containing earth's axis of rotation. If the angular speed of satellite is equal to that of earth, find the time it takes to move from a point above north pole to a point above equator.

Sol$^n$:

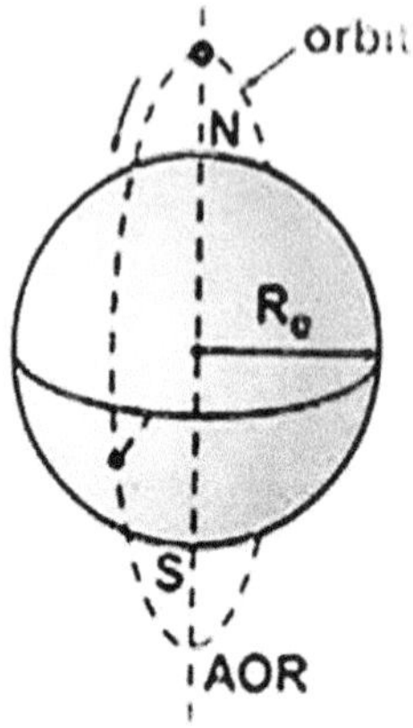


Ang speed of Sat is

$$\omega = \frac{2\pi}{T} = \frac{2\pi}{24} = \frac{\pi}{12} \text{ rad/hr}$$

time taken

$$t = \frac{\theta}{\omega} = \frac{\pi/2}{\pi/12} = 6 \text{ hrs } \underline{\text{Ans.}}$$

Scan for Video Explanation

Web Reference at www.physicsgalaxy.com

**Module-24: Solved Example-11**

**Ex :** If orbit radius of a geostationary satellite is $7R_0$, find the location having maximum latitude on earth which can receive signals from geostationary satellite.

Sol$^n$:

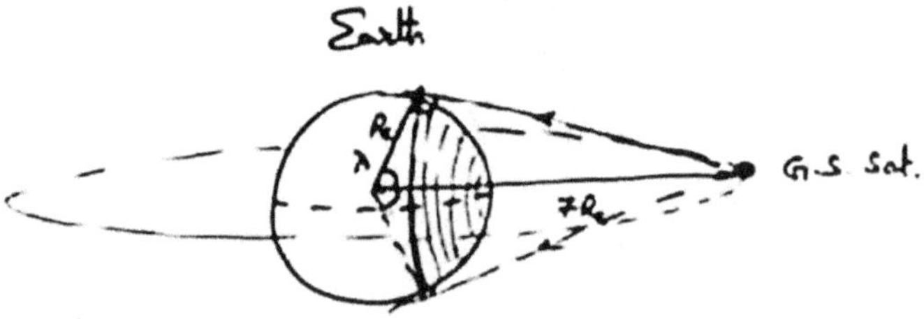


$$\cos\lambda = \frac{R_e}{7R_e}$$

$$\lambda = \cos^{-1}\left(\frac{1}{7}\right) \underline{\text{Ans.}}$$

Scan for Video Explanation

Web Reference at www.physicsgalaxy.com

**Module-25: Solved Example-12**

Ex : A satellite is revolving around the earth in an orbit of radius double that of the parking orbit. Find the time duration between two instants when this satellite is closest to a geostationary satellite.

Sol^n:

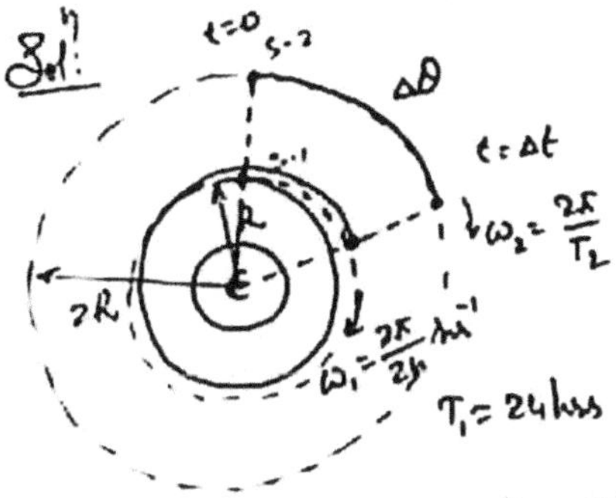


using Kepler's III law -

$$\frac{T_1}{T_2} = \left(\frac{R_1}{R_2}\right)^{3/2}$$

$$T_2 = T_1 \times \left(\frac{R_2}{R_1}\right)^{3/2} = 24 \times (2)^{3/2} \text{ hrs.}$$

$$= 67.88 \text{ hrs.}$$

$$\Delta t = \frac{\Delta\theta}{\omega_2} = \frac{2\pi + \Delta\theta}{\omega_1} = \frac{2\pi}{\omega_1 - \omega_2}$$

$$\Delta t = \frac{2\pi}{\frac{2\pi}{24} - \frac{2\pi}{67.88}} = \frac{67.88 \times 24}{67.88 - 24} = 37.13 \text{ hrs Ans.}$$

www.ingramcontent.com/pod-product-compliance
Lightning Source LLC
Chambersburg PA
CBHW080208170726
48282CB00030B/1047
* 9 7 8 9 3 8 8 1 8 2 4 4 7 *